AF411318

Novel Principles and Methods in Bacterial Cell Cycle Physiology: Celebrating the Charles E. Helmstetter Prize in 2022

Novel Principles and Methods in Bacterial Cell Cycle Physiology: Celebrating the Charles E. Helmstetter Prize in 2022

Editors

Vic Norris
Arieh Zaritsky

Basel • Beijing • Wuhan • Barcelona • Belgrade • Novi Sad • Cluj • Manchester

Editors
Vic Norris
University of Rouen
Mont Saint Aignan
France

Arieh Zaritsky
Ben-Gurion University of the Negev
Be'er-Sheva
Israel

Editorial Office
MDPI
St. Alban-Anlage 66
4052 Basel, Switzerland

This is a reprint of articles from the Special Issue published online in the open access journal *Life* (ISSN 2075-1729) (available at: https://www.mdpi.com/journal/life/special_issues/bacterial_cell_cycle).

For citation purposes, cite each article independently as indicated on the article page online and as indicated below:

Lastname, A.A.; Lastname, B.B. Article Title. *Journal Name* **Year**, *Volume Number*, Page Range.

ISBN 978-3-0365-9829-1 (Hbk)
ISBN 978-3-0365-9830-7 (PDF)
doi.org/10.3390/books978-3-0365-9830-7

Cover image courtesy of Philip C. Hanawalt and Charles E. Helmstetter
Some Giants at Cold Spring Harbor, 1968. See details in Philip C. Hanawalt's photo at the Editorial in this Special Issue and Charles E Helmstetter recording growth properties of newborn cells from a baby machine at Florida Institute of Technology

Contents

About the Editors

Vic Norris

Vic Norris is an Emeritus Professor in the Department of Biology in the University of Rouen (France). He has a bachelor's degree in Psychology and another in Biology and Mathematics. He obtained his PhD in Paris working on the cell cycle of *Escherichia coli*. Subsequently, he spent ten years in Leicester (UK) working again on *E. coli* but also on theoretical biology. He joined the Integrated Systems Institute at the University of Rouen in 1996 where he collaborated with scientists from across the disciplines on a variety of subjects that included plant memory and the bacterial cell cycle. He directed a group working on Secondary Ion Mass Spectrometry for a couple of years and also helped organize the Epigenomics Project in Evry (which ran for 14 years). Since retiring, he works on both neurodegeneration and the bacterial cell cycle. He has around 200 peer-reviewed publications.

Arieh Zaritsky

Arieh Zaritsky is an Emeritus Professor at the Life Sciences Department / Natural Sciences Faculty of Ben-Gurion University of the Negev (Be'er-Sheva, Israel). Dr. Zaritsky obtained a distinguished MSc in Genetics at the Hebrew University, Israel, in 1967. He completed his PhD from Leicester University, UK, in 1971 and post-doctorate training at the University Institute of Microbiology, Copenhagen, Denmark, in 1972. During his career, he has instructed over 50 graduate students/scientists and was awarded numerous research grants. He has visited higher education institutions around the world and delivered invited lectures at international meetings. Dr. Zaritsky is a recognized expert in bacterial and bacteriophage physiology, on which he has published about half of his 140 peer-reviewed articles and was awarded (1994) a Burroughs Wellcome/ASM Visiting Professorship. In 1989–1990, he served as a departmental chairman. At over 80 years of age, Dr Zaritsky is still running a laboratory with 3 prestigious grants, from the U.S.–Israel Binational Science Foundation (BSF), the Israel Science Foundation joint research program with the Chinese Academy of Sciences (ISF–NSFC), and from the Israel Agriculture Ministry, the latter to study the biological control of insect pests using specific toxin genes from Bacillus thuringiensis.

Preface

This Special Issue aims to celebrate the work of Charles E. Helmstetter following the awarding of the "Charles E. Helmstetter Prize for Groundbreaking Research in Bacterial Cell Cycle Physiology" to Moselio Schaechter, Philip C. Hanawalt, and Conrad L. Woldringh at the EMBO Workshop on "Bacterial Cell Biophysics" in Israel, December 2022. The contributions of these individuals are fundamental to our current understanding of the bacterial cell.

Science relies on history: new discoveries are based on earlier ones (as stated by Newton's "if I have seen further, it is by standing on the shoulders of giants"). All microbiologists will benefit from reading this Special Issue, which contains articles from several of the field's founders. These articles include those from Helmstetter and the prize laureates, along with those from Donachie, Nanninga, Kohiyama, Zaritsky, Leonard, Janniere, Liu, and their collaborators. The first decades of the "Copenhagen School" are described by Schaechter and Hanawalt; Helmstetter, Donachie, and Zaritsky follow suit; Nanninga and Woldringh outline the origins and roots of the so-called "Amsterdam School"; Leonard gives an insider's insight into the Helmstetter story; Kohiyama et al. and Holland et al. provide a French flavour, whilst Cao et al. exemplify the increasing importance of young Chinese scientists in the cell cycle field.

For funding, organization, and encouragement, we thank the participants in the EMBO Workshop, EMBO, and MDPI.

Vic Norris and Arieh Zaritsky
Editors

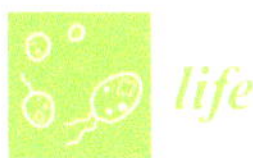

Editorial

Novel Principles and Methods in Bacterial Cell Cycle Physiology: Celebrating the Charles E. Helmstetter Prize in 2022

Vic Norris [1,*] and Arieh Zaritsky [2]

[1] Laboratory of Bacterial Communication and Anti-infection Strategies, EA 4312, University of Rouen, 76000 Rouen, France

[2] Life Sciences Department, Faculty of Natural Sciences, Ben-Gurion University of the Negev, Kiryat Bergman, HaShalom St. 1, Be'er-Sheva 8410501, Israel; ariehzar@gmail.com

* Correspondence: victor.norris@univ-rouen.fr

Citation: Norris, V.; Zaritsky, A. Novel Principles and Methods in Bacterial Cell Cycle Physiology: Celebrating the Charles E. Helmstetter Prize in 2022. *Life* **2023**, *13*, 2260. https://doi.org/10.3390/life13122260

Received: 25 October 2023
Accepted: 23 November 2023
Published: 27 November 2023

This Special Issue celebrates the creation of the Charles E. Helmstetter Prize for Groundbreaking Research in Bacterial Cell Cycle Physiology. The Prize was inaugurated at the EMBO Workshop "Bacterial cell biophysics: DNA replication, growth, division, size and shape", which was held in Ein Gedi, Israel, https://meetings.embo.org/event/22-bacteria-biophysics, accessed on 11–15 December 2022.

Understanding the cell cycle is fundamental to understanding the physiology of the bacterial cell and to all those fields that depend on it, ranging from clinical microbiology to industrial biotechnology, from microbial endocrinology to immunology, from the origins of life to xenobiology, and from environmental microbiology to synthetic biology. To take just one example, the cell cycle events of DNA replication and cell division underpin many approaches to dealing with the ever-growing threat of antibiotic resistance, including the phenotypic diversity that can lead to persister cells [1,2].

Our current understanding of the bacterial cell cycle owes a great deal to Charles Helmstetter, who has shaped the way microbiologists think about the bacterial cell for over fifty years. He has proved exceptionally gifted in formulating hypotheses for solving fundamental problems, in developing the experimental techniques to test them, in designing critical experiments, and in deriving paradigm-shifting ideas. In the early 1960s, well before the development of modern single-cell techniques, the main method of studying the bacterial cell cycle was through producing populations of cells belonging presumably to the same stage of the cell cycle—so-called synchronous cultures. The induction or selection methods used to produce such populations significantly perturbed the physiology of the cells and made it difficult to distinguish between real and artefactual changes in the cell cycle. Helmstetter's primary motivation was to solve this problem. After several years of making and correcting errors, trying different approaches, and never giving up, he eventually invented the famous "Baby Machine"—a simple device continuously producing newborn bacterial cells at the earliest stage of the division cycle (most importantly, with minimal environmental changes) (memorialized in [3]). Years later, this method was adapted for eukaryotic cells (see below), and was complemented by the sophisticated "Mother Machine", which combines microfluidics with fluorescence microscopy and single-cell analyses with big number statistics [4].

The Baby Machine possesses a feature that makes synchronous cultures unnecessary: the baby cells sequentially eluted from the machine are the descendants of cells in the parental culture in reverse age order, i.e., the first baby cells eluted are the products of the division of the oldest cells in the population. Using this 'backwards' method, the parental batch culture is pulse-labeled with radioactive thymidine before being transferred to the filter in the Machine; the radioactivity found in the eluted baby cells allows for the relative rate of DNA synthesis to be traced back to the division cycles of the cells growing in the unperturbed, steady state culture. The times of DNA replication's initiation and termination can then be deduced from the stepwise rise or fall in the rate of incorporated thymidine, respectively. The results at different growth rates were puzzling; whilst the time between the termination of replication and subsequent cell division was relatively consistent, the time between initiation and termination was highly variable: initiation could occur at any stage of the cycle—even after termination! These findings were impossible to reconcile with the G1-S-G2 m sequence of the eukaryotic cell cycle paradigm.

This apparent paradox was resolved by Helmstetter and Steve Cooper in 1968 by their paradigm-shifting proposal of the $I + C + D$ sequence for the bacterial cell cycle. Here, C is the chromosome replication time, D is the time between termination of replication and completion of division, and I is the time between consecutive initiations (equal to, but not coinciding with, the inter-division time, τ). Three major processes are responsible for a newborn cell growing and dividing to yield two daughters: the continuous increase in biomass (i.e., growth) and the discrete, cell cycle processes of chromosome replication (which is bounded by initiation and termination) and cell division. Helmstetter and Cooper considered these cell cycle processes to be highly coordinated and to overlap, albeit partly independently. Their proposed model solved the paradox of how a dynamic steady-state growing population can maintain a constant size distribution and yet have a mass doubling time τ that is shorter than the time it takes a cell to replicate its chromosome and then divide ($I < C + D$). Furthermore, if cells divide at the same rate as their mass increases, but C is longer ($I = \tau < C$), daughters that lack DNA may be created. These results led them to conclude that rounds of replication can also overlap so that the inter-initiation time I is shorter than C, consistent with the observed multi-forked replication. Once termination occurs, there are effectively two separate termini, i.e., chromosomes whose replication continues. Hence, these two structures in which replication is ongoing can be separated and segregated into the daughter cells, a fundamentally different picture from what happens in eukaryotes. The main characteristics of the bacterial cell cycle can be understood and predicted quite correctly for a wide range of values of C, D and I ($=\tau$), as well as for the various transitions between them, as is clarified by the $I + C + D$-based simulation program [5].

A few examples of subsequent work on the cell cycle by Helmstetter and his collaborators are briefly summarized here. In 1986, Leonard and Helmstetter reported that *oriC* plasmids (mini-chromosomes) initiated replication in coordination with the chromosome, irrespective of their copy number. With Olga Pierucci, he obtained the first evidence that chromosome segregation is not random. He and Leonard proposed an explanation for the observed non-random segregation pattern based on a mechanism that distinguishes between template strands of different ages. In 1997, he and Bogan found that the transcription of genes in the vicinity of *oriC*, namely *mioC*, *gidA* and *dnaA*, was affected by sequestration at the level of initiation rather than of elongation. Helmstetter and collaborators also found that transcription of several genes involved in the cell cycle, *ftsZ*, *dam*, *nrdA*, *mukB* and *seqA*, was reduced at a certain stage during chromosome replication, apparently coincident with the time the genes replicated. As part of a series of studies on plasmid replication, he and Leonard showed that F plasmid replication is not confined to a period of the cell cycle. In 1992, he and Zaritsky found that division was delayed after a nutritional shift down, and then its rate maintained for an extended period; these studies were followed up by other collaborators. With Thornton and Edward, he adapted the Baby Machine technique for cell cycle studies on hematopoietic mouse and human cells, and with collaborators, analyzed transcriptomics of cyclins and PCNA. Finally, in 2008, he used BrdU incorporation to study time zones of replication in the entire mouse genome.

It is well known that just as the importance of a prize enhances the reputation of its recipients, so the reputation of its recipients enhances the reputation of the prize. This is indeed the case for the Charles E. Helmstetter Prize, the first awardees of which are three distinguished specialists: Philip C. Hanawalt, Elio Schaechter, and Conrad L. Woldringh. Had the prize been named after someone else, Helmstetter himself would have been a prime recipient! We strongly advise those interested in the cell cycle to listen to their videos and read their papers in this Special Issue. Moreover, many of the other papers in this issue describe the contributions made by those who can fairly be said to have founded the bacterial physiology and the cell cycle field as we now know it (Figure 1).

Figure 1. Several scientists, among the founders of bacterial physiology and the cell cycle, at the Cold Spring Harbor Symposium on Quantitative Biology, "Ole Maaløe with Alumni", 1968. Front Row (sitting, Left-To-Right): KG Lark, M Schaechter, S Cooper, O Maaløe, Back Row: PC Hanawalt, DJ Clark, C Levinthal, J Watson *, P Kuempel, CE Helmstetter, D Glaser **. Nobel Prize Laureates in: * Physiology/Medicine (1962); ** Physics (1960) (From the personal collection of Philip C Hanawalt; https://web.stanford.edu/~hanawalt/), accessed on 22 November 2023. **"If I have seen further, it is by standing on the shoulders of giants"**—Sir Isaac Newton.

The study of DNA repair and the cell cycle in bacteria has direct implications for human health. Hanawalt details his discovery that the bacterial DNA replication cycle can be synchronized by temporarily inhibiting RNA and protein synthesis, and that an RNA synthesis step is required for the initiation but not for the completion of chromosome replication. His earlier graduate studies on the effects of UV on macromolecular synthesis in bacteria led to his discovery of repair replication and the co-discovery of excision-repair, which requires an undamaged complementary DNA strand. He postulates that excision-repair was essential for the persistence of life from its earliest forms. In summary, this paper offers insight into how fundamental discoveries can be made with relatively simple bacterial systems.

Schaechter [6] reminisces about his time in the laboratory of Ole Maaløe and his experience of the 'Copenhagen School', which was characterized by systematic approaches and careful measurements. This led to the discovery of the direct relationship between cell size and growth rate afforded by medium composition, and the constant rates of macromolecular syntheses in all media (at a given temperature). Back in America, he continued to make major discoveries about polysomes, and about the attachment of the chromosome to the membrane. Schaechter's discoveries have structured microbiology.

Woldringh [7] invokes polymer physics to help answer the question as to how bacteria segregate their chromosomes; in his hypothesis, the newly synthesized daughter strands, which are proposed to exist already as separate blobs in the early replication bubble, expand into large domains of the left and right chromosome arms, flanking the origin. One attractive advantage of this idea is that segregation only requires de novo DNA synthesis (though other players, like SMC-like proteins, are not excluded).

Donachie [8] asks "Why are *Escherichia coli* cells the size and shape that they are?" This is basically the question that he himself was asked by the late Kurt Nordström, many years ago. His paper argues that cell size and the geometry of cell growth are adapted to fit a growth rate-independent constant: the separation distance ("Unit Length") between sister nucleoids after the completion of each round of chromosome replication. He also suggests that (in *E. coli* and related Gram-negative rods) the mechanism of separation of sister chromosomes may be set by physics, not primarily by genes.

The metabolic control of replication couples the initiation and elongation phases of DNA replication to growth rate affected by nutrient richness; in their paper, Holland et al. provide evidence that this control, which is fundamental to genetic stability, depends on the dynamic recruitment of the glycolytic enzyme PykA by DnaE at sites of DNA synthesis.

Nanninga [9] gives his recollections of the origins of electron microscopy at Amsterdam University, and the importance of freeze-fracturing. He relates how the study of the cell cycle used electron microscopy to analyze bacteria grown in a steady state, and to show that the zonal growth of the envelope could not be responsible for the segregation of envelope-attached DNA. The combination of bacterial physiology, electron microscopy, and image analysis has been termed the Amsterdam School by Arthur Koch on several occasions. This work, among others, led to the idea that PBP3-independent peptidoglycan synthesis preceded a PBP3-dependent step. His experience with the 'mesosome' is instructive, and his review should be obligatory reading for those interested in cell division. Nanninga briefly describes the pioneering efforts to construct a confocal scanning light microscope in Amsterdam.

Kohiyama [10] tries, in his words, "to throw a cobblestone in the pond" (from the French 'jeter un pavé dans la mare'). He and his colleagues revisit the DnaA story from the point of view of hyperstructures; they propose the existence of a physico-chemical clock that simultaneously triggers the initiation of chromosome replication and of cell division.

Leonard [11] recounts his experience as a postdoctoral fellow with Helmstetter, focusing on construction of mini-chromosomes and using the 'backwards' baby machine to study their replication timing. He describes how his trials and tribulations eventually led to important discoveries about how mini-chromosomes replicate in synchrony with the host origin, how they segregate into daughter cells, and the important role of DNA supercoiling in mini-chromosome function. He discusses his experiences studying plasmid replication during the cell cycle.

Liu and his group [12] show the disconcerting extent to which the type of software and the values of settings can affect the quantification of cell size parameters using microscopic images. They argue persuasively that microscope-independent methods should be used to validate conclusions and provide a precious illustration in the case of the initiation mass. This caveat-focused paper should be read by all those interested in the cell cycle.

Finally, Zaritsky's review [13] summarizes his own contributions to the field, mostly related to the demonstration that DNA replication rate can be manipulated in thymine auxotrophic mutants by the external thymine concentration supplied, thus dissociating it from the rate of cell mass growth and duplication. This quantitative description complements Helmstetter's findings that in thymine prototrophs the replication time C is constant under a wide range of doubling times τ, as determined by the medium composition. This seemingly simple physiological manipulation enabled Zaritsky to discover the existence of Eclipse, namely a minimal distance possible between successive replisomes, which during thymine limitation causes the otherwise constant mass at initiation to gradually increase. Furthermore, cell width was found to be tightly related to nucleoid complexity (*NC*), which is the amount of DNA in genome equivalents associated per *terC* (i.e., the chromosome terminus). *NC* depends only on the number of replication positions n, where $n = C/\tau$, and can therefore be varied by changing τ or C.

Author Contributions: V.N. and A.Z.; writing—original draft preparation. All authors have read and agreed to the published version of the manuscript.

Funding: This investigation was supported by grants (to A.Z.) from the U.S.-Israel Binational Science Foundation (BSF, No. 2017004) and the ISF-NSFC joint research program (No. 3320/20), and facilities by BGU administration.

Conflicts of Interest: The authors declare no conflict of interest.

References

1. Balaban, N.Q.; Merrin, J.; Chait, R.; Kowalik, L.; Leibler, S. Bacterial Persistence as a Phenotypic Switch. *Science* **2004**, *305*, 1622–1625. [CrossRef]
2. Norris, V.; Kayser, C.; Muskhelishvili, G.; Konto-Ghiorghi, Y. The roles of nucleoid-associated proteins and topoisomerases in chromosome structure, strand segregation, and the generation of phenotypic heterogeneity in bacteria. *FEMS Microbiol. Rev.* **2022**. [CrossRef]
3. Helmstetter, C.E. A ten-year search for synchronous cells: Obstacles, solutions, and practical applications. *Front. Microbiol.* **2015**, *6*, 238. [CrossRef]
4. Wang, P.; Robert, L.; Pelletier, J.; Dang, W.L.; Taddei, F.; Wright, A.; Jun, S. Robust Growth of *Escherichia coli. Curr. Biol.* **2010**, *20*, 1099–1103. [CrossRef] [PubMed]
5. Zaritsky, A.; Wang, P.; Vischer, N.O.E. Instructive simulation of the bacterial cell division cycle. *Microbiology* **2011**, *157*, 1876–1885. [CrossRef] [PubMed]
6. Schaechter, M. The Birth of the Copenhagen School: Personal Recollections at the EMBO Workshop on Bacterial Growth Physiology. *Life* **2023**, *13*, 2235. [CrossRef]
7. Woldringh, C.L. The Bacterial Nucleoid: From Electron Microscopy to Polymer Physics—A Personal Recollection. *Life* **2023**, *13*, 895. [CrossRef]
8. Donachie, W.D. The Nordström Question. *Life* **2023**, *13*, 1442. [CrossRef] [PubMed]
9. Nanninga, N. Molecular Cytology of 'Little Animals': Personal Recollections of *Escherichia coli* (and *Bacillus subtilis*). *Life* **2023**, *13*, 1782. [CrossRef] [PubMed]
10. Kohiyama, M.; Herrick, J.; Norris, V. Open Questions about the Roles of DnaA, Related Proteins, and Hyperstructure Dynamics in the Cell Cycle. *Life* **2023**, *13*, 1890. [CrossRef] [PubMed]
11. Leonard, A.C. Recollections of a Helmstetter Disciple. *Life* **2023**, *13*, 1114. [CrossRef] [PubMed]
12. Cao, Q.; Huang, W.; Zhang, Z.; Chu, P.; Wei, T.; Zheng, H.; Liu, C. The Quantification of Bacterial Cell Size: Discrepancies Arise from Varied Quantification Methods. *Life* **2023**, *13*, 1246. [CrossRef] [PubMed]
13. Zaritsky, A. Extending Validity of the Bacterial Cell Cycle Model through Thymine Limitation: A Personal View. *Life* **2023**, *13*, 906. [CrossRef] [PubMed]

Editorial

The Birth of the Copenhagen School: Personal Recollections at the EMBO Workshop on Bacterial Growth Physiology, 2022

Moselio Schaechter

Division of Biological Sciences, University of California, San Diego, CA 92093, USA; elios179@gmail.com

Citation: Schaechter, M. The Birth of the Copenhagen School: Personal Recollections at the EMBO Workshop on Bacterial Growth Physiology, 2022. *Life* **2023**, *13*, 2235. https://doi.org/10.3390/life13122235

Received: 6 September 2023
Accepted: 3 November 2023
Published: 21 November 2023

The future of bacterial growth physiology is shining more brightly than ever. It is profiting from our increased ability to study individual bacteria, from advances in different forms of microscopy, and from the increased use of computational modeling. Bacterial physiology is merging with systems biology (a term that is perhaps less in vogue) and is now increasingly influenced by physical chemistry.

Our knowledge of bacteria continues to grow as we learn more and more about the interactions in living cells between and among micro- and macro- molecules, the happenings during the cell cycle, the variability of what were once thought to be homogeneous populations, and other such alluring subjects. These advances are due in no small measure to the return to biology of physicists, who introduce new ways of thinking via new technologies and new approaches to modeling and simulation.

That being said, the old, fundamental questions remain. Why do bacteria such as *Escherichia coli* have so many regulatory mechanisms that overlap? What are the underlying rules? How is it possible that in bacteria (as opposed to eukaryotes), the major macromolecular factories can, to some extent, operate independently of one another? In the broadest terms, questions related to morphology, chromosome replication, segregation, and cell division—and the couplings between them—are being posed in new ways, for example, in terms of polymer physics. This Special Issue follows the EMBO workshop in Israel "Bacterial cell biophysics: DNA replication, growth, division, size and shape" (2022) at which The Charles E. Helmstetter Prize for Groundbreaking Research in Bacterial Cell Cycle Physiology was inaugurated. I am a recipient of this prize, and am certain that had I been at the meeting, I would have found that the number of fascinating topics is increasing enormously, and that excitement in the field continues to mount.

The Birth of "The Copenhagen School"

I should know. I was there. But first, I arrived in Copenhagen in 1956 after a two-year stint at the Walter Reed Army Research Institute. Why Copenhagen and why Ole Maaløe's lab? A couple of good reasons. My wife at the time, Barbara, had taken a college course in Scandinavian Civilization and fell in love with both the topic and the teacher. Also, at the time, I found my barracks-mate, the soon-to-become famous phage geneticist Allan Campbell, sprawled on his Army bunk, reading a paper by Lark and Maaløe on synchronizing bacterial growth. He proclaimed aloud that "everything that has been done has to be done over". That was good enough for me although, uncharacteristically, Allan was wrong on this one. The reason, as was found out later in Ole's own lab, was that the temperature shifts used by Lark and Maaløe to induce synchronous divisions created artefacts. But I did not know this at the time, so I went there undeterred.

Ole had just returned from a year at Caltech, where he had been immersed in "Delbrückian" molecular biology. There, he published several papers on phage development with Jim Watson and Gunther Stent. Before that, Ole had attended medical school in Copenhagen. Upon his return to Denmark, he took a job in the department for the standardization of sera and vaccines of the State Serum Institute. As it turned out, this was a rather undemanding position. The nominal work was done almost entirely by Ole's assistant, the highly gifted Jens Ole Rostock. This left Ole with plenty of time to develop his interest in molecular biology. In time, Ole's lab (soon at the University of Copenhagen) became prominent, a place at which Americans and others could carry out interesting studies in molecular biology. Eventually, this led to the creation of the Copenhagen School of Bacterial Growth Physiology.

A few words about Ole as a person. In general terms, he was very much a Dane: somewhat reserved, yet sensitive to the needs for the well-being of his fellow men. I believe that this is indeed common in Denmark, a country which, by the way, rates No. 2 in the world in personal happiness. Ole had a special interest in oriental, read a lot about them, and collected mandalas. He was seldom seen without a cigar in his mouth which he temporarily removed, reluctantly, when mouth pipetting.

The birth of the Copenhagen School can be traced to the following modest event: Ole was looking for a way to detect synchronous growth faster than by counting cells via colony plate counts (which takes about a day). He had just acquired a new gadget, a Zeiss PMQII spectrophotometer. He imagined that with such a highly sensitive apparatus, he could detect small jumps in the optical density of a culture undergoing synchronous growth. But when we "shifted up" a culture in minimal medium by adding a rich medium to it, we saw no special jumps, just a steady increase in the growth rate. Abandoning his thoughts of synchronous division, Ole proposed that the speed at which bacteria grow may dictate their size (at the time, such thoughts were relegated to the messy corners of bacteriology). He had his post-doctoral students, Niels Ole Kjeldgaard and I, to test this notion by growing the bacterium *Salmonella typhimurium* in media of different richnesses, from a salt mixture with a single carbon source to a very rich broth. (By the way, with Ole, Jens Ole, and Niels Ole on board, I thought I should change my name to Elio Ole).

We were blessed by the Serum Institute, which had a most accommodating media kitchen. Whatever media we ordered, simple or complex, we received within 24 h. So, day after day, we grew our bacteria in a variety of different media. We expressed cell mass, ergo, cell size, as the ratio of the optical density of the culture vs. the number of cells (obtained via colony counts after plating). Sure enough, we soon found that Ole was correct: the faster the growth rate afforded by the medium, the larger the cells, in a direct relationship. Having measured the cells' content of nucleic acids, we could postulate that at a given temperature, the unit rate of macromolecular synthesis was constant in all media, something that was later demonstrated via direct measurements. Although this point was hinted at by previous work, ours was rendered more believable by our systematic approach and the detailed care with which we made the measurements. Additionally, Ole found convincing ways of making this business sound important.

I left Ole's lab in 1958 for my first academic job (at the University of Florida in Gainesville). Shortly thereafter, Ole was named the first professor of Microbiology at the University of Copenhagen. With the help of Jens Ole Rostock, he built up an exceptional research institution that attracted a large number of first-rate local and international investigators. Ole received many recognitions as a leading light in Danish science. A street in Copenhagen was named after him (Figure 1, from Anderson KB et al. [1]). He died in 1988.

Figure 1. Street sign (Ole Maaløe s' Way) and Ole's grandson.

My "After Copenhagen"

So, what did I do next? With a lab of my own and total freedom of action, should I follow up on the work in Copenhagen or try something different? I first tried a bit of both, but soon settled on the latter. Influenced by the developments in eukaryotic cell biology depending on cell fractionation techniques, I wanted to see what I would find when I opened up my bacterial cells. The conventional way to do this involves rather punitive methods, such as mangling the cells via sonication and grinding them with alumina. A gentle way to do it is to convert the bacteria into spheroplasts, cell-wall-less entities that can be lysed using a whiff of detergent. Sure enough, when we did this, we found that the ribosomes of the cell were linked together on nascent messenger RNA molecules in what soon became known as polyribosomes or polysomes. We visualized them using two methods, sucrose gradients and an analytical ultracentrifuge. This was a venerable and lumbering machine known as the Beckman Model E which, at that time, was the pride of any department that owned one (go see if you can find one anywhere now).

An unexpected event led us to study the attachment of DNA to the cell membrane. To lyse our spheroplasts, we decided to try a new detergent, sodium lauryl sarcosinate. To our surprise, after we centrifuged the gradients, we saw a thin white band about halfway down the tubes. This turned out to consist of the insoluble crystals of magnesium (present in our buffer) lauryl sarcosinate. Somehow, instead of abandoning the experiment, we analyzed the contents of the tubes and, to our surprise, found that all of the cells' DNA was in the white band, which was nicknamed the "M-band". About 30% of the cells' membranes were in the white band as well, but purified DNA was not. We reckoned that the DNA we found in this M-band was attached to the membrane. Probing further into it, we determined that the DNA is attached to the membrane at some 20 attachment sites.

At about that time, I closed my lab and moved from Boston to San Diego, where I have since been using social media to spread the microbiological word via a blog entitled "Small Things Considered" at http://schaechter.asmblog.org/ (accessed on 6 September 2023) or http://smallthingsconsidered.us (accessed on 6 September 2023) and a podcast, "This Week in Microbiology" (TWiM), at https://podcasts.google.com/search/%22This%20Week%20in%20Microbiology%22?hl=en-IL (accessed on 6 September 2023). A few selected publications relevant to this story are cited below [2–17].

Conflicts of Interest: The author declares no conflict of interest.

References

1. Anderson, K.B.; Atlung, T.; Bennett, P.M.; Cooper, S.; Dennis, P.; Diderichsen, B.; Edlin, G.; Eisenstark, A.; Fiil, N.; Friesen, J.; et al. Honoring Ole Maaløe. *Microbe* **2006**, *1*, 210–211. [CrossRef]
2. Schaechter, M.; Treece, E.L.; DeLamater, E.D. Studies on the cytochemistry of alkaline phosphatase in various bacteria. *Exp. Cell Res.* **1954**, *6*, 361–366. [CrossRef] [PubMed]
3. Schaechter, M.; Bozeman, F.M.; Smadel, J.E. Study on the growth of rickettsiae: II. Morphologic observations of living rickettsiae in tissue culture cells. *Virology* **1957**, *3*, 160–172. [CrossRef] [PubMed]
4. Schaechter, M.; Maaløe, O.; Kjeldgaard, N.O. Dependency on Medium and Temperature of Cell Size and Chemical Composition during Balanced Growth of *Salmonella typhimurium*. *Microbiology* **1958**, *19*, 592–606. [CrossRef] [PubMed]
5. Kjeldgaard, N.O.; Maaløe, O.; Schaechter, M. The Transition Between Different Physiological States During Balanced Growth of *Salmonella typhimurium*. *Microbiology* **1958**, *19*, 607–616. [CrossRef] [PubMed]
6. Schaechter, M.; Bentzon, M.W.; Maaløe, O. Synthesis of deoxyribonucleic acid during the division cycle of bacteria. *Nature* **1959**, *183*, 1207–1208. [CrossRef] [PubMed]
7. Hanawalt, P.C.; Maaløe, O.; Cummings, D.J.; Schaechter, M. The normal DNA replication cycle. II. *J. Mol. Biol.* **1961**, *3*, 156–165. [CrossRef] [PubMed]
8. Schaechter, M.; Williamson, J.P.; Hood, J.R.; Koch, A.L. Growth, cell and nuclear divisions in some bacteria. *Microbiology* **1962**, *29*, 421–434. [CrossRef] [PubMed]
9. Koch, A.L.; Schaechter, M. A model for statistics of the cell division process. *Microbiology* **1962**, *29*, 435–454. [CrossRef] [PubMed]
10. Earhart, C.F.; Tremblay, G.Y.; Daniels, M.J.; Schaechter, M. DNA replication studied by a new method for the isolation of cell membrane-DNA complexes. *Cold Spring Harb. Symp. Quant. Biol.* **1968**, *33*, 707–710. [CrossRef] [PubMed]
11. Tremblay, G.Y.; Daniels, M.J.; Schaechter, M. Isolation of a cell membrane-DNA-nascent RNA complex from bacteria. *J. Mol. Biol.* **1969**, *40*, 65–76. [CrossRef] [PubMed]
12. Dworsky, P.; Schaechter, M. Effect of Rifampin on the Structure and Membrane Attachment of the Nucleoid of *Escherichia coli*. *J. Bacteriol.* **1973**, *116*, 1364–1374. [CrossRef] [PubMed]
13. Slater, M.; Schaechter, M. Control of cell division in bacteria. *Bacteriol. Rev.* **1974**, *38*, 199–221. [CrossRef] [PubMed]
14. Leibowitz, P.J.; Schaechter, M. The attachment of the bacterial chromosome to the cell membrane. *Int. Rev. Cytol.* **1975**, *41*, 1–28. [PubMed]
15. Abe, M.; Brown, C.; Hendrickson, W.G.; Boyd, D.H.; Clifford, P.; Cote, R.H.; Schaechter, M. Release of *Escherichia coli* DNA from membrane complexes by single-strand endonucleases. *Proc. Nat. Acad. Sci. USA* **1977**, *74*, 2756–2760. [CrossRef]
16. Ogden, G.B.; Pratt, M.J.; Schaechter, M. The replicative origin of the *E. coli* chromosome binds to cell membranes only when hemi methylated. *Cell* **1988**, *54*, 127–135. [CrossRef]
17. Samitt, C.E.; Hansen, F.G.; Miller, J.F.; Schaechter, M. In vivo studies of DnaA binding to the origin of replication of *Escherichia coli*. *EMBO J.* **1989**, *8*, 989–993. [CrossRef] [PubMed]

Review

Fifty-Five Years of Research on *B*, *C* and *D* in *Escherichia coli*

Charles E. Helmstetter

Department of Biomedical and Chemical Engineering and Sciences, Florida Institute of Technology, Melbourne, FL 32901, USA; chelmste@fit.edu

Abstract: The basic properties of the *Escherichia coli* duplication process can be defined by two time periods: *C*, the time for a round of chromosome replication, and *D*, the time between the end of a round of replication and cell division. Given the durations of these periods, the pattern of chromosome replication during the cell cycle can be determined for cells growing with any doubling time. In the 55 years since these parameters were identified, there have been numerous investigations into their durations and into the elements that determine their initiations. In this review, I discuss the history of our involvement in these studies from the very beginning, some of what has been learned over the years by measuring the durations of *C* and *D*, and what might be learned with additional investigations.

Keywords: cell cycle; synchronous cells; chromosome replication; initiation age; *B* period; *C* period; *D* period

Citation: Helmstetter, C.E. Fifty-Five Years of Research on *B*, *C* and *D* in *Escherichia coli*. *Life* **2023**, *13*, 977. https://doi.org/10.3390/life13040977

Academic Editors: Vic Norris, Uri Gophna, Ron Elber, Itzhak Fishov and Arieh Zaritsky

Received: 7 March 2023
Revised: 31 March 2023
Accepted: 6 April 2023
Published: 10 April 2023

1. Introduction

Upon completion of the requirements for a PhD in biophysics at the University of Chicago in 1961, I had the honor to have the diploma handed to me by George Wells Beadle, the newly installed president of the university, recent Nobel laureate and one of the first acclaimed researchers in molecular genetics. The work of Beadle and Edward Tatum, and their demonstration of the one gene–one enzyme hypothesis [1], was a prominent topic in genetics courses at the time. Since that time, I have had the opportunity to read several essays by and about George Beadle and have, as a consequence, become interested in his approach to the study of science. As is described below, my scientific journey followed a similar path, although with a decidedly less consequential outcome. I suspect that many of us may have followed similar paths during the course of our work. However, before discussing that issue, another aspect of Beadle's thinking about science should be mentioned. He began his contribution to Phage and the Origins of Molecular Biology [2] by writing: "I have often thought how much more interesting science would be if those who created it told how it really happened, rather than reported it logically and impersonally, as they often do in scientific papers. This is not easy, because of normal modesty and reticence, reluctance to tell the whole truth, and protective tendencies toward others". I have attempted to follow his suggestion in this personal review.

My primary interest in Beadle's work concerns the story behind his long-term effort to investigate the gene–enzyme paradigm [3–5] due to its correspondence with our similar search for a way to determine the relationship between chromosome replication and cell division in *E. coli*. He and Tatum studied the fruit fly, *Drosophila*, in an attempt to understand the gene–enzyme relationship by searching for enzymatic reactions controlled by known genes. This continued for about five years without success. Then, one day, while listening to a lecture by Tatum, he apparently suddenly realized that the best and easiest approach was to conduct the experiments in reverse, that is, to search for genes that controlled known chemical reactions. He realized that experiments of this type could be conducted quite easily by X- or UV-irradiating *Neurospora*, with which he was already familiar, and then by simply selecting and characterizing those mutants that could no longer grow on a minimal medium.

It only took 5 months to obtain the first set of nutritional mutants. It was not long before hundreds of mutants with requirements for specific nutrients were generated, thereby producing evidence for the "one gene–one enzyme" hypothesis. Recently, Strauss [5] suggested that the true significance of this work by Beadle and Tatum may be overlooked. They were awarded their Nobel Prize "for their discovery that genes act by regulating definite chemical events". However, that finding was made possible by Beadle's simple, but inspired idea to mutagenize, and then select for a required nutrient. It was a revolutionary approach to experimental biology that many in our field have used successfully for decades. As I explain later, our work to identify the *E. coli* replication–division relationship also stemmed directly from a comment made by a colleague, followed by a decision to conduct cell cycle experiments in reverse order.

2. Synchronous Cell Growth

2.1. Bacterial Cells

My interest in the cell cycle began while I was a graduate student in the mid-1950s, not long after the report by Howard and Pelc defining the stages of the eukaryotic cell cycle was published [6]. I started by performing some radiobiological studies on newt heart cells in mitosis. In an adjacent laboratory, Aaron Novick and Leo Szilard continued to develop and study chemostats [7]. Novick, who had recently returned from the Pasteur Institute, encouraged me to read the magnificent work that had been published by researchers at the Pasteur Institute at that time (e.g., [8,9]). They were the most exciting papers I had read at that young age, and I still feel that way. Between reading about that work and the excitement in the Novick–Szilard lab, I decided to switch my focus to the bacterial cell cycle and study something related to DNA. Based on the thinking at the time, it appeared that the only way to study the cycle was to obtain synchronously dividing cultures. The only bacteria with which I had any familiarity were the strains being used by Novick and Szilard, primarily *E. coli* strain B, and occasionally, B/r. After looking at their cells using a microscope, strain B did not appear to be suitable for synchronization since the size distribution was broad, with many filamentous cells. Strain B/r however looked very promising since the size distribution was narrower, there did not appear to be any filaments, and they were nonmotile. I thought I might be able to synchronize the division of that strain, so I asked my major professor at the time, Robert Uretz, to purchase *E. coli* B/r 12407 from the American Type Culture Collection. That strain was later called B/r A.

Back then, there were basically three different approaches to bacterial synchronization: single or multiple temperature shifts, single or multiple nutritional deprivations, and size selection by filtration or centrifugation [10]. All of these methods were found to cause some growth disturbances, as assessed by the requirements that the cells undergo at least two cycles of detectably synchrony, that whatever is measured in the first cycle repeats in the second cycle, and that the fundamental properties of synchronous cells, such as sizes and growth rates, mimic the initial exponential phase population. The method that appeared to cause the least disturbance involved filtration of a culture through a stack of Whatman cellulose filter papers, which enabled the smaller newborn cells to pass through into the effluent, while retaining the larger, older cells in the stack [11]. A modification of this technique, involving the use of pressure rather than vacuum filtration, enabling me to perform a few simple experiments on *E. coli* B/r A cells at very low concentrations.

Everything changed in 1963 during a meeting of the Biophysical Society in New York. One afternoon, I was involved in a discussion with a small group talking about synchronizing cells. I believe Philip Hanawalt was among the group and am certain another participant was David Friefelder because he asked me a question that completely changed the course of the work. After I described the technique I was using, Friefelder asked a number of questions including how long filtration took. I said, "A few minutes". He then said, as I recall, "Well then, the cells must be growing while in the filter stack". That turned out to be the comment that led to the eventual generation of a very simple method to determine the DNA–cell cycle relationship, which did not involve synchronizing cells at all.

The idea was that if growing cells became attached to a substrate while a culture medium passed through, the only cells that could possibly be released from the substrate would be newborn cells originating from the portion of the bound cells that was not involved in their original attachment. Some of the dividing cells might not release a progeny at all, or some progenies might reattach, but in the ideal case in which all attachments are permanent, only newborns will be released. Realistically, strong attachment plus good flushing ought to yield a highly pure population of newborn cells. The best part is that this process might be able to yield a minimally disturbed, synchronously dividing population because it simply involved collecting and incubating cells released from a culture growing under ideal conditions. The theory seemed to work perfectly on the first try. The development and testing of this new approach rapidly progressed, with the final configuration consisting of filtering cells of the strain B/r A using a nitrocellulose membrane filter, inverting the filter, pumping the medium backwards through the filter, and collecting the cells that fell out [12]. It eventually became known as the "baby machine".

Although the device was clearly designed as a means to obtain minimally disturbed synchronous cells, this is not the application that was eventually used in most studies, including those that yielded information on replication–division coordination. In the very early experiments, nucleic acid synthesis during the cell cycle was analyzed by collecting newborn cells released from the instrument, growing them synchronously, and then pulse labeling them with radioactive precursors. That approach was laborious and resulted in cells that showed some evidence of growth disturbance. The growth disturbance turned out to be related to incorrect media preparation, which was soon corrected, but that issue forced me to rethink the whole idea of synchronous growth studies. Then, in 1966, I realized that cell cycle studies with the baby machine could best be conducted with minimal disturbance by performing the experiments in reverse. In this approach, the exponentially growing cells were pulse labeled just prior to attachment in the baby machine, and radioactivity of the newborn cells was determined as they were eluted. Each generation of newborn cells sequentially released during the growth and division of attached cells in the instrument were the daughters of the oldest cells, then the youngest cells in the original exponential phase culture, respectively. Thus, the radioactivity of newborn cells reflected the reversed incorporation during the cell cycle. From this point on, cell cycle studies were stunningly easy to perform and the data produced were very clear. After many years of trying, it only took a few weeks to determine the pattern of DNA replication during the cycle of slow-growing *E. coli* B/r A cells with that procedure using radioactive thymidine [13]. In essence, our work was based on a comment originally made by David Friefelder in 1963 plus the realization that we must conduct the experiments in the most effective, easiest manner.

The idea to use the baby machine in reverse stemmed directly from an experiment Steve Cooper and I conducted while we were both postdocs working with Ole Maaløe in Copenhagen in 1963–1964. We were interested in the process of chromosome segregation and decided that the baby machine technique would be an ideal method to determine if there were nonrandom aspects of chromosome segregation. So, we performed an experiment such as the type described above with radioactive thymidine and observed that the radioactivity per released newborn cell decreased essentially two-fold in each generation grown with the instrument, suggesting the random segregation of the labelled DNA strand between attached and released daughter cells. Cooper and I did not think our finding was very interesting at the time, and we both moved on; Cooper took up another postdoctoral position, and I took up a position in what is now the Roswell Park Comprehensive Cancer Center. Then, about two years later, upon belatedly arriving at the proper interpretation of an experiment of this type, I wrote to Cooper to tell him about it. He was euphoric, and it just happened that he was finishing a postdoctoral position and was moving to a new facility that was not quite ready for his arrival. As a result, we thought it would be a great idea for us to work together in my laboratory at Roswell Park for a while and investigate the cell cycle properties of rapidly growing cells. Our work progressed rapidly because it only involved loading the device with labelled cells every day and simply measuring the radioactivity of the cells that poured out. Resultant data showing radioactivity per newborn cell versus elution time required analysis [14–16], but even that task was simplified by the critical prior work of Schaechter et al. [17], which showed us how to conduct the experiments, and the findings by Sueoka and colleagues of the existence of multi-forked chromosome replication [18].

2.2. Mammalian Cells

Many years later, the baby machine concept was extended to mammalian cells. It was found to be applicable to cells that did not normally adhere to or spread on surfaces such as lymphoid cell lines [19]. The performance of the technique, in terms of both the purity of the released newborn cells and the longevity of their production, proved to be superior to the technique used for bacterial cells. In fact, it was possible to operate a rather complex mammalian cell baby machine for at least five generations with minimal degradation of the purity or concentration of newborn cells, thus achieving a nearly steady-state growth condition [20]. Unfortunately, mammalian cell cycle studies rarely employ techniques designed to produce minimally disturbed cells, such as the baby machine or mitotic shake-off [21]. Although the baby machine technique is very simple in principle, it can be labor-intensive. During operation, samples of newborn cells must be collected over lengthy periods of time before the actual experiment can begin.

Many mammalian cell cycle experiments employ an inhibitor to align cells at specific stages in the cycle [21]. These procedures have the advantage of being easier to perform, while producing considerably more aligned cells. A frequently used technique involves treatment with inhibitors of DNA replication, which are intended to align cells at the beginning of S phase, followed by the initiation of inhibition to produce synchronized growth. It has been suggested that this treatment produces synchronized cells that the reflect processes of the normal cell cycle [22]. That argument depends on the definition of S phase. If it is defined as merely the period of chromosomal DNA synthesis, such as the C period in *E. coli*, then the cells would certainly be aligned at the start of S and reflect the processes associated with that alignment. However, if the S phase is defined as a stage or interval of the cycle, then all aspects of that interval would not be aligned. Thus, if there were unique events in the cycle interval during which chromosome replication took place that were not governed, in some fashion, by chromosome replication itself, those events would not be aligned or detected. This is why Cooper and I decided to define the *E. coli* cell cycle in terms of the time periods C and D, rather than phases such as S and possibly G2 [15,16]. C and D can only be considered to define phase s in the cycle when the doubling time is equal to or longer than $(C + D)$.

3. Cell Age at the Initiation of Chromosome Replication

3.1. E. coli K-12 and B/r

The average durations of C and D, as well B, the time between cell division and the initiation of replication by the single chromosome in slow-growing cells, have been measured in numerous strains of *E. coli*. I have used some of these data to examine the timing of the initiation of chromosome replication as a function of generation time (τ). Although this may be an unusual method for analyzing the durations of cell cycle periods, it yields interesting information that might not be obvious otherwise. This was accomplished by calculating the "set number", defined as the generations between the start of a round of replication and the division after the end of that round [16], given by $(C + D)/\tau$. The set numbers were then used to determine average age at initiation (a_i) by subtracting the fractional part of the set number from 1.0 [16]. Figure 1A shows a collation of calculated set numbers and average ages at the initiation of replication for several K-12 and B/r strains grown in batch and chemostat cultures at 37 °C using data for C and D determined by flow cytometry [23–27]. As expected, the set numbers, and thus, the ages changed continuously throughout the period of rapid growth when the generation time was less than ($C + D$) for all strains, and a B period appeared when it was longer than ($C + D$). During very slow growth, there was minimal change in age at the initiation of replication. An interesting observation is that the initiation age immediately after $\tau = C + D$ varied considerably in different strains. ($C + D$) increased in all strains after τ became longer than the duration of ($C + D$) during rapid growth, but this occurred at different rates. In some strains, it increased in correlation with τ, such that the initiation age remained close to 0 for a time before a significant B period appeared (Figure 1B, inset). Ascribing any significance to the strain differences in durations of ($C + D$) requires additional data. Fortunately, some early studies with strain B/r are useful in this regard.

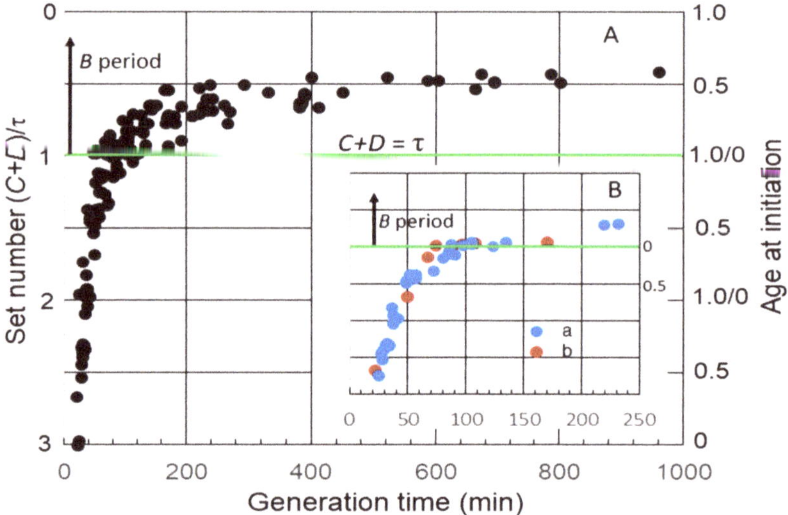

Figure 1. Set numbers and cell ages at the initiation of chromosome replication in *E. coli* growing with various generation times. (**A**) Collation of both set numbers and ages at the initiation of replication for seven strains of *E. coli* K-12 and three strains of B/r [23–27]. The green horizontal line in this and the subsequent figures shows age at initiation of replication when $C + D$ is equal to the generation time, i.e., the initiation of replication at division in daughter cells containing a single chromosome. At longer generation times, there is a gap between division and initiation, relating to the B period above the green line. (**B**) Age at the initiation of replication versus generation time for: a: K-12 NJ24 [26] and b: K-12 GM1655 [27].

3.2. Age at the Initiation of Chromosome Replication in E. coli B/r A, F and K

As indicated above, the baby machine technique was developed using *E. coli* B/r A as the experimental organism. Although we were eventually able to use K-12 strains in the technique [28], the original procedure functioned poorly with K-12. If I had not stumbled upon B/r A in the beginning, the technique might not have been developed, and our work might have taken an entirely different direction. For the first 15 years, all of our work was performed exclusively with B/r A, including the work Cooper and I published in 1968 [14–16]. Eventually two additional B/r strains were used, B/r F from Ole Maaløe via James Friesen, and B/r K from Herbert Kubitschek. My aim in the remaining sections of this review is to summarize some of the interesting findings reported over the years using strain B/r and to introduce the possibility that continuing investigations with this strain might prove to be enlightening.

The relationships between average ages at the initiation of chromosome replication and generation times for *E. coli* B/r A, F, and K observed in our experiments [29] are shown in Figure 2. *B* periods began to appear in strains B/r F and K when the generation time became equal to $(C + D)$ min, or shortly thereafter. On the other hand, a B period was not detected in B/r A, at least between the generation times of 65 and 120 min. During this interval, it appeared that age at initiation remained basically constant just before cell division, at or near the end of the D period, meaning that $(C + D)$ and τ were increasing at nearly the same rate.

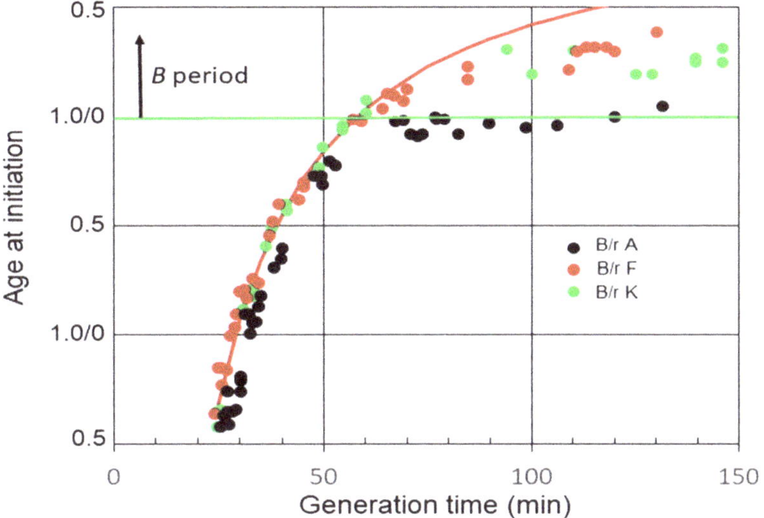

Figure 2. Cell age at the initiation of chromosome replication in *E. coli* B/r A, F, and K growing with various generation times. The red curve is theoretical age at the initiation of replication for B/r F, assuming $(C + D)$ is constant and equal to 58 min.

The average C period at 37 °C was found to be 42 min during rapid growth ($\tau \leq C + D$) for all three strains [29]. On the other hand, the D periods differed, with the average D periods equaling 22, 16, and 14 min for B/r A, F, and K, respectively. The red curve in the figure shows the theoretical age at the initiation of replication for B/r F if $(C + D)$ was constant at 58 min ($C = 42$ min; $D = 16$ min) at all the generation times. It is evident that the measured values for $(C + D)$ in B/r F were constant until the generation time equaled $(C + D)$. Thereafter, $(C + D)$ began to increase, since the measured initiation ages lie below the curve, but it increased at a slower rate than the generation time did, unlike B/r A. In the same analysis, $(C + D)$ also appeared to be constant in B/r A and K during rapid growth, but the generation time at which $(C + D)$ began to increase was less certain. It is possible that $(C + D)$ always begins to increase once the generation time is longer than the duration of $(C + D)$ during rapid growth, but that is unclear from these data. The important point to note here is simply that B/r A differs from F and K during slow growth.

Figure 3 shows additional calculations of the average age at the initiation of replication in strain B/r A using data for C and D determined by researchers in other laboratories [23,26,30]. Calculations of the initiation age using data reported by Skarstad et al. [23] and Michelsen et al. [26] yielded values for slow-growing B/r A, which were very similar to those for F and K, as shown in Figure 2, and very different from our findings for B/r A; in other words, there was a significant early B period. However, another set of experiments performed with B/r A by Koppes et al. [30] were more consistent with our findings. As seen in Figure 3, they reported B periods in B/r A of 5 min and 3 min at generation times of 109 min and 135 min, respectively. They also found a B period of 39 min for B/r K growing with a generation time of 100 min, which is in basic agreement with our findings for this strain (Figure 2). The explanation for this odd disparity seen only with B/r A remains unknown, but it does not appear to be related to the techniques employed or the carbon sources used in the culture media since there were no consistent differences. The one consistent difference may relate to the culture media components other than carbon sources. The studies that found a small or no *B* period when the experiment was performed in C medium, which consists of M9 salts plus MgSO$_4$ [29]. The experiments detecting significant *B* periods in B/r A when the experiment was performed in either M9 or AB medium, which contain additional salts, such as CaCl$_2$ and FeCl$_3$ [31]. Possible implications of these findings are described in Section 4.

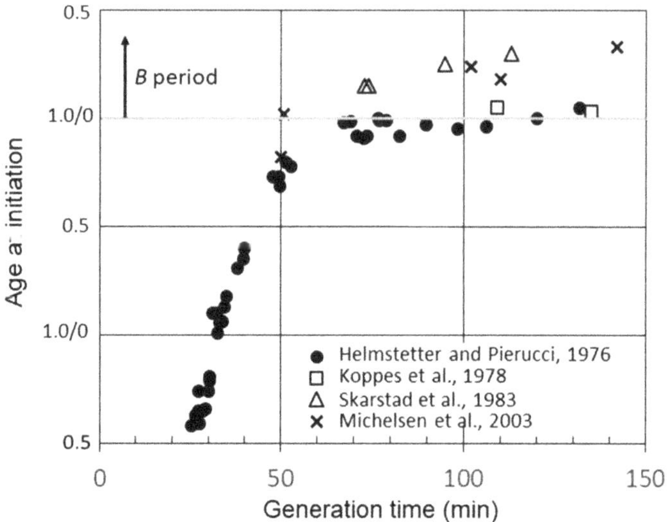

Figure 3. Cell age at the initiation of chromosome replication versus generation time for *E. coli* B/r A determined from measurements of *B*, *C*, and *D* durations in batch cultures by Helmstetter and Pierucci [28], Koppes et al. [30] and Michelson et al. [26], and in chemostat cultures by Skarstad et al. [23].

3.3. Effects of Protein Synthesis Inhibition on Initiation of Replication

Another interesting property of *E. coli* B/r grown in C medium relates to the response of slow-growing cells to exposure to chloramphenicol or chloramphenicol plus rifampicin [32,33]. When the inhibitors were added to B/r A cells growing with generation times between approximately 60 and 120 min, that is, cells that normally initiated replication with two chromosomal origins during the *D* period, the cells continued to progress to initiation of replication. For example, when chloramphenicol was added to B/r A cells growing with a doubling time of 120 min, the cells continued to progress to initiation of replication for over 30 min (Figure 4). The triangles in the figure indicate the ages of the youngest cells that were able to continue to the initiation of replication in the presence of inhibitors. The cells that initiated replication were the only cells that divided between the two origins. However, division per se was not required since initiation of replication also takes place in the presence of penicillin [33]. This phenomenon was only observed in slow-growing cells that normally initiated during the *D* period with two origins. However, it was not unique to B/r A because when chloramphenicol and rifampicin were added to B/r F cells growing with a generation time of 60 min, i.e., cells that normally initiated toward the end of *D* [32], they also continued to initiate, but for only 10 min (Figure 4), presumably because they only divided for 10 min. If this behavior is true for all strains of *E. coli*, then determinations of *C*, as well as the average number of replication origins per cell and (*C* + *D*), in strains growing with generation times between *C* min and (*C* + *D*) min obtained by measuring the extent of DNA replication in the presence of rifampicin or chloramphenicol could yield erroneously large values. This would be especially true for cells growing with generation times slightly shorter than (*C* + *D*).

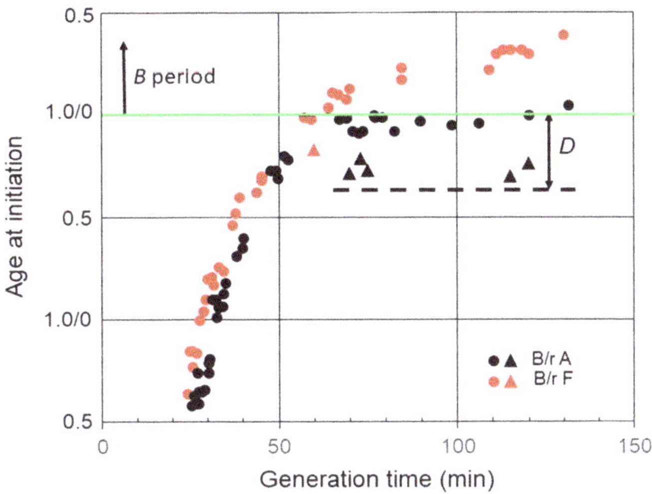

Figure 4. Effect inhibition of RNA and protein synthesis on the initiation of replication in slow-growing B/r A and F. The triangles indicate the ages of the youngest cells that were capable of progressing to initiation of replication in the presence of chloramphenicol or rifampicin plus chloramphenicol. The dashed line shows the approximate cell age at the start of the *D* period during slow growth of B/r A.

3.4. Relationship between Durations of C and D

It is well known that the average *C* period has been found to be constant in *E. coli* growing with generation times less than 60 70 min at 37 °C [26]. In *E. coli* B/r, the average duration of (*C* + *D*) has also been found to be indistinguishable from the constant during rapid growth [29], and that may also be true for some K-12 strains [27]. As a consequence, the average ratios of *C/D* during rapid growth of B/r A, F, and K are 1.9, 2.6, and 3.0, respectively [29]. To examine whether this relationship between *C* and *D* might continue during slow growth, the measured values for *D* during the slow growth of three strains (29) were multiplied by their respective ratios during rapid growth and compared to *C*, as shown in Figure 5. Since the values for *D* × (*C/D*) appear to superimpose on the values for *C* in B/r A and K during slow growth, the ratios *C/D* seem to be invariant at all growth rates examined in these strains and probably in B/r F as well.

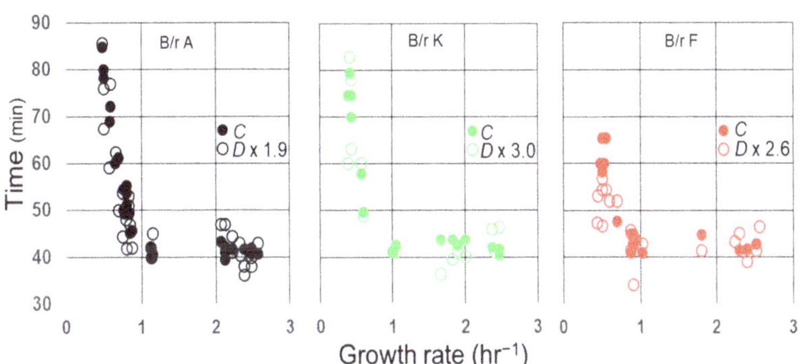

Figure 5. Correlation between *C* and *D* in *E. coli* B/r as a function of growth rate. Measurements of *C* and *D* and the average *C/D* ratios during rapid growth [τ ≤ (*C* + *D*)] were taken from Helmstetter and Pierucci [29] and used to plot *C* (closed circles) and *D* × *C/D* (open circles) versus growth rate.

3.5. Relative Sizes of B/r A, K, and F

A final interesting aspect of the B/r strains relates to their sizes and shapes. Woldringh et al. [34] have shown that the shapes of newborn B/r A and K cells grown in C medium are very similar during rapid growth, but they diverge considerably in shape during slow growth. At slower growth rates, B/r A becomes more spherical, whereas B/r K maintains a rod-like shape. However, newborn cell volumes calculated using cell dimensions were found to be similar for all three B/r strains when they were grown at the same rate. Referring back Figure 2, since in C medium B/r A initiates replication at an earlier age than B/r F or K do at all growth rates, if newborn cells have similar volumes, the cell volume per chromosomal origin at the initiation of replication must be smaller in A than those in F or K. Indeed, that is what has been reported [30], and it can be found using known values of $(C + D)$ to calculate cell volume at the time of the initiation of replication from the cell dimension measurements. Perhaps the difference in initiation volumes is related to the different shapes of strains during slow growth. However, that cannot explain the differing initiation volumes during rapid growth. When B/r A and K were grown rapidly in C medium at a rate of about 2.5 doublings/hr, the newborn cells were very similar in shape [34]. Thus, at this rapid growth rate, the newborn cell volumes, cell shapes, C period durations (42 min), and times for cell constriction (T $\approx$ 10 min [34]) were found to be essentially the same. The one consistent difference is in the D duration, along with a concomitant difference in initiation age. These findings raise questions regarding the relationship between initiation timing and D period duration.

4. Discussion

The preceding information suggests, but it certainly does not prove, that the comparative cell cycle properties of *E. coli* B/r A, F, and K may depend on the culture conditions. In one set of conditions, the findings, albeit limited, suggest that the three strains may be very similar at all growth rates with regard to the newborn cell size, the cell age/volume at the initiation of chromosome replication, the durations of C and D periods, and the cell shape. Under an alternative culture condition, during growth in C medium, B/r F and K appear to maintain the same cell cycle properties, but the key properties of B/r A differ, namely, the initiation volume is smaller, and the D period is longer at all growth rates. Given this striking difference in these properties in one strain grown at the same rate in different media, it is interesting to enquire about the connection between cell size at the initiation of chromosome replication and D duration. There are a few possible explanations. The first and least interesting one is the possibility that the observed differences arise due to differing techniques and procedures used in different laboratories. Although this explanation seems to be unlikely, it would be decidedly more reassuring if these studies were performed simultaneously by one research group.

A likely scenario is that the longer duration of D in B/r A grown in C medium is a direct consequence of the smaller size at initiation, resulting in the initiation of replication at an earlier cell age. There are several observations that, taken together, support this conclusion. Based on current views of the cell cycle and assuming that the cell volume and cell mass are interchangeable, if the cell volumes (V) and C periods are the same in B/r A and K grown at the same rate, but the initiation volume (V_i) of B/r A is smaller, then B/r A must, out of necessity, have a longer D period since $V_i = V/\ln 2e^{(C+D)/\tau}$ [35]. Additionally, the duration of D seems to be uninfluenced by the time for cell constriction and the T period, since during rapid growth, T averages about 10 min in both B/r A and K [34], whereas the average D periods are 22 and 14 min, respectively [29]. The T period also increases considerably more in B/r A than it does in K during slow growth [34], but in spite of that, the D/C ratio remains essentially unchanged in both strains. As plausible as it may be that the initiation size could set the D duration, it does not account for the smaller initiation volume in this B/r strain in one culture medium. The explanation could simply be that there is an increased relative concentration of DnaA or DnaA-ATP in B/r A when it is grown in C medium, but information on that issue is not currently available.

Another scenario is that the smaller initiation size is a direct consequence of the longer D in B/r A grown in C medium. On the surface, the involvement of D in the process of replication initiation would be surprising since it has been reported that increasing the duration of D by altering the cell shape does not alter the initiation mass [36–38]. However, in the studies on *E. coli* B/r referenced here, the cell shapes during rapid growth were basically the same, in spite of the differing D durations. On the other hand, the observation of a dramatic shape change in B/r A as it grew slower in C medium might suggest that some aspect of surface growth could be aberrant in B/r A at all growth rates, possibly yielding a longer D period when it is grown in C medium. How might a longer D result in a smaller initiation volume? We know that B/r cells that were normally initiated in D were able to progress to the initiation of replication in the absence of protein synthesis as long as they progressed toward division between two chromosomal origins. That phenomenon cannot be a direct consequence of the increased relative concentration of DnaA protein that has been observed during slow growth [27,39,40] since initiation of replication during the inhibition was not seen in cells that normally initiated replication just before or just after the D period. It seems to be associated with some specific aspect of D, such as envelope formation/division progression. There is evidence that domains of acidic phospholipids may form during D [41], which might be capable of DnaA-ADP-to-DnaA-ATP conversion and the consequent initiation of replication. Perhaps this process could augment the progression to initiation during the cycle of cells with longer D periods, resulting in a smaller initiation size.

Lastly, since cell age at the initiation of replication is relatively unchanged, at or near cell division, in B/r A for an extended period once the generation time becomes longer than the average value of $(C + D)$ during rapid growth (Figure 2), as may be the case with other strains as well (Figure 1), a question arises as to the size at initiation during that interval. In the case of B/r A, cell dimensions determined by Woldringh et al. [34] indicated that the initiation volume decreased as the generation time increased from 65 to 120 min. This is also a time when D becomes increasingly longer which, as discussed above, could be a potential explanation for the smaller initiation size in these cells. Alternatively, the observed increase in relative DnaA concentration during slow growth [27,40] could also be a reasonable explanation for the decreasing initiation size. Even if these conjectures have some level of validity, an attempt to explain the basis for the apparent lack of significant change in cell age at the initiation of replication during an interval of slow growth would be premature in the absence of additional information.

Much of the preceding data are speculative and may turn out to be of minimal significance. What is significant is that I feel it is possibility that a more comprehensive analysis of the properties of *E. coli* B/r, and especially the seemingly unusual B/r A, might yield some interesting new information on cell cycle controls, and hence, the validity of current cell cycle control models.

Funding: This research received no external funding.

Institutional Review Board Statement: Not applicable.

Informed Consent Statement: Not applicable.

Data Availability Statement: A video with additional information can be viewed at https://drive.google.com/file/d/1r7mrP6E46-fam7wKsjHYritBGWMgJhJC/view.

Acknowledgments: Thanks to Arieh Zaritsky, Vic Norris, Itzhak Fishov and Conrad Woldringh for help with the preparation of this review and the accompanying data, and especially for the organization of this Special Issue. It has been an incredible honor.

Conflicts of Interest: The author declares no conflict of interest.

References

1. Beadle, G.W.; Tatum, E.L. Genetic Control of Biochemical Reactions in Neurospora. *Proc. Natl. Acad. Sci. USA* **1942**, *27*, 499–506. [CrossRef] [PubMed]
2. Beadle, G.W. Biochemical Genetics: Some Recollections. In *Phage and the Origins of Molecular Biology*; Cairns, J., Stent, G.S., Watson, J.D., Eds.; Cold Spring Harbor Laboratory of Quantitative Biology: Cold Spring Harbor, NY, USA, 1966; pp. 23–32.
3. Beadle, G.W. Recollections. *Annu. Rev. Biochem.* **1974**, *43*, 1–14. [CrossRef]
4. Singer, M.; Berg, P. George Beadle: From Genes to Enzymes. *Nat. Rev. Gen.* **2004**, *5*, 949–954. [CrossRef]
5. Strauss, B.S. Biochemical Genetics and Molecular Biology: The Contributions of George Beadle and Edward Tatum. *Genetics* **2016**, *293*, 13–20. [CrossRef] [PubMed]
6. Howard, A.; Pelc, S.R. Synthesis of Nucleoprotein in Bean Root Cells. *Nature* **1950**, *167*, 599–600. [CrossRef] [PubMed]
7. Novick, A.; Szilard, L. Experiments with the Chemostat on Spontaneous Mutants of Bacteria. *Proc. Natl. Acad. Sci. USA* **1950**, *36*, 708–719. [CrossRef]
8. Wollman, E.L.; Jacob, F. Mechanism of the Transfer of Genetic Material During Recombination in *Escherichia coli* K-12. CR Hebd. *Seances Acad. Sci.* **1955**, *240*, 2449–2451.
9. Wollman, E.L.; Jacob, F.; Hayes, W. Conjugation and Genetic Recombination in *Escherichia coli* K-12. Cold Spring Harb. Symp. *Quant. Biol.* **1956**, *21*, 141–162. [CrossRef]
10. Helmstetter, C.E. Methods for Studying the Microbial Division Cycle. In *Methods in Microbiology*; Norris, J.R., Ribbons, D.W., Eds.; Academic Press: New York, NY, USA, 1969; pp. 327–363.
11. Maruyama, Y.; Yanagita, T. Physical Methods for Obtaining Synchronous Culture of *Escherichia coli*. *J. Bacteriol.* **1956**, *71*, 542–546. [CrossRef]
12. Helmstetter, C.E.; Cummings, D.J. An Improved Method for Selection of Bacterial Cells at Division. *Biochim. Biophys. Acta* **1964**, *82*, 608–610. [CrossRef]
13. Helmstetter, C.E. Rate of DNA Synthesis During the Division Cycle of *Escherichia coli* B/r. *J. Mol. Biol.* **1967**, *24*, 417–427. [CrossRef]
14. Helmstetter, C.E.; Cooper, S. DNA Synthesis During the Division Cycle of Rapidly Growing *Escherichia coli* B/r. *J. Mol. Biol.* **1968**, *31*, 507–518. [CrossRef] [PubMed]
15. Cooper, S.; Helmstetter, C.E. Chromosome Replication and the Division Cycle of *Escherichia coli* B/r. *J. Mol. Biol.* **1968**, *31*, 519–540. [CrossRef]
16. Helmstetter, C.E.; Cooper, S.; Pierucci, O.; Revelas, E. On the Bacterial Life Sequence. Cold Spring Harbor Symp. *Quant. Biol.* **1968**, *33*, 809–822. [CrossRef] [PubMed]
17. Schaechter, M.; Maaløe, O.; Kjeldgaard, N.O. Dependency on Medium and Temperature of Cell Size and Chemical Composition during Balanced Growth of *Salmonella typhimurium*. *J. Gen. Microbiol.* **1958**, *19*, 592–606. [CrossRef] [PubMed]
18. Yoshikawa, H.; O'Sullivan, A.; Sueoka, N. Sequential Replication of the *Bacillus subtilis* Chromosome. III. Regulation of Initiation. *Proc. Natl. Acad. Sci. USA* **1964**, *52*, 973–980. [CrossRef]
19. Thornton, M.; Eward, K.L.; Helmstetter, C.E. Production of Minimally Disturbed Synchronous Cultures of Hematopoietic Cells. *Biotechniques* **2002**, *32*, 1098–1105. [CrossRef]
20. LeBleu, V.S.; Thornton, M.; Gonda, S.R.; Helmstetter, C.E. Technology for Cell Cycle Research with Unstressed Steady-State Cultures. *Cytotechnology* **2006**, *51*, 149–157. [CrossRef]
21. Banfalvi, G. Overview of Cell Synchronization. *Methods Mol. Biol.* **2017**, *1524*, 3–27.
22. Spellman, P.T.; Sherlock, G. Final Words: Cell Age and Cell Cycle Are Unlinked. *Trends Biotech.* **2004**, *22*, 277–278. [CrossRef]
23. Skarstad, K.; Steen, H.B.; Boye, E. Cell Cycle Parameters of Slowly Growing *Escherichia coli* B/r Studied by Flow Cytometry. *J. Bacteriol.* **1983**, *154*, 656–662. [CrossRef] [PubMed]
24. Skarstad, K.; Steen, H.B.; Boye, E. *Escherichia coli* DNA Distributions measured by Flow Cytometry and Compared with Theoretical Computer Simulations. *J. Bacteriol.* **1985**, *163*, 661–668. [CrossRef]
25. Allman, R.; Schjerven, T.; Boye, E. Cell Cycle Parameters of *Escherichia coli* K-12. *J. Bacteriol.* **1991**, *173*, 7970–7974. [CrossRef]
26. Michelsen, O.; de Mattos, M.J.T.; Jensen, P.R.; Hansen, F.G. Precise Determination of C and D Periods by Flow Cytometry in *Escherichia coli* K-12 and B/r. *Microbiology* **2003**, *149*, 1001–1010. [CrossRef] [PubMed]
27. Zheng, H.; Bai, Y.; Jiang, M.; Tokuyasu, T.A.; Huang, X.; Zhong, F.; Wu, Y.; Fu, C.; Kleckner, N.; Hwa, T.; et al. General Quantitative Relations Linking Cell Growth and the Cell Cycle in *Escherichia coli*. *Nature Microbiol.* **2020**, *5*, 995–1001. [CrossRef] [PubMed]
28. Helmstetter, C.E.; Eenhuis, C.; Theisen, P.; Grimwade, J.; Leonard, A.C. Improved Bacterial Baby Machine: Application to *Escherichia coli* K-12. *J. Bacteriol.* **1992**, *174*, 3445–3449. [CrossRef]
29. Helmstetter, C.E.; Pierucci, O. DNA Synthesis during the Division Cycle of Three Substrains of *Escherichia coli* B/r. *J. Mol. Biol.* **1976**, *102*, 477–486. [CrossRef]
30. Koppes, L.J.H.; Woldringh, C.L.; Nanninga, N. Size Variations and Correlation of Different Cell Cycle Events in Slow-Growing *Escherichia coli*. *J. Bacteriol.* **1978**, *134*, 423–433. [CrossRef]
31. Clark, D.J.; Maaløe, O. DNA Replication and the Division Cycle in *Escherichia coli*. *J. Mol. Biol.* **1967**, *23*, 99–112. [CrossRef]
32. Helmstetter, C.E. Initiation of Chromosome Replication in *Escherichia coli* I. Requirements for RNA and Protein Synthesis at Different Growth Rates. *J. Mol. Biol.* **1974**, *84*, 1–19. [CrossRef]
33. Helmstetter, C.E. Initiation of Chromosome Replication in *Escherichia coli* II. Analysis of the Control Mechanism. *J. Mol. Biol.* **1974**, *84*, 21–36. [CrossRef] [PubMed]

34. Woldringh, C.L.; de Jong, M.A.; van den Berg, W.; Koppes, L. Morphological Analysis of the Division Cycle of Two *Escherichia coli* Substrains During Slow Growth. *J. Bacteriol.* **1977**, *131*, 270–279. [CrossRef] [PubMed]
35. Churchward, G.; Estiva, E.; Bremer, H. Growth Rate-Dependent Control of Chromosome Replication Initiation in *Escherichia coli*. *J. Bacteriol.* **1981**, *145*, 1232–1238. [CrossRef]
36. Hill, N.S.; Kadoya, R.; Chattoraj, D.K.; Levin, P.A. Cell Size and Initiation of DNA Replication in Bacteria. *PloS Genet.* **2012**, *8*, e1002549. [CrossRef]
37. Zheng, H.; Ho, P.; Jiang, M.; Tang, B.; Liu, W.; Li, D.; Yu, X.; Kleckner, N.E.; Amir, A.; Liu, C. Interrogating the *Escherichia coli* Cell Cycle by Cell Dimension Perturbations. *Proc. Natl. Acad. Sci. USA* **2016**, *113*, 15000–15005. [CrossRef] [PubMed]
38. Si, F.; Li, D.; Cox, S.E.; Sauls, J.T.; Azizi, O.; Sou, C.; Schwartz, A.B.; Erickstad, M.J.; Jun, Y.; Li, X.; et al. Invariance of Initiation Mass and Predictability of Cell Size in *Escherichia coli*. *Curr. Biol.* **2018**, *27*, 1278–1287. [CrossRef]
39. Hansen, F.G.; Atlung, T.; Braun, R.E.; Wright, A.; Hughes, P.; Kohiyama, M. Initiator (DnaA) Protein Concentration as a Function of Growth Rate in *Escherichia coli* and *Salmonella typhimurium*. *J. Bacteriol.* **1991**, *173*, 5194–5199. [CrossRef]
40. Flåtten, I.; Morigen, M.; Skarstad, K. DnaA Protein Interacts with RNA Polymerase and Partially Protects it from the Effect of Rifampicin. *Mol. Microbiol.* **2009**, *71*, 1018–1030. [CrossRef]
41. Mileykovskaya, E.; Dowhan, W. Cardiolipin Membrane Domains in Prokaryotes and Eukaryotes. *Biochim. Biophys. Acta* **2009**, *1788*, 2084–2091. [CrossRef]

Perspective

Unbalanced Growth, the DNA Replication Cycle and Discovery of Repair Replication

Philip C. Hanawalt

Department of Biology, Stanford University, Stanford, CA 94305, USA; hanawalt@stanford.edu

Abstract: This article recounts my graduate research at Yale University (1954–1958) on unbalanced growth in *Eschericia coli* during thymine deprivation or following ultraviolet (UV) irradiation, with early evidence for the repair of UV-induced DNA damage. Follow-up studies in Copenhagen (1958–1960) in the laboratory of Ole Maaløe led to my discovery that the DNA replication cycle can be synchronized by inhibiting protein and RNA syntheses and that an RNA synthesis step is essential for initiation of the cycle, but not for its completion. This work set the stage for my subsequent research at Stanford University, where the repair replication of damaged DNA was documented, to provide compelling evidence for an excision-repair pathway. That universal pathway validates the requirement for the redundant information in the complementary strands of duplex DNA to ensure genomic stability.

Keywords: bacterial cell cycle; DNA replication cycle; unbalanced growth; thymineless death; DNA repair replication; excision-repair

Citation: Hanawalt, P.C. Unbalanced Growth, the DNA Replication Cycle and Discovery of Repair Replication. *Life* **2023**, *13*, 1052. https://doi.org/10.3390/life13041052

Academic Editors: Ron Elber and Bruce J. Nicholson

Received: 21 March 2023
Revised: 13 April 2023
Accepted: 14 April 2023
Published: 20 April 2023

1. Graduate Studies at Yale

The ground-breaking Watson/Crick model for the base-paired double-helical structure of DNA emerged in 1953–54, just as I completed an undergraduate physics major at Oberlin College and joined the Biophysics Department at Yale University for graduate study. I initiated my primary research project in the laboratory of Richard Setlow, a spectroscopist who was studying the effects of ultraviolet light (UV) on enzyme activities and on the inhibition of bacterial cell division. I was interested in focusing upon the effects of UV on growth and macromolecular synthesis in bacteria, and the molecular mechanisms for these effects. My eventual PhD thesis was titled "Macromolecular synthesis in *Escherichia coli* during conditions of unbalanced growth" [1].

Iverson and Giese had used colorimetric assays, (Indole for DNA and Orcinal for RNA), to follow the synthesis of DNA and RNA over an 8 h period after UV irradiation of *E. coli* [2]. To analyze in detail the early impact of UV upon DNA synthesis, I wanted to label the nascent DNA with radioactive thymine and use a thymine-deficient mutant to achieve maximum labeling efficiency with exogenous thymine. Seymour Cohen had studied a mutant strain, *E. coli* 15T-, deficient in thymidylate synthase, and had observed an exponential loss of viability with time in the absence of thymine. He named this phenomenon thymineless death (TLD) and attributed it to unbalanced growth, since protein synthesis and other metabolic activities continued unabated for a short period in the absence of DNA synthesis [3]. Balanced growth was defined by Allan Campbell as "over a time interval, if during that interval, every extensive property of the growing system increases by the same factor" [4]. Cohen provided the thymine-deficient mutant for my studies, and I used ^{14}C-thymine to follow DNA synthesis. To follow other macromolecular syntheses for comparisons at the same time, I developed a sensitive microassay to resolve the incorporation of ^{32}P, from ^{32}PO$_4$, into DNA, RNA, and phospholipids [1,5]. The procedure, fully detailed in my thesis and publication, employs the precipitation of trichloroacetic acid-treated cells on collodion membrane filters, ethanol to remove phospholipids, and quantitative hydrolysis of RNA to mononucleotides by KOH, without loss of DNA The procedure was validated by comparison with colorimetric determinations.

I compared unbalanced growth during thymine deficiency to that following UV irradiation. I confirmed that UV inhibited DNA synthesis in *E. coli* and that it resumed after low UV doses, with a lag that increased as a function of the dose, while the synthesis of RNA and protein was less affected [1]. This was evidence that DNA synthesis could recover from the damage inflicted by UV, but we did not know what kind of damage was produced. Albert Kelner had discovered a phenomenon, termed photoreactivation, in which bacterial survival was greater if UV irradiation was followed by exposure to visible light [6]. I found that visible light exposure following UV shortened the lag in recovery of DNA synthesis, and I suggested that "the visible light facilitates repair of DNA integrity" [1,7]. That was my first use of the term, repair, but again, at that time we did not know what damage was being repaired, or even if the cellular lethality was primarily due to damage in DNA. I used monochromatic UV at 265 nm for the dose/response studies reported in the publication of my thesis work [1,8].

2. Postdoctoral Research at the University of Copenhagen

Upon completion of my graduate thesis, I was interested in a postdoctoral period to study synchronous growth in bacterial cultures so that I could look more closely at the time course of macromolecular syntheses during the cell cycle. (I also wanted the experience of living for a few years in another country!) Conveniently, it turned out that the acknowledged expert in synchronizing bacterial growth was Ole Maaløe, then at the State Serum Institute in Copenhagen. I applied and was accepted with an NIH postdoctoral fellowship. Just before I departed for Denmark, Seymour Cohen generously provided a further mutated strain of *E. coli* 15T-, not only deficient in the synthesis of thymine but also arginine and uracil. Thus, it was possible that I might compare the effects of RNA and protein syntheses on the phenomenon of TLD in the triple mutant strain.

Upon my arrival in Denmark, Ole Maaløe had just been appointed Professor and Director of the Institute of Microbiology at the University of Copenhagen, in a magnificent new building adjacent to the botanical garden overlooking Rosenberg Castle. He did not offer an immediate suggestion for my project, so I simply continued my studies on TLD, using the new triple mutant, which allowed me to look at the effects of protein and RNA synthesis on that process. Very few cells survived in the absence of thymine while protein and RNA synthesis continued for a limited period. However, when arginine and uracil were also deficient, TLD leveled off at about 3% survival. In the presence of thymine during the absence of protein and RNA synthesis, there was roughly a 40% increase in DNA followed by a plateau. If DNA synthesis was allowed for different periods, followed by thymine starvation, during the continued absence of protein synthesis, the survival level was increased until all the cells survived. This led to our hypothesis that the DNA replication cycle could be completed in the absence of protein and RNA synthesis and that thymine starvation only killed the cells that were actively carrying out DNA synthesis. The initiation of the cycle evidently required protein and RNA synthesis.

I needed to learn single-cell autoradiography for further analysis, so I visited my Yale graduate colleague, Robert van Tubergen. Bob was an expert in this technique and was currently in a postdoctoral position with Roy Markham in the UK. We carried out several preliminary experiments for me to master the approach and then I returned to Copenhagen to document the fraction of the population of individual cells that were synthesizing DNA in the absence of protein and RNA synthesis. This approach confirmed our model in which the number of cells actively carrying out DNA synthesis decreased with time under conditions in which protein and RNA synthesis were inhibited.

A new postdoc, Don Cummings, arrived in 1959 from Lloyd Kozloff's laboratory at the University of Chicago to join Maaløe's group, bringing experience in ultracentrifugation and the analysis of DNA replication by density labeling, as had been employed by Meselson and Stahl to prove the semiconservative mode of chromosomal DNA synthesis [9]. This approach validated our expectation that normal semiconservative DNA synthesis continued to completion in the absence of arginine and uracil, but that no new cycles of

DNA replication were initiated under these conditions. The density labeling established that none of the initially labeled DNA molecules had begun a second cycle of replication. Thus, we had succeeded in synchronizing the bacterial DNA replication cycle.

Professor Maaløe presented the results from our studies in an invited lecture on "Control of normal DNA replication in bacteria" at the 1960 Cold Spring Harbor Symposium on "Cellular Regulatory Mechanisms" [10]. Unfortunately, the meeting was over-subscribed and I learned about it too late to participate. Several papers on my postdoctoral work were published later [11,12], and one of those papers [11] was selected as a Citation Classic by *Current Contents*, having been cited in over 530 publications between 1961 and 1964.

3. Second Postdoc, at the California Institute of Technology

I was interested in pursuing the density-labeling approach to study further details of DNA replication, and I received a fellowship from the American Cancer Society for a year with Robert Sinsheimer at Caltech, where I also became acquainted with Max Delbrook and Matt Meselson. We discussed the relationships of DNA molecules to the structure of the bacterial chromosome, and we postulated that the chromosome may consist of short segments of DNA, of about 7 million daltons, connected to each other by protein linkers. (It turned out that this was simply the characteristic size of DNA molecules obtained inadvertently upon shearing larger molecules during the loading of the ultracentrifuge rotor cell with a small-bore needle syringe.) John Cairns then showed by autoradiography in 1963 that the chromosome in *E. coli* consists of a single closed-circular duplex DNA molecule [13].

During my year at Caltech, Delbruck taught an exciting course on photobiology, in which he highlighted the recent discovery by Beukers and Berends, that UV irradiation causes co-valent dimerization of thymine molecules [14]; and it was soon shown that cyclobutane pyrimidine dimers could be formed between adjacent pyrimidines in DNA. So, there was finally a documented candidate for a responsible DNA lesion that caused growth inhibition and lethality in UV-irradiated bacterial cells. In 1962, Wulff and Rupert reported that photoreactivation involves the direct reversal of pyrimidine dimers to free thymine in situ, without affecting the phosphodiester backbone of DNA [15]. That finding supported my earlier speculation that photoreactivating light facilitated the repair of damage that was hindering the resumption of DNA synthesis in the UV-irradiated cells. I obtained further experience in density-labeling technology, with mentorship from Sinsheimer and Jerry Vinograd, while also searching unsuccessfully for physical changes in DNA during thymine deprivation.

4. Faculty Appointment at Stanford University

In September 1961, I joined the Biophysics Laboratory at Stanford University as a Research Biophysicist and Lecturer. I initiated a new graduate course in Molecular Biophysics and began undergraduate teaching in Arthur Giese's Cell Physiology course, remembering that my graduate research had followed upon his early studies! In 1965, I was promoted to Associate Professor with tenure in the Department of Biological Sciences.

Two incoming biophysics graduate students, Dan Ray and David Pettijohn, joined me in 1962 to compare DNA replication under normal growth conditions with that following UV irradiation. Dan and I were able to isolate partially replicated growing fork DNA fragments in *E. coli* through ^{32}P pulse labeling of DNA during bacterial growth in a medium containing 5-bromouracil, replacing thymine to density-label replicating DNA, and analysis by density-gradient equilibrium sedimentation. We found that the fork-containing DNA fragments were selectively sensitive to shearing into replicated and unreplicated sub-fragments [16].

David and I focused on the qualitative nature of DNA replication in UV-irradiated bacteria. We had expected to isolate DNA segments in which the replication fork was blocked by pyrimidine dimers, and that these would appear as partially replicated fragments during density labeling. We confirmed that semiconservative DNA synthesis was inhibited, but surprisingly, a new mode of non-conservative DNA synthesis appeared, in which much of the density-labeled nascent DNA was in very short segments, too short to significantly shift the density of DNA fragments containing them. Intentional shearing did not resolve them into fragments of different densities, leading to our conclusion that the nascent DNA was in short "patches" embedded in the parental DNA fragments [17].

Meanwhile, Richard Setlow had been continuing the studies on the inhibition of DNA synthesis that I had initiated at Yale; but now, with the knowledge that UV exposure induced pyrimidine dimers and with the availability of a UV-sensitive mutant bacterial strain, *E. coli* B_{S-1}, isolated by Ruth Hill [18], Setlow identified very short oligomers containing pyrimidine dimers that were excised from the DNA in the UV-irradiated parental bacterial strain, but not in the UV-sensitive mutant. He proposed an excision-repair mechanism to explain his results [19]. Paul Howard-Flanders obtained the same results using wild-type and UV-sensitive mutants of *E. coli* K12 strains [20]. Of course, the excision of those damaged DNA oligomers would leave another lesion, a gap in the parental DNA strand.

David Pettijohn and I had discovered the second step in excision-repair, the filling of those gaps by repair replication, a non-conservative mode, using the undamaged parental strand as a template [21]. Unlike the control of chromosomal DNA replication, repair replication was not affected by the cell cycle. It was just as efficient in cells that had completed the cycle as in those undergoing chromosomal DNA synthesis [22]. We concluded that the Watson/Crick duplex DNA structure was needed, not only for sequential replication of the genome but also that it was absolutely essential for the repair of damage in the respective DNA strands. My colleague, Robert Haynes, and I provided a short history of the emerging field of DNA repair in *Scientific American* [23]. The studies in my laboratory on repair replication were then presented in the 1968 Cold Spring Harbor Symposium [24], and I published a broader review on cellular responses to photochemical damage in a chapter in a book on Photophysiology, edited by Arthur Giese [25].

5. Reflections on the Primordial Genome and the Need for DNA Repair

DNA is not unusually stable, as had been originally assumed, and we now know that multiple repair pathways are needed to maintain the integrity of the genome. The genome is subject to endogenous oxidative damage, and DNA is also intrinsically unstable due to spontaneous depurination, which generates a basic site that must be repaired. In contrast, the depurination of RNA is very much slower, while RNA is more prone than DNA to strand breaks and degradation (see the classic review by Tomas Lindahl [26]). For the existence of living cells, there must have been informational nucleic acids of sufficient lengths to encode the needed proteins/enzymes and ribozymes. In abiogenic synthesis experiments, only relatively short DNA molecules have been generated. Perhaps longer informational nucleic acid chains were formed by the ligation of short ones. An undergraduate physics student, Roger Lewis, and I carried out a proof-of-principle study in which we demonstrated that UV irradiation of duplex DNA, in which short thymine-containing oligomers were paired with a long adenine-containing strand, could result in the thymine–dimer linkage of the oligomers, to leave a strand break at each linkage site without any loss of continuity in the other strand [27].

Nucleic acid repair would have been essential for the origin of life as well as for its persistence, and the redundant double-strand genome must have been required at the very beginning of life. (This of course assumes that the original informational genome consisted of nucleic acids!) I think it likely that the first functional genomes were composed of a mixture of ribonucleotides and deoxyribonucleotides, since both were present in the primordial "stew", and that this combination may have helped to ensure both the informational and the structural stability of the first genomes. The primordial earth was bombarded by a high flux of UV light, not attenuated by an ozone layer. It is likely that sunlight photochemistry played an important role in the origin of life, while paradoxically, it was also one of the principal threats to its persistence [28].

Funding: The early studies in my laboratory at Stanford were supported by an NIH grant (GM09901) from the Institute of General Medical Sciences and a contract with the Atomic Energy Commission.

Institutional Review Board Statement: Not applicable.

Acknowledgments: I am indebted to a long list of talented students and colleagues from 36 different countries for the scientific contributions from my group. The complete list with more details can be found on my website: https://web.stanford.edu/~hanawalt/.

Conflicts of Interest: The author declares no conflict of interest.

References

1. Hanawalt, P.C. Macromolecular Synthesis in Escherichia coli during Conditions of Unbalanced Growth. Ph.D. Thesis, Yale University, New Haven, CT, USA, September 1958.
2. Iverson, R.M.; Giese, A.C. Synthesis of nucleic acids in ultraviolet-treated Escherichia coli. *Biochim. Biophys. Acta* **1957**, *25*, 62–68. [CrossRef] [PubMed]
3. Cohen, S.S.; Barner, H.D. Studies on unbalanced growth in *E. coli. Proc. Natl. Acad. Sci. USA* **1954**, *40*, 885–893. [CrossRef]
4. Campbell, A. Synchronization of cell division. *Bacteriol. Rev.* **1957**, *21*, 263–272. [CrossRef]
5. Hanawalt, P.C. Use of phosphorus-32 in microassay for nucleic acid synthesis in *Escherichia coli. Science* **1959**, *130*, 386–387. [CrossRef]
6. Kelner, A. Effect of visible light on the recovery of *Streptomyces griseus* conidia from ultraviolet irradiation injury. *Proc. Natl. Acad. Sci. USA* **1949**, *52*, 73–79. [CrossRef] [PubMed]
7. Hanawalt, P.C.; Buehler, J. Photoreactivation of macromolecular synthesis in Escherichia coli. *Biochim. Biophys. Acta* **1960**, *37*, 141–143. [CrossRef]
8. Hanawalt, P.C.; Setlow, R.B. Effect of monochromatic UV on macromolecular synthesis in *E. coli. Biochim. Biophys. Acta* **1960**, *41*, 283–294. [CrossRef]
9. Meselson, M.; Stahl, F. The replication of DNA in *Escherichia coli. Proc. Natl. Acad. Sci. USA* **1958**, *44*, 671–682. [CrossRef]
10. Maaløe, O. The control of normal DNA replication in bacteria. *Cold Spring Harb. Symp. Quant. Biol.* **1960**, *26*, 45–52. [CrossRef] [PubMed]
11. Maaløe, O.; Hanawalt, P.C. Thymine deficiency and the normal DNA replication cycle I. *J. Mol. Biol.* **1961**, *3*, 144–155. [CrossRef]
12. Hanawalt, P.C.; Maaløe, O.; Cummings, D.; Schaecter, M. The normal DNA replication cycle II. *J. Mol. Biol.* **1961**, *3*, 156–165. [CrossRef]
13. Cairns, J. The bacterial chromosome and its manner of replication as seen by autoradiography. *J. Mol. Biol.* **1963**, *6*, 208–213. [CrossRef]
14. Beukers, R.; Berends, W. Isolation and identification of the irradiation product of thymine. *Biochim. Biophys. Acta* **1960**, *41*, 550–551. [CrossRef] [PubMed]
15. Wulff, D.L.; Rupert, C.S. Disappearance of thymine photodimers in ultraviolet irradiated DNA upon treatment with a photoreactivating enzyme from baker's yeast. *Biochem. Biophys. Res. Commun.* **1962**, *7*, 237–240. [CrossRef] [PubMed]
16. Hanawalt, P.C.; Ray, D.S. Isolation of the growing point in the bacterial chromosome. *Proc. Natl. Acad. Sci. USA* **1964**, *52*, 125–132. [CrossRef] [PubMed]
17. Pettijohn, D.; Hanawalt, P.C. Deoxyribonucleic acid replication in bacteria following ultraviolet irradiation. *Biochim. Biophys. Acta* **1963**, *72*, 127–129. [CrossRef]
18. Hill, R.F. A radiation sensitive mutant of *E. coli. Biochim. Biophys. Acta* **1958**, *30*, 636–637. [CrossRef]
19. Setlow, R.B.; Carrier, W.L. The disappearance of thymine dimers from DNA: An error-correcting mechanism. *Proc. Natl. Acad. Sci. USA* **1964**, *51*, 226–231. [CrossRef]
20. Boyce, R.P.; Howard-Flanders, P. Release of ultraviolet light-induced thymine dimers from DNA in E. coli K-12. *Proc. Natl. Acad. Sci. USA* **1964**, *51*, 293–300. [CrossRef]
21. Pettijohn, D.; Hanawalt, P.C. Evidence for repair-replication of ultraviolet damaged DNA in Bacteria. *J. Mol. Biol.* **1964**, *9*, 395–410. [CrossRef]
22. Hanawalt, P.C. The U.V. Sensitivity of Bacteria: Its Relation to the DNA Replication Cycle. *Photochem. Photobiol.* **1966**, *5*, 1–12. [CrossRef] [PubMed]
23. Hanawalt, P.; Haynes, R. The Repair of DNA. *Sci. Am.* **1967**, *216*, 36–43. [CrossRef] [PubMed]
24. Hanawalt, P.C.; Pettijohn, D.E.; Pauling, C.E.; Brunk, C.F.; Smith, D.W.; Kanner, L.C.; Couch, J.L. Repair replication of DNA in vivo. *Cold Spring Harb. Symp. Quant. Biol.* **1968**, *36*, 187–194. [CrossRef]
25. Hanawalt, P.C. Cellular Recovery from Photochemical Damage. In *Photophysiology*; Giese, A.C., Ed.; Academic Press: New York, NY, USA, 1968; Volume IV, Chapter 12, pp. 203–251.
26. Lindahl, T. Instability and decay of the primary structure of DNA. *Nature* **1993**, *362*, 709–715. [CrossRef] [PubMed]

27. Lewis, R.J.; Hanawalt, P.C. Ligation of oligonucleotides by pyrimidine dimers–a missing "link" in the origin of life? *Nature* **1982**, *298*, 393–396. [CrossRef] [PubMed]
28. Ganesan, A.; Hanawalt, P. Photobiological origins of the field of genomic maintenance. *Photochem. Photobiol.* **2016**, *92*, 52–60. [CrossRef]

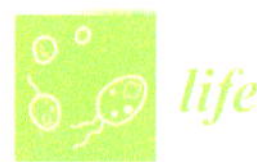

Review

The Bacterial Nucleoid: From Electron Microscopy to Polymer Physics—A Personal Recollection

Conrad L. Woldringh

Bacterial Cell Biology, Swammerdam Institute for Life Sciences (SILS), University of Amsterdam, 1098 XH Amsterdam, The Netherlands; c.woldringh@icloud.com

Abstract: In the 1960s, electron microscopy did not provide a clear answer regarding the compact or dispersed organization of the bacterial nucleoid. This was due to the necessary preparation steps of fixation and dehydration (for embedding) and freezing (for freeze-fracturing). Nevertheless, it was possible to measure the lengths of nucleoids in thin sections of slow-growing *Escherichia coli* cells, showing their gradual increase along with cell elongation. Later, through application of the so-called agar filtration method for electron microscopy, we were able to perform accurate measurements of cell size and shape. The introduction of confocal and fluorescence light microscopy enabled measurements of size and position of the bacterial nucleoid in living cells, inducing the concepts of "nucleoid occlusion" for localizing cell division and of "transertion" for the final step of nucleoid segregation. The question of why the DNA does not spread throughout the cytoplasm was approached by applying polymer-physical concepts of interactions between DNA and proteins. This gave a mechanistic insight in the depletion of proteins from the nucleoid, in accordance with its low refractive index observed by phase-contrast microscopy. Although in most bacterial species, the widely conserved proteins of the ParABS-system play a role in directing the segregation of newly replicated DNA strands, the basis for the separation and opposing movement of the chromosome arms was proposed to lie in preventing intermingling of nascent daughter strands already in the early replication bubble. *E. coli*, lacking the ParABS system, may be suitable for investigating this basic mechanism of DNA strand separation and segregation.

Keywords: electron microscopy; phase-contrast microscopy; bacterial nucleoid; DNA polymer physics; protein depletion; chromosome arms; replication bubble; active and passive DNA segregation

Citation: Woldringh, C.L. The Bacterial Nucleoid: From Electron Microscopy to Polymer Physics—A Personal Recollection. *Life* **2023**, *13*, 895. https://doi.org/10.3390/life13040895

Academic Editors: Ron Elber and Paolo Mariani

Received: 2 March 2023
Revised: 22 March 2023
Accepted: 24 March 2023
Published: 28 March 2023

1. Electron and Light Microscopy

In 1966, the preparation of T2-phages that I purified during my master's degree program was photographed at the Laboratory of Electron Microscopy in Amsterdam by Nanne Nanninga (Figure 1a,b). The preparation was pure and satisfied the director, Dr. Woutera van Iterson, who consequently accepted me later as a Ph.D. student. I also remember how van Iterson and I looked together at the pictures of highly magnified photographs of thin sections of *Escherichia coli* cells, fixed by the Ryter–Kellenberger method [1], in which aggregated DNA threads and poly-ribosomes can be distinguished. As in the case of mesosomes in *Bacillus subtilis* (see below), van Iterson saw in the continuation of DNA threads through the cytoplasm towards the plasma membrane (arrow in Figure 1c), a confirmation of the first model for bacterial DNA segregation, in which Jacob, Brenner and Cuzin [2] proposed a connection between DNA and the plasma membrane.

Nanninga, however, was skeptical. At that time, around 1970, he was involved in applying the freeze-fracture technique to *B. subtilis* cells which were expected to contain mesosomes. These membranous organelles had no clear function, but in the thin sections studied by van Iterson, they were often seen in contact with nucleoids [3]. The shadowed replicas of unfixed, freeze-fractured *B. subtilis* cells, however, did not show any sign of mesosomes; these only appeared in the freeze-fractures when cells were previously fixed (causing permeabilization of the plasma membrane); for instance, with osmium tetroxide used for thin sectioning. A similar phenomenon seemed to occur with the variable visibility in freeze fractures of the *E. coli* nucleoid: the latter could not always be distinguished, probably due to ice crystal formation. These problems and, in addition, the difference in nucleoid appearance between osmium tetroxide and glutaraldehyde fixed [4] led Nanninga to stimulate the development of a confocal scanning light microscope [5,6], which promised to bridge the gap in resolution between electron and light microscopy.

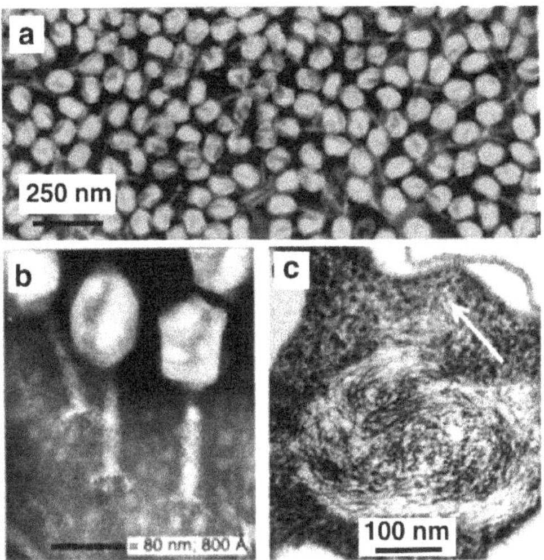

Figure 1. (a,b) Electron micrographs of T2r-bacteriophages, taken by N. Nanninga, 28 March 1966. The preparations were negatively stained with phosphotungstic acid and photographed at an instrumental magnification of 40,000×. (c) Electron micrograph of a thin section of *E. coli* K-12, fixed with osmium tetroxide according to the Ryter–Kellenberger conditions [1]. White arrow points to a presumed DNA-membrane connection.

2. Cell Size, Shape and Growth Models

While in the lab, the interpretation of electron micrographs of fixed or frozen cells led to emotional and unsolvable discussions about mesosomes [7], I had found, in the small library of the institute, the book: "Control of macromolecular synthesis", by Maaløe and Kjeldgaard [8]. Especially intriguing, there was a scheme of the nucleoid and cytoplasm (see their Figure 7-1, at the end of the book [8]). During my Ph.D. program, I also tried to understand the Helmstetter–Cooper model published in 1968 [9]. There was nobody in my surroundings who knew about this model, but there was interest in my study of thin-sectioned nucleoids showing that replication and segregation went hand in hand during the cell cycle [10]. After obtaining my Ph.D. in 1974, I had the opportunity to visit the laboratory of Charles Helmstetter in Buffalo (New York), where I also met Olga Pierucci. Travelling for the first time in the US, and also meeting scientists such as Herb Kubitschek, Arthur Koch and Elio Schaechter, was an impressive and stimulating experience.

Another important stimulating event occurred when I participated at the Lunteren Lectures on Molecular Genetics of 1974. There, I showed measurements of the size and shape of *E. coli* mutant cells [11], prepared by the agar filtration method developed by Kellenberger [12,13]. After my presentation, Arieh Zaritsky approached me with a clear message: "We have to meet and talk about cell shape!". Having already received his Ph.D. at the laboratory of Bob Pritchard (Leicester, UK), Arieh seemed to understand the recent physiological experiments of Maaløe and Kjeldgaard, as well as the Helmstetter–Cooper model that described the coordination between chromosome replication and cell division [14,15]. This was the beginning of a still-lasting cooperation [16] that started with learning how to culture *E. coli* cells under steady state conditions while analyzing shape changes during a nutritional shift-up of cells prepared by agar filtration and understanding the distinction between the two completely different physiological states of "thymine starvation" and "thymine limitation" [17].

Together with Nanne Nanninga, Arieh Zaritsky, Bob Rosenberger, Norman Grover, Wim Voorn and Luud Koppes, the electron microscope measurements of fixed and air-dried cells were compared to growth models that predicted the observed shape of length distributions; we discussed cell elongation modes (linear with a rate doubling or exponential), shape changes and correlations between cell cycle events such as initiation of DNA replication (derived from radio-autograms) and initiation of cell constriction, the so-called *C+D-T* period. In 1993, Voorn, Koppes and Grover, remarked in a short paper [18] that a newly developed "incremental-size model" could not be rejected. Previously, the occurrence of "a constant size increment" during the *C+D-T* period was mentioned in Figure 6 of Koppes et al. [19] and Figure 1 of Koppes and Nanninga [20], suggesting a strong positive correlation between the events of initiation of DNA replication and initiation of cell constriction.

More than 20 years later, the same model was going to basically cause an explosion of studies [21,22]. This revival of the model can be ascribed to Suckjoon Jun, who gave it the name "adder", writing: "The beauty of this "adder" is that it automatically ensures size homeostasis" (see also the Movie S1, "Size convergence by adder principle, related to Figure 3" in [23]). According to this now widely accepted adder model, based on measurements of living cells, often grown in microfluidic devices, cells do not sense their size (sizer model) nor their age (timer model), but add a constant size, between birth and division, that is independent of birth size. Whether and how cells could "sense" a constant size increment in large and small newborn cells is still unknown. However, measuring the amount of DNA in large and small prospective daughter cells in fast-growing *E. coli* cells [24] showed an increased amount of DNA (20% higher) in large siblings. This observation is in agreement with the prediction that large newborns initiate DNA replication earlier [25]. In addition, nucleoid segregation was found to be advanced in these larger prospective daughter cells, allowing them to divide earlier, as to be expected from the adder model. Confirmation of this adder-like behavior based on DNA replication and segregation has to await visualization of differently sized siblings in quantitative time-lapse experiments, as performed by the group of Jaan Männik [26].

3. Nucleoid Occlusion and Transertion

During his short-term EMBO-fellowship visit to Amsterdam in 1977, Arieh Zaritsky proposed to organize together with Nanne Nanninga the first EMBO workshop on bacterial duplication. It was held in 1980 in Noordwijkerhout (The Netherlands) with leaders in the field of bacterial physiology, such as Donachie, Grover, Helmstetter, Koch, Kubitschek, Maaløe, Messer, Pierucci, Pritchard and Schwarz.

Arieh organized the second workshop in Sede Boqer (Israel) in 1984, which I attended after enjoying a sabbatical leave in the lab of Jim Walker at the University of Texas at Austin. While continuing our cooperation, the study of populations of cell division mutants in Amsterdam was greatly facilitated by Norbert Vischer, who listened to our wishes for measuring cell properties and who translated them into practical software for image analysis and visualization of results [27]. This also enabled us to develop an interactive cell cycle simulation (CCS) program [28], which was used for decades to predict behavior of emerging cell-cycle mutants and to teach students the Helmstetter–Cooper model [29].

It was also during this period that, together with Nanninga, Wientjes and Zaritsky and Ph.D. students (Egbert Mulder, Marko Roos, Peter Taschner, Frank Trueba, Jacques Valkenburg and Joop van Helvoort), the concept of "nucleoid occlusion" was developed [30]. The term was coined by Larry Rothfiel and originally applied to the idea that transcriptional activity around the nucleoid occludes the increased rate of peptidoglycan synthesis necessary to initiate constriction [31]. Along with the ideas of Vic Norris [32], our observations on

E. coli nucleoids and quantitative measurements by Evelien Pas, Peter Huls and Norbert Vischer resulted in the formulation of the "transertion model". Observations of an expansion of non-replicating nucleoids by active protein synthesis [33], their compaction and fusion by inhibition of protein synthesis with chloramphenicol [34] and re-segregation after release from inhibition that occurred faster than cell elongation [35] led to the proposal that coupled transcription–translation–translocation of envelope proteins (transertion) could play an active role in DNA segregation (Figure 2).

However, about 20 years later, this idea could be falsified with the help of constructs made by Flemming Hansen (Denmark). Because the positioning of the left (L) and right (R) chromosome arms during replication showed a similar ordering pattern in either growing cells (e.g., L-ori-R L-ori-R or L-ori-R-R-ori-L), or during run-off DNA replication in protein-synthesis inhibited cells, transertion could not play a role in the mere movement of the chromosome arms [36]. This movement was proposed to be the passive result of DNA synthesis itself rather than of active protein synthesis (see Section 5). It should be noted, however, that active transertion influences the ordering pattern of the left and right chromosome arms and is still required for separation and movement of the entire daughter nucleoids into the prospective daughter cells [36].

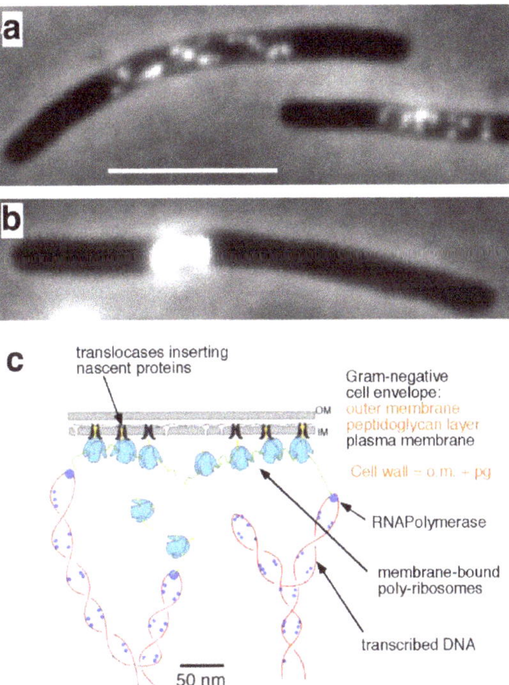

Figure 2. (**a**,**b**) Filaments of *E. coli dnaX* (Ts) grown at restrictive temperature (42 °C) for several mass doublings, fixed with 0.1% osmium tetroxide and stained with DAPI. (**a**) While DNA synthesis stops immediately, cells continue to grow, forming SOS-filaments. During elongation, the original nucleoid is pulled apart into small lobules. (**b**) Upon growth inhibition with 300 μg/mL chloramphenicol the DNA lobules re-compact into a confined region [34]. Bar in (**a**) also holds for (**b**) and represents 5 μm. (**c**) Schematic representation of the "transertion model" [37].

4. Physical DNA Model

However, what about the remaining controversy concerning the dispersed or compact organization of DNA in the bacterial nucleoid? During and after my Ph.D., I remained fascinated by Figure 7-1 of Maaløe and Kjeldgaard [8] and I was glad that, with the help of Theo Odijk, we could make a similar figure, based on our measurements of nucleoid volume [38], on recent data of macromolecular concentrations in *E. coli* cells [39] and on Odijk's free-energy approach of calculating the excluded-volume interactions between soluble proteins and DNA [40]. This so-called depletion theory (see explanation in [41]) formulates the free energy of the system that tends to reach equilibrium by minimizing its total free energy. The theory considers the free energy of self-interactions between DNA supercoils and of cross-interactions between DNA and soluble, cytoplasmic proteins and predicts a phase separation between nucleoid and cytoplasm as described in Figure 3.

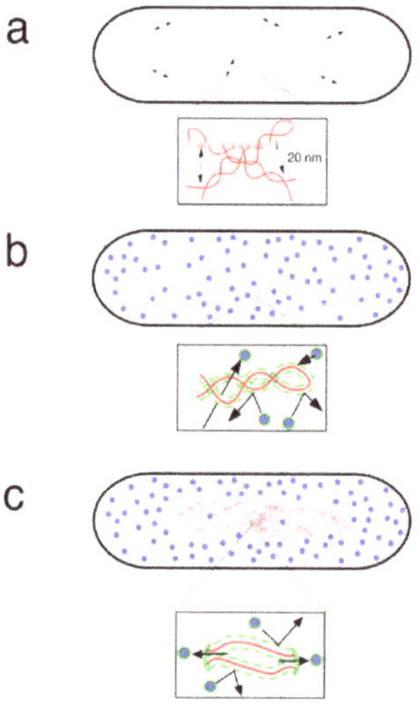

Figure 3. Description of DNA–DNA and DNA–protein interactions that contribute to a phase separation between nucleoid and cytoplasm. (**a**) When the supercoiled DNA is introduced into an empty cell, free energy is stored because the expansion of the network due to the colliding supercoil segments (indicated by double arrows) has to be overcome. (**b**) When, in addition, many proteins are introduced into the cell, a free energy increase occurs that is associated with the cross-interactions between proteins and the DNA helix, both exhibiting an electrostatic depletion radius, indicated by the green zone around the DNA chains. This cross-interaction energy overwhelms the self-interaction energy in (**a**), leading to an unstable situation. (**c**) To minimize the free energy of the total system, a phase separation is established in which the DNA is compacted in a smaller volume with decreased protein-DNA cross-interactions. The latter is obtained because overlapping depletion zones (green areas) between the DNA strands in the compacted nucleoid, squeeze-out (deplete) proteins, resulting in a lower protein concentration and in ~30% reduction in nucleoid density as shown in phase-contrast images [38,42].

Together with colleagues such as the late Michiel Meijer [43], Paul Sloof [44] and, subsequently, with Suckjoon Jun [45], we finally succeeded in reproducibly liberating nucleoids from *E. coli* spheroplasts by osmotic shock and in measuring the size of free-floating nucleoids under different crowding conditions (e.g., PEG; see Figure 9C in [46]). We also calculated the very small diffusion coefficient of a DNA region near *oriC* in isolated nucleoids [47] and, with Steve Elmore, Michiel Müller, Norbert Vischer and Theo Odijk, also in living cells [48].

Finally, a model for the bacterial nucleoid could be developed [49] (see [45] for microfluidics experiments). In the model, the DNA is represented by branched supercoils, partly relaxed through association with DNA-binding proteins and cross-linked by a substantial number of physical entanglements and/or proteins into a homogeneous, core-less network, without any sign of a "highly ordered structure", as often proposed. While during osmotic shock of the spheroplasts, the nucleoids enlarge about 100 fold in volume, the liberated and DAPI-stained nucleoids expand further under continued UV irradiation (Figures 2 and 4 in ref. [49]).

In all our experiments, a distinctive substructure of granules (diameter ~ 2 μm) became visible during this expansion, showing Brownian motion. It is tempting to speculate that these granules correspond with the uncrosslinked blobs calculated by Odijk to have a radius of gyration of ~0.9 μm (see Appendix B in [49]).

5. Segregation of Chromosome Arms

However, how do daughter strands, newly synthesized in such a seemingly homogeneous network (see Figure 5 in [49]), separate and remain unmixed? Although *E. coli* lacks the ParABS system, measurements of fluorescently tagged gene loci showed [50,51] that the two newly synthesized chromosome arms end up as individual domains in different halves of the two daughter nucleoids (Figure 4a, panel 4). This arrangement could be explained by assuming that already at initiation of DNA replication, each of the two replisomes in the replication bubble synthesizes daughter strands that do not mix because of their physical differences (Figure 4b,c). These differences could arise because the leading strands become supercoiled, while in the lagging strands the Okazaki fragments have to first be ligated (Figure 4b). It is proposed that the four nascent strands exclude each other and fold into four individual blobs, screened-off from each other. Their intermingling would require extra excluded-volume interactions and thus, extra free energy (loss of entropy); as a result, the four nascent strands will remain separated in a minimum energy situation. During continued *de novo* DNA synthesis, the blobs may fold into four enlarging and separate domains stabilized by newly recruited nucleoid-associated proteins (NAPs; see review [52]) required for gene expression (Figure 4c).

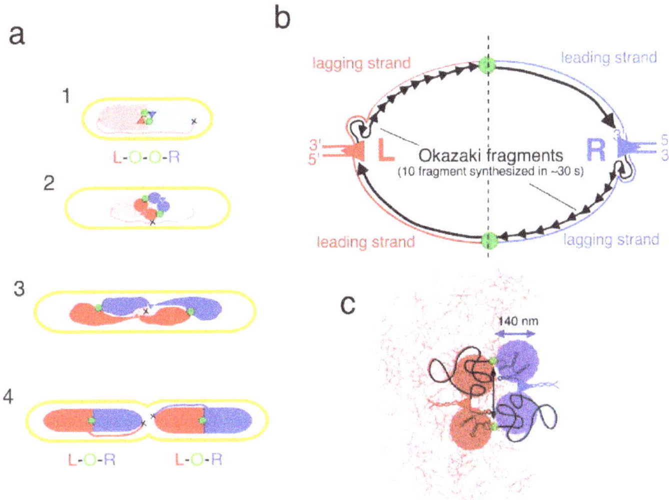

Figure 4. Segregation of chromosome arms in *E. coli*. (**a**) In newborn cells, the two replichores (red and blue) occupy separate regions within the nucleoid as documented in [50,51]. Colored triangles are the replisomes duplicating the left (L; red) and right (R; blue) chromosome arms. The duplicated origins

(green circles) lie in between, giving the pattern L-O-O-R (panel 1). Note the replacement of the unreplicated, parental nucleoids (light colors) by the newly synthesized DNA (dark colors). The two pairs of replichores that are synthesized in the replication bubble are assumed not to become mixed or entangled, but to form four individual domains (dark-colored circles in panel 2; see text) in a transversal arrangement [53,54]. De novo DNA synthesis expands and rearranges the domains (panel 3); they end up as four separate domains in each half of the two daughter nucleoids (panel 4). (**b**) Schematic representation of the replication bubble or orisome [55], ~30 s after initiation, when 10 Okazaki fragments have been synthesized. Physical differences between leading and lagging strands are proposed to prevent the mixing of the nascent strands (see text). (**c**) Hypothetical representation of the same replication bubble as in (**b**), consisting of four enlarging blobs (filled circles) that do not mix. The length of DNA (~10 μm) synthesized in 30 s (see (**b**)) is drawn as a spherical blob with a volume of 0.0014 μm^3 (cf. volume non-replicating nucleoid of 0.24 μm^3; calculation to be published elsewhere). Due to tethering of the replisomes to parental DNA, the origins between the nascent domains are pushed apart (indicated by double arrow). During continued DNA synthesis the blobs fold into enlarging domains (see (**a**), panel 2 and 3).

By comparing the time of initiation and the time of duplication of fluorescent *oriC*-GFP spots in *E. coli*, it became evident that newly synthesized origins separate soon after their duplication [48], without a significant period of "cohesion". It should be noted, however, that in several laboratories, data were obtained that were interpreted to indicate a period of cohesion [56–58]. An early separation, not necessarily incompatible with a transient cohesion period, is to be expected if the replicated origin-DNA in the replication bubble is more mobile than the two replisomes. This could be the case if the replisomes remain tethered to the compact mass of unreplicated parental DNA which they are reeling in. Tethering of the replisomes will force the duplicated origins to move apart (double arrow in Figure 4c). The expanding domains, enlarging through de novo DNA synthesis, will rearrange themselves in the long axis of the rod-shaped cell towards the two halves of the daughter nucleoids in a segregation process that requires no other driving force than continued replication (Figure 4a, panel 3). Similar ideas were expressed by Suckjoon Jun [45,59,60]. The hypothesis that segregation is merely driven by the process of de novo DNA synthesis and accumulation was previously proposed by Alan Grossman [61].

6. Conclusions

Studies of bacterial DNA organization and segregation exhibit two different views: either resolution and movement of replicated daughter strands is performed by a dedicated, active process based on DNA loop extrusions through structural maintenance of chromosome (SMC) complexes [62], or by the passive process of de novo DNA synthesis, as described here. If, in the replication bubble (Figure 4b), initial intermingling of the newly synthesized DNA strands would occur, it is to be expected that the entanglements could only be resolved with an elaborate mechanism of topoisomerases and SMC proteins [62]. However, the different physical properties of the nascent leading and lagging daughter strands (Figure 4b), together with different gene expression activities between the two replichores, could prevent the mixing of the four daughter strands right from the beginning. In that case, the secret of segregation lies in the build-up of the replication bubble: if no initial mixing occurs due to their different physical properties, they will become confined in four individual blobs (Figure 4c) that expand into individual domains (Figure 4a). A similar build-up of replication bubbles and early separation of strands could occur in eukaryotic chromosomes [63].

When a more detailed quantification of the number of proteins involved in the replication bubble will become available, calculations of the free energy state of the proposed four domains, as performed by Odijk for the whole nucleoid (compare Figure 3), could become possible. Such calculations might support the above proposal of passive DNA strand exclusion and formation of the four domains (Figure 4c) that gradually replace the parental nucleoid.

So far, microscopic observations have not given any indication for the existence of these domains. Further developments in spatial light interference microscopy [64], or digital holographic microscopy combined with optical diffraction tomography [65] and improved labeling techniques for nascent DNA strands [66] will be necessary to evaluate the above hypothesis of the four blobs initially created in the early replication bubble and developing into the four domains that end up in different halves of the two daughter nucleoids (Figure 4a).

Funding: This research received no external funding.

Institutional Review Board Statement: Not applicable.

Informed Consent Statement: Not applicable.

Data Availability Statement: Not applicable.

Acknowledgments: Nanne Nanninga, Arieh Zaritsky, Theo Odijk, Norman Grover, Vic Norris, Charles Helmstetter, Jim Walker and Suckjoon Jun are gratefully acknowledged for their encouragements and cooperations throughout the years. In addition, I thank Norbert Vischer, Peter Huls and Sebastian Robalino for discussions and help in preparing this manuscript. Leendert Hamoen is thanked for his hospitality in the laboratory of Bacterial Cell Biology of the University of Amsterdam.

Conflicts of Interest: The author declares no conflict of interest.

References

1. Ryter, A.; Kellenberger, E.; Birch-Andersen, A.; Maaløe, O. Etude au microscope électronique de plasmas contenantde l'acide déoxyribonucléique. I. Les nucléoides des bactéries en croissance active. *Z. Naturf.* **1958**, *13*, 597–605. [CrossRef]
2. Jacob, F.; Brenner, S.; Cuzin, F. On the regulation of DNA replication in bacteria. *Cold Spring Harb. Symp. Quant. Biol.* **1963**, *28*, 329–348. [CrossRef]
3. van Iterson, W. Symposium on the fine structure of bacteria and their parts. II. Bacterial cytoplasm. *Bacteriol. Rev.* **1965**, *29*, 290–325. [CrossRef] [PubMed]
4. Woldringh, C.L.; Nanninga, N. Structure of nucleoid and cytoplasm in the intact cell. In *Molecular Cytology of Escherichia coli*; Nanninga, N., Ed.; Academic Press: New York, NY, USA, 1985; pp. 161–197.
5. Brakenhoff, G.J.; Blom, P.; Barends, P.J. Confocal scanning light microscopy with high aperture lenses. *J. Microsc.* **1979**, *117*, 219. [CrossRef]
6. Valkenburg, J.A.C.; Woldringh, C.L.; Brakenhoff, G.J.; van der Voort, H.T.M.; Nanninga, N. Confocal scanning light microscopy of the *Escherichia coli* nucleoid: Comparison with phase-contrast and electron microscope images. *J. Bacteriol.* **1985**, *161*, 478–483. [CrossRef]
7. Nanninga, N. *Het Vertekende Beeld*; Amsterdam University Press: Amsterdam, The Netherlands, 2007; pp. 123–132.
8. Maaløe, O.; Kjeldgaard, N.O. *The Control of Macromolecular Synthesis*; W.A. Benjamin Inc.: New York, NY, USA, 1966.
9. Helmstetter, H.E.; Cooper, S.; Pierucci, O.; Revelas, E. On the bacterial life sequence. *Cold Spring Harb. Symp. Quant. Biol.* **1968**, *33*, 809–822. [CrossRef] [PubMed]
10. Woldringh, C.L. Morphological analysis of nuclear separation and cell division during the life cycle of *Escherichia coli*. *J. Bacteriol.* **1976**, *125*, 248–257. [CrossRef] [PubMed]
11. Wijsman, H.J.W. A genetic map of several mutations affecting the mucopeptide layer of *Escherichia coli*. *Genet. Res.* **1972**, *20*, 65–74. [CrossRef]
12. Kellenberger, E.; Arber, W. Electron microscopical studies of phage multiplication: I. A method for quantitative analysis of particle suspensions. *Virology* **1957**, *3*, 245–255. [CrossRef]
13. Woldringh, C.L.; de Jong, M.A.; van den Berg, W.; Koppes, L. Morphological analysis of the division cycle of two *Escherichia coli* substrains during slow growth. *J. Bacteriol.* **1977**, *131*, 270–279. [CrossRef]
14. Zaritsky, A.; Pritchard, R.H. Changes in cell size and shape associated with changes in the replication time of the chromosome of *Escherichia coli*. *J. Bacteriol.* **1973**, *114*, 824–837. [CrossRef]
15. Zaritsky, A. On dimensional determination of rod-shaped bacteria. *J. Theor. Biol.* **1975**, *54*, 243–248. [CrossRef] [PubMed]
16. Zaritsky, A.; Woldringh, C.L. Chromosome replication, cell growth, division and shape: A personal perspective. *Front. Microbiol.* **2015**, *6*, 756. [CrossRef]
17. Zaritsky, A.; Woldringh, C.L.; Einav, M.; Alexeeva, S. Use of thymine limitation and thymine starvation to study bacterial physiology and cytology. *J. Bacteriol.* **2006**, *188*, 1667–1679. [CrossRef] [PubMed]
18. Voorn, W.J.; Koppes, L.J.H.; Grover, N.B. Mathematics of cell division in *Escherichia coli*: Comparison between sloppy-size and incremental-size kinetics. *Curr. Top. Mol. Gen.* **1993**, *1*, 187–194.
19. Koppes, L.J.H.; Woldringh, C.L.; Nanninga, N. Size variations and correlation of different cell cycle events in slow-growing *Escherichia coli*. *J. Bacteriol.* **1978**, *134*, 423–433. [CrossRef]

20. Koppes, L.J.H.; Nanninga, N. Positive correlation between size at initiation of chromosome replication in *Escherichia coli* and and size at initiation of cell constriction. *J. Bacteriol.* **1980**, *143*, 89–99. [CrossRef] [PubMed]

21. Amir, A. Cell size regulation in bacteria. *Phys. Rev. Lett.* **2014**, *112*, 208102. [CrossRef]

22. Jun, S.; Taheri-Araghi, S. Cell-size maintenance: Universal strategy revealed. *Trends Microbiol.* **2014**, *23*, 4–6. [CrossRef]

23. Taheri-Araghi, S.; Bradde, S.; Sauls, J.T.; Hill, N.S.; Levin, P.A.; Paulsson, J.; Vergassola, M.; Jun, S. Cell-size control and homeostasis. *Curr. Biol.* **2015**, *25*, 385–391.

24. Huls, P.G.; Vischer, N.O.E.; Woldringh, C.L. Different amounts of DNA in newborn cells of *Escherichia coli* preclude a role for the chromosome in size control according to the "adder" model. *Front. Microbiol.* **2018**, *9*, 664. [CrossRef] [PubMed]

25. Hansen, F.G.; Atlung, T. The DnaA tale. *Front. Microbiol.* **2018**, *9*, 319. [CrossRef] [PubMed]

26. Tiruvadi-Krishnan, S.; Männik, J.; Kar, P.; Lin, J.; Amir, A.; Männik, J. Coupling between DNA replication, segregation, and the onset of constriction in *Escherichia coli*. *Cell Rep.* **2022**, *38*, 110539. [CrossRef]

27. Vischer, N.O.E.; Huls, P.G.; Woldringh, C.L. Object-image: An interactive image analysis program using structured point collection. *Binary* **1994**, *6*, 160–166.

28. Zaritsky, A.; Wang, P.; Vischer, N.O.E. Instructive simulation of the bacterial division cycle. *Microbiology* **2011**, *157*, 1876–1885. [CrossRef] [PubMed]

29. Helmstetter, C.E.; Cooper, S. DNA synthesis during the division cycle of rapidly growing *Escherichia coli* B/R. *J. Mol. Biol.* **1968**, *31*, 507–518.

30. Mulder, E.; Woldringh, C.L. Actively replicating nucleoids influence the positioning of division sites in DNA-less cell forming filaments of *Escherichia coli*. *J. Bacteriol.* **1989**, *171*, 4303–4314. [CrossRef] [PubMed]

31. Woldringh, C.L.; Mulder, E.; Valkenburg, J.A.C.; Wientjes, F.B.; Zaritsky, A.; Nanninga, N. Role of nucleoid in toporegulation of division. *Res. Microbiol.* **1990**, *141*, 39–49. [CrossRef] [PubMed]

32. Norris, V.; Madsen, M.S. Autocatalytic gene expression occurs via transertion and membrane domain formation and underlies differentiation in bacteria: A model. *J. Mol. Biol.* **1995**, *253*, 739–748. [CrossRef]

33. Woldringh, C.L.; Zaritsky, A.; Grover, N.B. Nucleoid partitioning and the division plane in *Escherichia coli*. *J. Bacteriol.* **1994**, *176*, 6030–6038. [CrossRef]

34. van Helvoort, J.M.L.M.; Kool, J.; Woldringh, C.L. Chloramphenicol causes fusion of separated nucleoids in *Escherichia coli* K-12 cells and filaments. *J. Bacteriol.* **1996**, *178*, 4289–4293. [CrossRef] [PubMed]

35. van Helvoort, J.M.L.M.; Huls, P.G.; Vischer, N.O.E.; Woldringh, C.L. Fused nucleoids resegregate faster than cell elongation in *Escherichia coli pbpB*(Ts) filaments after release from chloramphenicol inhibition. *Microbiology* **1998**, *144*, 1309–1317. [CrossRef] [PubMed]

36. Woldringh, C.L.; Hansen, F.G.; Vischer, N.O.E.; Atlung, T. Segregation of chromosome arms in growing and non-growing cells. *Front. Microbiol.* **2015**, *6*, 448. [CrossRef]

37. Woldringh, C.L. The role of co-transcriptional translation and protein translocation (transertion) in bacterial chromosome segregation. *Mol. Microbiol.* **2002**, *45*, 17–29. [CrossRef] [PubMed]

38. Valkenburg, J.A.C.; Woldringh, C.L. Phase separation between nucleoid and cytoplasm in *Escherichia coli* as defined by immersive refractometry. *J. Bacteriol.* **1984**, *160*, 1151–1157. [CrossRef]

39. Bremer, H.; Dennis, P.D. Modulation of chemical composition and other parameters of the cell at different exponential growth rates. In *EcoSal Plus*; Slauch, J.M., Ed.; ASM Press: Washington, DC, USA, 2008. [CrossRef]

40. Odijk, T. Osmotic compaction of supercoiled DNA into a bacterial nucleoid. *Biophys. Chem.* **1998**, *73*, 23–30. [CrossRef]

41. Woldringh, C.L.; Odijk, T. Structure of DNA within the bacterial cell: Physics and physiology. In *Organization of the Prokaryotic Genome*; Charlebois, R.L., Ed.; American Society for Microbiology: Washington, DC, USA, 1999; Chapter 10; pp. 171–187.

42. Mason, D.J.; Powelson, D.M. Nuclear division as observed in live bacteria by a new technique. *J. Bacteriol.* **1956**, *71*, 474–479. [CrossRef]

43. Meijer, M.; de Jong, M.A.; Woldringh, C.L.; Nanninga, N. Factors affecting the release of folded chromosomes from *Escherichia coli*. *Eur. J. Biochem.* **1976**, *63*, 469–475.

44. Sloof, P.; Maagdelijn, A.; Boswinkel, E. Folding of prokaryotic DNA: Isolation and characterization of nucleoids from *Bacillus licheniformis*. *J. Mol. Biol.* **1983**, *163*, 277–297. [CrossRef]

45. Pelletier, J.; Halvorsen, K.; Ha, B.-Y.; Paparcone, R.; Sandler, S.J.; Woldringh, C.L.; Wong, W.P.; Jun, S. Physical manipulation of the *Escherichia coli* chromosome reveals its soft nature. *Proc. Natl. Acad. Sci. USA* **2012**, *109*, E2649–E2656. [CrossRef]

46. Cunha, S.; Woldringh, C.L.; Odijk, T. Polymer-mediated compaction and internal dynamics of isolated *Escherichia coli* nucleoids. *J. Struct. Biol.* **2001**, *136*, 53–66. [CrossRef]

47. Cunha, S.; Woldringh, C.L.; Odijk, T. Restricted diffusion of DNA segments within the isolated *Escherichia coli* nucleoid. *J. Struct. Biol.* **2005**, *150*, 226–232. [CrossRef] [PubMed]

48. Elmore, S.; Müller, M.; Vischer, N.O.E.; Odijk, T.; Woldringh, C.L. Single-particle tracking of oriC-GFP fluorescent spots during chromosome segregation in *Escherichia coli*. *J. Struct. Biol.* **2005**, *151*, 275–287. [CrossRef] [PubMed]

49. Wegner, A.S.; Alexeeva, S.; Odijk, T.; Woldringh, C.L. Characterization of *Escherichia coli* nucleoids released by osmotic shock. *J. Struct. Biol.* **2012**, *178*, 260–269. [CrossRef]

50. Nielsen, H.J.; Ottesen, J.R.; Youngren, B.; Austin, S.J.; Hansen, F.G. The *Escherichia coli* chromosome is organized with the left and right chromosome arms in separate cell halves. *Mol. Microbiol.* **2006**, *62*, 331–338. [CrossRef]

51. Wang, X.; Liu, X.; Possoz, C.; Sherratt, D.J. The two *Escherichia coli* chromosome arms locate to separate cell halves. *Gen. Dev.* **2006**, *20*, 1727–1731. [CrossRef]

52. Norris, V.; Kayser, C.; Muskhelishvili, G.; Konto-Ghiorghi, Y. The roles of nucleoid-associated proteins and topoisomerases in chromosome structure, strand segregation, and the generation of phenotypic heterogeneity in bacteria. *FEMS Microbiol. Rev.* **2022**, fuac049. [CrossRef]

53. Reyes-Lamothe, R.; Wang, X.; Sherratt, D. *Escherichia coli* and its chromosome. *Trends Microbiol.* **2008**, *16*, 238–245. [CrossRef]

54. Wang, X.; Rudner, D.Z. Spatial organization of bacterial chromosomes. *Curr. Opin. Microbiol.* **2014**, *22*, 66–72. [CrossRef]

55. Leonard, A.C.; Grimwade, J.E. The orisome: Structure and function. *Front. Microbiol.* **2015**, *6*, 545. [CrossRef] [PubMed]

56. Bates, D.; Kleckner, N. Chromosome and replisome dynamics in *E. coli*: Loss of sister cohesion triggers global chromosome movement and mediates chromosome segregation. *Cell* **2005**, *121*, 899–911. [CrossRef] [PubMed]

57. Joshi, M.C.; Magnan, D.; Montminy, T.P.; Lies, M.; Stepankiw, N.; Bates, D. Regulation of sister chromosome cohesion by the replication fork tracking protein SeqA. *PLoS Genet.* **2013**, *9*, e1003673. [CrossRef] [PubMed]

58. Wang, X.; Reyes-Lamothe, R.; Sherratt, D.J. Modulation of *Escherichia coli* sister chromosome cohesion by topoisomerase IV. *Genes Dev.* **2008**, *22*, 2426–2433. [CrossRef]

59. Jun, S.; Mulder, B. Entropy-driven spatial organization of highly confined polymers: Lessons for the bacterial chromosome. *Proc. Natl. Acad. Sci. USA* **2006**, *103*, 12388–12393. [CrossRef] [PubMed]

60. Jun, S. Chromosome, cell cycle and entropy. *Biophys. J.* **2015**, *108*, 785–786. [CrossRef]

61. Lemon, K.P.; Grossman, A.D. The extrusion-capture model for chromosome partitioning in bacteria. *Genes Develop.* **2001**, *15*, 2031–2041. [CrossRef]

62. Gogou, C.H.; Japaridze, A.; Dekker, C. Mechanisms for chromosome segregation in bacteris. *Front. Microbiol.* **2021**, *12*, 685687. [CrossRef]

63. Hirano, T. Condensin-based chromosome organization from bacteria to vertebrates. *Cell* **2016**, *164*, 847–857. [CrossRef]

64. Mir, M.; Babacan, S.D.; Bednarz, M.; Do, M.N.; Golding, I.; Popescu, G. Visualizing *Escherichia coli* Sub-Cellular Structure Using Sparse Deconvolution Spatial Light Interference Tomography. *PLoS ONE* **2012**, *7*, e39816. [CrossRef]

65. Oh, J.; Sung Ryu, J.; Lee, M.; Jung, J.; Han, S.Y.; Jung Chung, H.; Park, Y. Three-dimsional label-free observation of individual bacteria upon antibiotic treatment using optical diffraction tomography. *Biomed. Opt. Express* **2020**, *11*, 1257–1267. [CrossRef]

66. Spahn, C.H.; Glaesmann, M.; Grimm, J.B.; Ayala, A.X.; Lavis, L.D.; Heilemann, M. A toolbox for multiplexed super-resolution imaging of the *E. coli* nucleoid and membrane using novel PAINT labels. *Sci. Rep.* **2018**, *8*, 14768. [CrossRef] [PubMed]

Hypothesis

The Nordström Question

William D. Donachie

Institute of Cell and Molecular Biology, University of Edinburgh, Edinburgh EH9 3FD, Scotland, UK;
uilleam4@gmail.com

Abstract: It is suggested that the absolute dimensions of cells of *Escherichia coli* may be set by the separation distance between newly completed sister nucleoids.

Keywords: Initiation volume (*Vi*); unit length; partition

1. Introduction

A bit of background.

In 1967 and 1968, the "*C+D*" model of Steven Cooper and Charles Helmstetter provided a framework via with which to interpret measurements on the patterns of chromosomal DNA synthesis in cultures of *Escherichia coli* B/r cells, at different growth rates (at 37 °C) [1–3].

A possible explanation for the constancy of "C" at moderate to high growth rates was fairly clear: the rate of DNA synthesis at each replication fork could be independent of substrate concentration (i.e., "substrate saturation") and therefore of growth rate. At that time, no explanation could be given for the growth rate independence of "*D*", the number of minutes between the completion of the "*C*" period and the completion of cell division. Although so much is now known about the mechanism of division, the constancy of "*D*" remains a bit of a mystery (although I would guess that it has something to do with the mechanism of growth of the septal ring, and with the polymerisation of FtsZ in particular). However, what was not obvious was that which cells had in common when new rounds of chromosome replication were initiated, thus apparently initiating the cell duplication cycle. Therefore, the question to be asked was, "What do all of these cells have in common at the time of initiation of rounds of DNA replication?".

The answer was in the library. I found the clue in a paper by Schaechter, Maaløe and Kjeldgaard [4], in which they showed that the mean dry weight/cell increased exponentially with the growth rate in asynchronous cultures of *Salmonella typhimurium* growing in "balanced" exponential growth at 37 °C on a variety of different carbon sources.

The next step was easy (having ascertained from Bacteriologist colleagues that *Salmonella typhimurium* was really much the same as *E. coli*!). Assuming that cells grow exponentially in mass over each cell cycle, and using Powell's age distribution equation for ideal asynchronous logarithmic-phase populations [5] with Elio's graph [4], the dry weight/cell could be calculated for cells of all growth rates at the time of the initiation of chromosome replication in each cell cycle. As it turned out, the dry weight/cell was in fact *not* the same at the time of initiation in cells at every different growth rate; however, what *was* the same for cells with doubling times between 60 min and 20 min (the shortest known doubling time for *E. coli* B/r at 37 °C) was the ratio of the cell mass to the number of chromosome origins (*oriC*) at the time every new round of chromosome replication was initiated (r.o.r.). Eureka! . . . sort of.

Therefore, I called this ratio (mass/*oriC*) at the start of each r.o.r., "Initiation Mass (*Mi*)" and submitted a short paper to *Nature* [6]. This paper is no more than an exercise in logic. Noone would have been less surprised than me if the measured values of cell mass at initiation proved to vary from this model, because, if nothing else, the experimental

Citation: Donachie, W.D. The Nordström Question. *Life* **2023**, *13*, 1442. https://doi.org/10.3390/life13071442

Academic Editor: Lluís Ribas de Pouplana

Received: 5 May 2023
Revised: 14 June 2023
Accepted: 15 June 2023
Published: 26 June 2023

measurements of the mean cell dry weight in cultures growing at different rates were scattered around the best-fit regression line. The measured values of the *C* and *D* times were also variable. The composition of cells also changes with the growth rate. In particular, the proportion of cell mass due to ribosomes also changes dramatically with the growth rate, as Maaløe and Kjeldgaard [7] have shown; how could a cell measure its mass anyway? Models for the initiation of chromosome replication, being based on chemical interactions, of course implied that the concentration of reactants, rather than the absolute amounts, was what was important. For this reason (and assuming that cell mass and cell volume were proportional to one another), I later preferred to refer to "Initiation Volume (*Vi*)" rather than Mi.

2. Kurt Nordström's Killer Question

In 2000, I retired and became interested in other things; however, annoyingly, even after 32 years, there was still no satisfactory explanation as to *how* the initiation of chromosomal DNA replication is linked to cell size. For one last try, and because there were lots of interesting new experiments, Garry Blakely and I devised yet another model for periodic initiation [8]. As everyone seemed to agree, it was clear that a critical event was the binding of the DnaA protein to *oriC*; however, although reducing the rate of production of DnaA protein does indeed delay initiation, alas increasing the production of DnaA has only a marginal effect on initiation timing. Therefore, something else must also be limiting for initiation. Our model [8] proposed that competition between DnaA-ATP protein (active form) and DnaA-ADP (inactive) for binding to *oriC*, taken together with the post-initiation inactivation of all DnaA-ATP and a transient block to reinitiation at *oriC*, could provide part of the explanation for the periodic initiation of chromosomal DNA replication linked to cell growth and size. Our model predicted a peak in the DnaA-ATP/DnaA-ADP ratio at every doubling in cell volume.

In 2002, I was invited to give that year's Nordström Lecture in Uppsala. I presented our model and thought that the lecture had been well enough received. However, in his office afterward, Kurt Nordström asked me a question, something like this: "Tell me, Willie, what do you think determines the *actual* volume of Vi?"

I was so glad that Kurt had not asked that question during the seminar, because it left me completely flummoxed. The model (like similar previous models) did not specify the actual value of *Vi*; indeed, it would work for any arbitrary cell volume and give peaks in DnaA-ATP/DnaA-ADP at each doubling of whatever volume that was.

In June 2022, I gave a short "Zoom" presentation at a symposium in Copenhagen ("Major Ideas in Quantitative Microbial Physiology: Past, Present and Future", organised by Suckjoon Jun) because I thought that I had finally come up with a (partial) answer to Kurt Nordström's question. You will decide whether I have!

3. "Unit Length" (*L* microns) Extension between Completion of r.o.r. Defines *Vi* (Cubic Microns)

(The following idealised description of the geometry of *E. coli* cells applies only under conditions in which C + D = 40 + 20 min.)

One of the many attractions of an *E. coli* cell is its simple geometry and consistent mode of growth. To a first approximation, the cell is a cylinder with hemispherical poles. When growing under constant conditions, the cell elongates without change in the width of the cylinder. Again to a first approximation, the rate of cell elongation appears to be exponential.

This is the mode of cell growth for most if not all steady growth rates in different media. However, both the average length and average width of the growing cylinders change with the growth rate, such that the average (length/width) is constant at both low and high growth rates, as shown by Zaritsky [9]. Thus, cells growing exponentially with a growth rate of three doublings/h are, on average, twice as long and twice as wide as cells growing at one doubling/h, and the fast growing cell therefore has a volume at initiation that is four times that of the slow-growing cell).

Figure 1 shows the relative proportions (length and width) of such an ideal cell at 20 min intervals while growing exponentially at either one doubling/h (top) or three doublings/h (bottom).

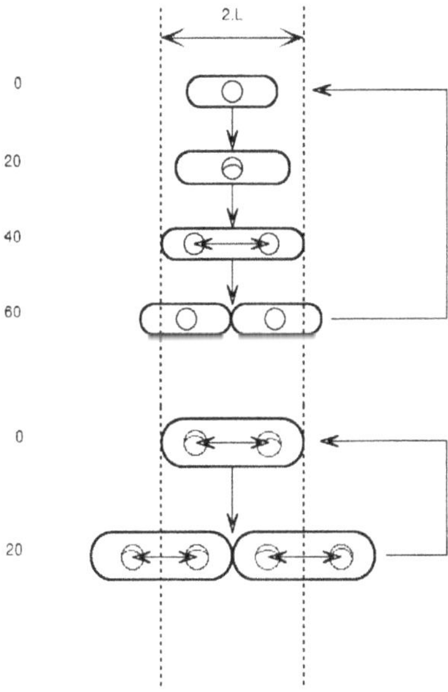

Figure 1. Cell size and shape, chromosome replication, and position during normal cell cycles of E. coli are discussed. The top section shows the cycle for a cell growing in balanced growth with a doubling time of 60 min, and the lower section shows the cycle for a doubling time of 20 min. Cells are drawn to scale (where 2.L, or two unit lengths, is about 2.8 microns). In balanced growth at any constant growth rate, a cell grows exponentially only by elongation, without change in width. However, average cell volume increases exponentially with growth rate [4] although the relative proportions of length to width remain constant [9]. The chromosome is drawn as a small circle (with replication forks when present), but this is not to the same scale. However, the circle is drawn with a diameter roughly equal to that of the nucleoid [10]. The time taken for each replication fork to travel from *oriC* (top of circle) to the terminus (bottom of circle) is 40 min (independent of growth rate) [1–3]. At all growth rates, completion of each round of chromosome replication takes place as the cell reaches a fixed length (2.L) [11,12] and sister nucleoids then separate by a fixed distance, 1.L, which positions them in the centres of the incipient sister cells [10]. Cell division also commences in the cell centre at this cell length (2.L) and takes 20 min to complete, independent of growth rate [1–3].

Despite the differences in the *average* length, width and volume, it is a curious fact that there is a point in the growth cycle when all cells have the same length ("2L"), independent of the growth rate or cell volume [11,12]. We may therefore ask whether there is anything else that takes place only at this stage in the cell growth cycle.

The figure shows three events that occur only at around this cell length:

1. Completion of a round of chromosome replication (r.o.r.) [1–3].
2. Relocation of sister chromosomes from the cell centre to the cell quarters [10,13].
3. Initiation of the septal ring at the mid-point between sister chromosomes [14].

If cells obey these growth rules, then it follows from geometry that Vi (cell volume/origin) must also be constant at the time of the initiation of each round of chromosome replication (i.e., C minutes earlier), independent of growth rate. Therefore, when we ask a question about some aspect of the cell's geometry, we are also asking questions about all the others: volume is defined by length and width, and vice versa. Asking what determines *Vi* is the same as asking what determines *L*. There is little doubt in my mind that *L*, or the unit length, is a real constant independent of growth rate; it is the fixed amount by which cells extend their length between the end of one r.o.r. and the end of the next. Considered this way, we could answer the Nordström question by stating that the constancy and value (microns) of *L* implies the constancy and value (cubic microns) of *Vi*.

4. Does Partition Distance Govern Unit Length?

Our laboratory, like many others, searched for cell cycle genes for over 20 years. From scores of mutants, the genes and proteins specific for cell division have been successfully uncovered and yet no one (to my doubtless outdated knowledge) has found proteins in *E. coli* that are specific for chromosome partition. Mutations in a few proteins (e.g., MukB, DNA gyrase, FtsK) may interfere with sister chromosome separation but none of these proteins provide an actual partition mechanism. Yet, sister nucleoids clearly do (and logically must) move apart after replication. Moreover, they move apart, seemingly quite suddenly, by about half a cell length [10,13] (as they must for an efficient growth and division cycle); therefore, what is the hidden mechanism of partition? Alas, I do not know the answer. However, I remember discussing this problem with Vic Norris in Paris (in a gym!) sometime in the late 1980s. Vic's suggestion was along the lines that, "If sister DNA molecules each have a net negative charge, then they ought to repel one another". In a narrow cylinder such as an *E. coli* cell, the direction of movement would necessarily be in the long axis, while the inverse square law plus the viscosity of the cytoplasm would determine the average distance moved. Could this distance correspond to "*L*"? If it did, then Kurt Nordström's question would be answered: "The invariant increase in cell length/cycle (*L* microns), and hence the value of *Vi* (cubic microns/*oriC*) would have evolved by Natural Selection to fit the partition distance (*L* microns) which in turn is determined by Physics."

It has not escaped my notice that the above suggestion raises any number of questions! e.g.,

1. How is a single replicating molecule of DNA kept together as a single nucleoid (by MukB and Gyrase), but suddenly changed to two mutually repellent nucleoids at the completion of replication? (If it is).
2. How is the extent of cell elongation between the completion of each r.o.r. fixed independent of the rate of volume growth?
3. Am I describing the "adder" phenomenon in another way?
4. Gram-positive rods, such as *Bacillis subtilis*, which do not change width with the growth rate, would require a different process to ensure that sister chromosomes are located in sister cell centres. These organisms, unlike *E. coli*, possess *par* genes.
5. And so on.

5. Thanks and Apologies

When I retired, I deliberately stopped reading about the cell cycle, because by that time, it looked as if most cell-division-specific genes and proteins had been identified and that the main outline of the *E. coli* cell cycle was clear enough. Job done! I blush as I write that but it is what I did. Therefore, all the marvellous new technologies and amazing new insights passed me by, like Rip van Winkle. Nearly 20 years, later I received a phone call from Johan Elf, who kindly thought that I might like to know that *Vi* really is constant. That woke me up briefly, but I went back to sleep almost at once (albeit with a smile on my face!); this was until Suckjoon Jun and his colleagues sent me a draft of their great review [15] and the extent of the whole rejuvenated world of *E. coli* cell cycle research was revealed to me! And then I had to think of something new to say in my little talk for the Copenhagen meeting, and I remembered what Kurt Nordström had asked me 20 years before.

I hope I have not made any embarrassing errors in fact or logic, or tediously reinvented the wheel, but I have enjoyed trying to answer Kurt's question. The regret of course is that

I cannot discuss it with him (although I am sure that he would have had some other killer question!). However, I am glad that I can thank Johan Elf, Ariel Amir (who visited me here, after I missed his seminar!), Arieh Zaritsky (who, amongst other kindnesses, asked me for this contribution) and Suckjoon Jun, for really waking me up!

My final thanks go to Vic Norris, not only for providing the electrifying idea about partition, but also for offering to get this little essay into some sort of publishable form.

Funding: This research received no external funding.

Conflicts of Interest: The author declares no conflict of interest.

References

1. Helmstetter, C.E. Rate of DNA synthesis during the division cycle of *Escherichia coli* B/r. *J. Mol. Biol.* **1967**, *24*, 417–427. [CrossRef]
2. Cooper, S.; Helmstetter, C.E. Chromosome replication and the division cycle of *Escherichia coli* Br. *J. Mol. Biol.* **1968**, *31*, 519–540. [CrossRef]
3. Helmstetter, C.E. DNA synthesis during the division cycle of rapidly growing *Escherichia coli* Br. *J. Mol. Biol.* **1968**, *31*, 507–518. [CrossRef]
4. Schaechter, M.; Maaloe, O.; Kjeldgaard, N.O. Dependency on medium and temperature of cell size and chemical composition during balanced grown of salmonella typhimurium. *J. Gen. Microbiol.* **1958**, *19*, 592–606. [CrossRef]
5. Powell, E.O. Growth rate and generation time of bacteria, with special reference to continuous culture. *J. Gen. Microbiol.* **1956**, *15*, 492–511. [CrossRef]
6. Donachie, W.D. Relationship between cell size and time of initiation of DNA replication. *Nature* **1968**, *219*, 1077–1079. [CrossRef]
7. Maaloe, O.; Kjeldgaard, N.O. *Control of Macromolecular Synthesis: A Study of DNA, RNA, and Protein Synthesis in Bacteria*; Benjamin: New York, NY, USA, 1966.
8. Donachie, W.D.; Blakely, G.W. Coupling the initiation of chromosome replication to cell size in *Escherichia coli*. *Curr. Opin. Microbiol.* **2003**, *6*, 146–150. [CrossRef]
9. Zaritsky, A. On dimensional determination of rod-shaped bacteria. *J Theor Biol.* **1975**, *54*, 243–248. [CrossRef]
10. Begg, K.J.; Donachie, W.D. Experiments on chromosome separation and positioning in *Escherichia coli*. *New Biol.* **1991**, *3*, 475–486.
11. Donachie, W.D.; Begg, K.J.; Vicente, M. Cell length, cell growth and cell division. *Nature* **1976**, *264*, 328–333. [CrossRef]
12. Grover, N.B.; Woldringh, C.L.; Zaritsky, A.; Rosenberger, R.F. Elongation of rod-shaped bacteria. *J. Theor. Biol.* **1977**, *67*, 181–193. [CrossRef]
13. Sargent, M.G. Nuclear segregation in *Bacillus subtilis*. *Nature* **1974**, *250*, 252–254. [CrossRef]
14. Bi, E.F.; Lutkenhaus, J. FtsZ ring structure associated with division in *Escherichia coli*. *Nature* **1991**, *354*, 161–164. [CrossRef]
15. Jun, S.; Si, F.; Pugatch, R.; Scott, M. Fundamental principles in bacterial physiology—History, recent progress, and the future with focus on cell size control: A review. *Rep. Prog. Phys.* **2018**, *81*, 056601. [CrossRef]

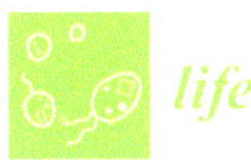

Review

Open Questions about the Roles of DnaA, Related Proteins, and Hyperstructure Dynamics in the Cell Cycle

Masamichi Kohiyama [1], John Herrick [2] and Vic Norris [3,*]

[1] Institut Jacques Monod, Université Paris Cité, CNRS, 75013 Paris, France; masamichi.kohiyama@univ-paris-diderot.fr
[2] Independent Researcher, 3 rue des Jeûneurs, 75002 Paris, France; jhenryherrick@yahoo.fr
[3] CBSA UR 4312, University of Rouen Normandy, University of Caen Normandy, Normandy University, 76000 Rouen, France
*   Correspondence: victor.norris@univ-rouen.fr

Abstract: The DnaA protein has long been considered to play the key role in the initiation of chromosome replication in modern bacteria. Many questions about this role, however, remain unanswered. Here, we raise these questions within a framework based on the dynamics of hyperstructures, alias large assemblies of molecules and macromolecules that perform a function. In these dynamics, hyperstructures can (1) emit and receive signals or (2) fuse and separate from one another. We ask whether the DnaA-based initiation hyperstructure acts as a logic gate receiving information from the membrane, the chromosome, and metabolism to trigger replication; we try to phrase some of these questions in terms of DNA supercoiling, strand opening, glycolytic enzymes, SeqA, ribonucleotide reductase, the macromolecular synthesis operon, post-translational modifications, and metabolic pools. Finally, we ask whether, underpinning the regulation of the cell cycle, there is a physico-chemical clock inherited from the first *protocells*, and whether this clock emits a single signal that triggers both chromosome replication and cell division.

Keywords: Charles E. Helmstetter Prize; *E. coli*; ribonucleotide reductase; sequestration; *oriC*; macromolecular crowding; differentiation; macromolecular synthesis operon; integrative suppression; L-form

Citation: Kohiyama, M.; Herrick, J.; Norris, V. Open Questions about the Roles of DnaA, Related Proteins, and Hyperstructure Dynamics in the Cell Cycle. *Life* **2023**, *13*, 1890. https://doi.org/10.3390/life13091890

Academic Editor: Tao Huang

Received: 18 July 2023
Revised: 29 August 2023
Accepted: 6 September 2023
Published: 10 September 2023

1. Introduction

The coordination of cell growth and chromosome replication is achieved by mechanisms that are still being uncovered. One approach to investigating this coordination is genetics and, over the last half century, this has led to the isolation of conditional lethal mutants of cell division or DNA synthesis. As part of these investigations, in early 1960, Kohiyama started to isolate mutants of *Escherichia coli* K12 that fail to grow at a high temperature. With his collaborators, he found that nearly 1% of colonies obtained at 30 °C from a mutagenized culture failed to grow at 42 °C but, on examining each clone for DNA or protein syntheses and morphological changes after transfer to 42 °C, he found only a few mutants affected in DNA synthesis, with the majority being those defective in protein synthesis, such as valyl-sRNA synthetase [1], or in cell division, without identification of the mutated genes [2]. Isolation of temperature-sensitive (*ts*) mutants continued at the Pasteur Institute and the resulting strain collection has been beneficial to studies on the cell cycle such as the discovery of the FtsZ ring, which is essential for division [3], and to studies on metabolism such as those on the ribonucleotide reductase [4].

The first priority was the elucidation of the regulatory mechanism of chromosome replication as hypothesized in the Replicon Theory [5], according to which DNA replication starts from the genetically defined point (*oriC*) by the action of an initiator. Kohiyama, therefore, sought mutants that failed to initiate replication at high temperatures and found two [6]. These mutations were mapped to the same locus and the gene was called *dnaA* [7].

Further characterization of these *dnaA* mutants demonstrated a close connection between initiation of replication and cell cycle control: at a non-permissive temperature, a *dnaA* mutant temporarily stops dividing and forms filamentous cells [2]; division later resumes towards one end of the filament to produce normal-sized cells that lack DNA [8]. The fact that the size of these anucleate cells is relatively constant (but see [9]) is consistent with the idea that DnaA is involved, directly or indirectly, in the positioning of the division site.

In fact, the possibility that DnaA protein acts as a regulator of gene expression was raised by Hansen a few years after the isolation of the first mutant [10], and DnaA was subsequently shown to regulate many operons [11]. These observations, therefore, help make DnaA a candidate for the role of coordinator of the cell cycle.

To explore this proposal, it is essential to characterise the biochemical properties of the DnaA protein. A large part of this was done by Kornberg and his collaborators using genetic engineering only 20 years after the first isolation of a *dnaA* mutant [12]. They found that the DnaA protein is an ATPase possessing a high affinity for the replication origin (*oriC*) via DnaA boxes constituted of nine bases. The consequence of this interaction is the opening of *oriC*, which allows the insertion of DNA helicase into *oriC* in order to start DNA synthesis after the loading of DNA polymerase III. This interaction between DnaA and DnaA boxes seems to be important in most of the processes in which DnaA is involved [13,14]. DnaA has four domains [15,16] (Figure 1). Domain I binds both the helicase, DnaB, and the DiaA protein that may help link DnaA dimers and monomers; it also has a site for a low-affinity domain I–domain I interaction to generate dimers. Domain II is a linker. Domain III has an AAA+ motif that binds ATP or ADP; this binding of ATP leads to head-to-tail homo-oligomers in a helix; it also binds ssDNA and has a region that interacts with membrane. Domain IV is the dsDNA-binding domain that recognizes DnaA boxes (for references, see [11]).

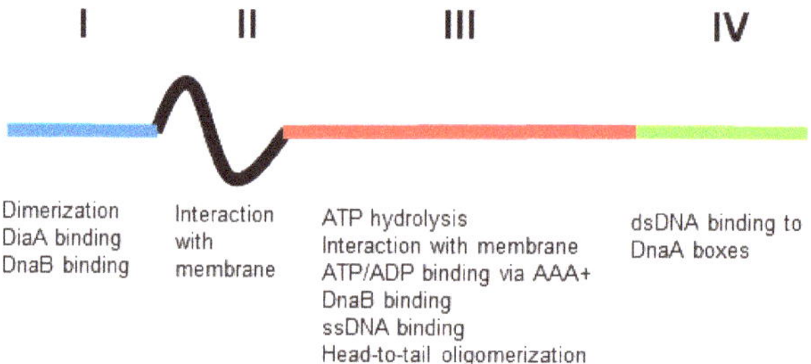

Figure 1. Functional structure of DnaA. The four domains (I, II, III and IV, blue, black, red and green, respectively) have different functions (see text for explanation). Not to scale.

Hyperstructures are large assemblies of molecules and macromolecules that have particular functions; they constitute a level of organization intermediate between the macromolecule and the cell. We have refrained from defining the concept too precisely because it is still "under development" and a prematurely narrow definition might limit its usefulness. In doing this, we adopt the principle of using a "generous Darwinian fog" [17].

Here, we raise questions that need to be addressed to clarify the roles of DnaA and related proteins in the cell cycle. We do this in the theoretical framework of hyperstructure dynamics and in the context of an origins-of-life scenario in which the early cells or *protocells* had a cell cycle regulated by a cellular clock or timer based on physical chemistry. We consider the possibility that DnaA—and the hyperstructures with which it is associated—are the heirs to the original clock and that they integrate different sorts of information in order to coordinate the cell cycle with the environment.

2. The Theoretical Framework for the Questions

We propose that the precursors of cells, *alias* protocells, had a physico-chemical "clock" that emitted a single signal to trigger, simultaneously, the processes of both DNA replication and cell division. If so, it is conceivable that this clock still functions in modern cells but that this functioning now involves sophisticated macromolecules and the complex structures (hyperstructures) into which these macromolecules assemble in order to function. In this framework, we ask how the different hyperstructures involving DnaA and related proteins might be central to the operation of this clock. Along with its partners, the DnaA-based hyperstructure integrates environmental information via the structures of the membrane and chromosome, and via metabolites and ions (Figure 2). The output of this integration is the signal for the cell to make the transition from the non-replicating to the replicating state; the latter state ends with the division that produces daughter cells with possibly different phenotypes as modulated by DnaA. This integration has characteristics shared with actions typical of both AND and OR logic gates insofar as some, but not all, of the information integrated by the DnaA-based initiation hyperstructure is required for it to trigger replication.

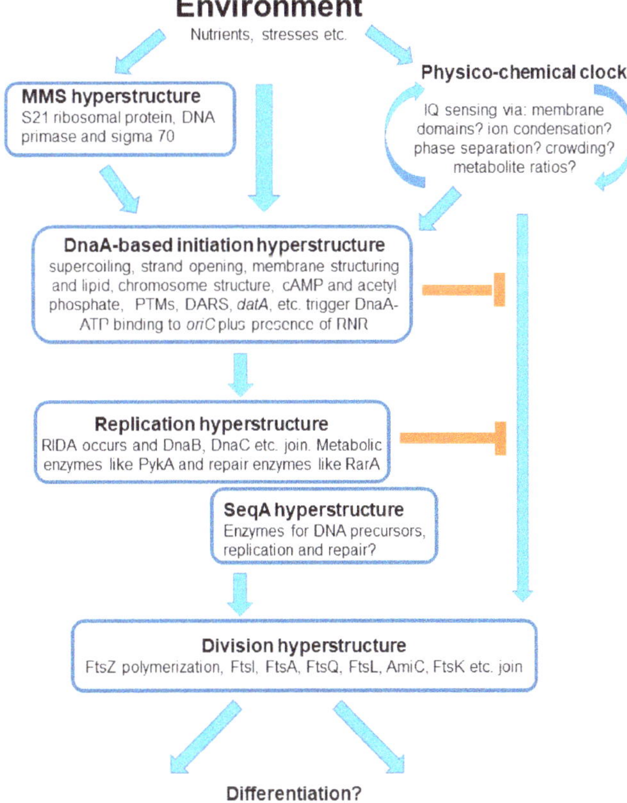

Figure 2. Control of the bacterial cell cycle by hyperstructures and DnaA. The boxes represent hyperstructures. A clock or timer based on physical–chemical processes provides the intensity and quantity (IQ) sensing needed for a successful cell cycle. This clock sends a signal to initiate both chromosome replication (via a DnaA-based initiation hyperstructure) and cell division. The initiation and replication hyperstructures can inhibit the cell division process (orange lines).

3. Crowding, Phase Separation, and the Cell Cycle

Biomolecular condensates are a class of hyperstructures that play major roles in prokaryotic and eukaryotic physiology (for a review see [18]). They form because of phase separation, which is due to transient, low-affinity, cohesive interactions between their constituent polymers, whilst stereo-specific, macromolecular interactions only play a secondary role [19]. These interactions are determined by molecular and macromolecular crowding [20]. Different crowding agents can have different effects, as illustrated by the compaction and DNA binding of HU [21]. This paves the way for hyperstructures to respond in different ways in transducing information from inside and outside the cell. In the case of the initiation of chromosome replication, a low membrane occupancy is needed for the activation or rejuvenation of DnaA [22,23], whilst macromolecular crowding is required for the replication of *oriC* in vitro [24]; in the case of chromosome segregation, phase separation is proposed to underpin the segregation of newly replicated chromosomes [25]; in the case of cell division, macromolecular crowding is required for the formation of FtsZ droplets in vitro [26]. Fundamental questions here include: Are crowding and phase separation the primary determinants of hyperstructure dynamics? Do they determine the assembly and disassembly of hyperstructures? Do they control the fusion and fission of hyperstructures? In short, are they the essence of the clock that controls the cell cycle?

4. Is There a Chromosomal DnaA Hyperstructure?

The chromosomal *datA* site is a 1 kb region that contains binding sites for DnaA and that helps in its inactivation [27]. An excess of *datA* sites results in a delay to the initiation of replication, whereas the lack of *datA* results in extra initiations, and it was originally suggested that the binding of DnaA to newly duplicated *datA* during *oriC* sequestration could help prevent premature reinitiation when *oriC* is desequestered [27]. When *datA* is negatively supercoiled, DnaA-ATP oligomers are stabilised, and *datA*-IHF interactions and DnaA-ATP hydrolysis are promoted [28]. These results are consistent with a *datA*-based chromosomal hyperstructure helping regulate initiation. The two chromosomal intergenic regions, DnaA-reactivating sequence 1 (DARS1) and DnaA-reactivating sequence 2 (DARS2), each contain a cluster of DnaA binding sites; these sites promote regeneration of DnaA-ATP from DnaA-ADP by nucleotide exchange, and thereby help to promote the initiation of replication [29]. (Note that DnaA-ADP is mainly monomeric and unable to go from high-affinity binding sites to nucleate polymerisation at the low-affinity binding sites in the origin of replication.) The reactivation of DnaA by DARS2 is coordinated by the site-specific binding to DARS2 of IHF and Fis; this binding of IHF is temporally regulated during the cell cycle [30], as is the binding of Fis, which occurs specifically prior to initiation [31]. It is thought that DARS1 is mainly involved in maintaining the origin concentration, whereas DARS2 is also involved in maintaining single cell synchrony [32].

After initiation, the ATPase activity of DnaA is stimulated by the regulatory inactivation of the DnaA (RIDA) complex composed of the Hda protein interacting with the DNA-loaded β-clamp [33]. Recently, it has been found that only the disruption of RIDA has a major effect on initiation (since DARS and *datA* can compensate for one another) [34]. All of this raises the question of whether the inactivation of DnaA takes place within a chromosomal DnaA hyperstructure.

DnaA is also a sequence-specific transcriptional regulator. Such regulators bind to sites that are distributed on the chromosome with a periodicity consistent with a solenoidal-type organization [35]; this organization would bring together a regulator and its sites into a hyperstructure. If this hyperstructure does indeed exist, what is its relationship with the *datA* and the DARS sites? Do the above constitute separate hyperstructures and, if so, how do they interact? Or are they all part of an initiation hyperstructure?

5. Is There a Membrane DnaA Hyperstructure?

The involvement of the membrane in the initiation of replication has long been known [36–38], and it is tempting to speculate that the initiation hyperstructure may also contain acidic phospholipids such as cardiolipin, and even that the hyperstructure is physically associated with the bilayer itself, possibly in a fluid state. Indeed, a significant proportion of the DnaA (10% or more) in the cell is associated with the membrane [39,40]. In this association, both domain II and the hydrophobic domain III of DnaA are important [41]. Mutations and deletions affecting the membrane interaction sequence in domain III of DnaA restored growth to cells with a lowered content of acidic phospholipids [42]. DnaA association with the membrane may dissociate ADP from DnaA depending on the degree of protein crowding on the membrane [22,23]. Insertion into or association with the membrane is fundamental to *transertion* (the coupled transcription, translation, and insertion of proteins into a membrane) and to the existence of transertion hyperstructures [43–45]; this raises the question of whether a DnaA hyperstructure based on transertion might exist for part of the cell cycle and, moreover, whether such transertion might be important for the existence and/or operation of a chromosomal DnaA or other hyperstructures.

6. Does the Initiation Hyperstructure Contain Glycolytic Enzymes?

Metabolism is coupled to DNA synthesis through nutrient richness and growth rate in a variety of ways. One way this occurs is via (p)ppGpp [46], with the transcription of *dnaA* in *E. coli* [47] and the level of DnaA protein in *C. crescentus* [48] being lowered by (p)ppGpp. Another way is via a central carbon metabolism that, in *E. coli*, can: (1) promote DnaA to its active DnaA-ATP form and its binding to *oriC* by cAMP (a regulator of this part of metabolism) [49]; (2) suppress the defects of the *dnaA46* mutant by changes in pyruvate and acetate metabolism [50]; (3) inhibit DnaA conversion to DnaA-ATP and its binding to *oriC* (via acetylation of DnaA, with acetyl-CoA and acetyl-phosphate as donors) (for references see [51]). Evidence for the involvement of the central carbon metabolism in DNA replication in *Bacillus subtilis* includes: (1) subunits of pyruvate dehydrogenase (PdhC) and related enzymes bind the origin of replication region, DnaC and DnaG inhibit the initiation of replication; (2) mutations in the genes of central carbon metabolism suppress initiation and elongation defects in *dnaC*, *dnaG*, and *dnaE* mutants; (3) mutations in *gapA* (which encodes glyceraldehyde 3-phosphate dehydrogenase) perturb the metabolic control of replication; (4) pyruvate kinase (PykA) can both inhibit initiation and stimulate elongation via proposed interactions with DnaC, DnaG, and DnaE as modulated, perhaps, by phosphorylation [51,52].

Given that metabolic enzymes can exist as hyperstructures in their own right, a question that arises here is whether metabolic enzymes are also part of an initiation hyperstructure. And a related question is whether the metabolites themselves are directly part of initiation and/or replication hyperstructures, not just by binding to the constituent proteins (as in the case of DnaA and cAMP [49]) or by being used to modify the protein (as in the case of DnaA and acetyl-CoA), but also by binding directly to RNA or DNA. If metabolites were indeed to bind the origin region, this would make the connection with the *Ring World*, an origins-of-life scenario in which small, double-stranded DNA rings were selected firstly because they catalysed the reactions of central carbon metabolism [53].

7. Does the DnaA-Initiation Hyperstructure Contain SeqA?

E. coli avoids multiple reinitiations of chromosome replication by a sequestration mechanism that depends on the SeqA protein binding preferentially to newly replicated, hemi-methylated GATC sites, many of which are clustered in *oriC*. This sequestration, which involves the membrane, occurs only when *oriC* is hemi-methylated [54] and when SeqA, which has an affinity for the membrane, is present [55,56]; the result is an inhibition of initiation [57]. SeqA forms multimers and SeqA-DNA complexes can cover 100 kb of DNA and are close or integral to the replication hyperstructure(s), and have a bidirectional movement that differs from that of the origins (which goes to the poles) [58,59]. GATC sites are clustered not only in the *oriC* region but also in many genes involved in the replication and repair of DNA (such as *dnaA, dnaC, dnaE, gyrA, topA, hepA, lhr, parE, mukB, recB, recD,* and *uvrA*), as well as genes involved in the synthesis of the precursors of DNA (such as *nrdA, purA, purF, purL, pyrD* and *pyrI*), consistent, perhaps, with the presence of these genes and their products in a replication hyperstructure [60]. One related question is whether the DnaA-initiation hyperstructure in its earliest form contains SeqA and, reciprocally, a second question is whether the replication hyperstructure contains DnaA?

8. What Is the Relationship between Strand Opening and DnaA Binding?

Kornberg's group first showed that DnaA opens *oriC* in vitro depending on the presence of ATP [61]. The opening of *oriC* was detected by P1 nuclease sensitivity in this case, and repeatedly shown by other techniques such as $KMnO_4$ modification [62] and DMS footprinting [63]. *oriC* contains (1) the DnaA-ATP-Oligomerization Region (DOR), with twelve DnaA boxes, and (2) the neighbouring Duplex Unwinding Element (DUE), which contains three AT-rich 13-mer repeats along with DnaA binding motifs. DnaA-ATP assembles into a pentamer via its binding to DnaA boxes in one half of the DOR and, progressively, via its binding to the single-stranded motifs in the DUE, which stabilizes the unwound DUE. This unwinding is promoted by the nucleoid-associated protein (NAP), IHF, or indeed by HU [14]. Katayama's group analysed, also by P1 nuclease sensitivity, the opening of M13 *oriC* DNA by DnaA in vitro at the DUE using various types of mutated *oriC* and IHF; the results obtained were consistent with those obtained in vivo [64]. That said, it is difficult to follow the kinetics of *oriC* opening with these techniques.

Strick's group performed a single molecule analysis on DnaA–*oriC*(2kb) interaction using an optical magnetic tweezer to follow the rapid kinetics of double-stranded DNA opening. They observed formation of stable complexes between supposedly DnaA-ATP oligomers and *oriC* with different degrees of positive supercoiling. The formation of these complexes occurred using an *oriC* that lacked the DUE, raising the question of whether they were studying a non-canonical reaction. Other questions include why the kinetics of the complex formation was not studied, why the formation of the complex did not occur constantly [65], and whether DnaA was actually present in the complex. It should be pointed out that the use of optical magnetic tweezers is technically demanding: it requires the attachment of *oriC* DNA to a magnetic bead followed by the selection of intact *oriC*-containing beads (which are easily damaged and consequently in a minority).

Techniques based on minicircles of DNA facilitate the detection of fine-scale modifications to the DNA structure. Using an *oriC* minicircle of 641 bp with three negative supercoils, Landoulsi and Kohiyama found that around 80% of this substrate was positively twisted three times during incubation with DnaA and that the efficiency of unwinding was affected by the degree of negative superhelicity of the minicircle (three negative turns proved more effective in causing unwinding than two or four negative turns). Unwinding of this *oriC* minicircle by DnaA was verified by Bal31 sensitivity (rather than by P1 nuclease sensitivity), whilst the presence of DnaA on the unwound minicircle was confirmed by an anti-DnaA antiserum. The problem raised by this work is that the unwinding did not require ATP [66]. It should also be noted that the above work on *oriC* minicircles depends on a sophisticated technique that requires the formation of circles from a linear 641 bp *oriC* fragment that can only be achieved in a glass capillary after overnight incubation in the presence of DNA ligase and ethidium bromide, which introduces superhelicity; modification of the superhelicity of minicircles resulting from DnaA action is scored after Topo I treatment and is not directly measured.

This work raised the question of whether or not the ATP-dependent opening of *oriC* by DnaA (as demonstrated by P1 nuclease sensitivity) is the unique pathway for the initiation of replication. The fact that the mutant isolated first, *dnaA46*, which has lost the ATP binding site, can grow normally at a low temperature indicates the existence of an alternative pathway whereby *oriC* can fire without ATP. Consistent with this, the growth of *dnaA46* is more sensitive than the wild type to gyrase inhibitors [67], whilst the opening of *oriC* minicircles by DnaA is sensitive to negative supercoiling densities [66] (see above). Another explanation, offered by Kaguni's group, is that the DnaA46 protein, with the aid of DnaK, can form a structure similar to that of DnaA-ATP [68]. Although no data are presently available from X-ray crystallography of the whole molecule of DnaA or from cryoEM analysis of DnaA-*oriC*, significant advances have been made by the Berger group using DnaA from *Aquifex aeolicus* along with the nonhydrolyzable ATP analog AMP-PCP [15], and by Katayama and collaborators using a combination of biochemistry and computer simulations to model the central part of the *oriC*-DnaA-IHF complex [16].

9. Does DnaA Participate in Differentiation?

In the strand segregation hypothesis, a coherent phenotypic diversity is generated by the segregation of certain hyperstructures with only one of the parental DNA strands [69]; candidate hyperstructures for such asymmetric segregation include those containing the NAPs and the topoisomerases. An asymmetric segregation of a chromosomal DnaA hyperstructure is another seductive possibility: could DnaA play a particular role in generating phenotypic diversity (e.g., in preparing a population to confront stresses via its role in modulating gene expression) or in connecting different phenotypes with different patterns of the cell cycle—or indeed both?

10. What Modifications Does DnaA Undergo and What Are Their Roles?

It has been proposed that a hyperstructure might be assembled if enzymes (such as protein kinases and acetyltransferases) and their NAP substrates were to associate with one another in a positive feedback loop in which, for example, the modification of an NAP by its cognate enzyme increases the probability of colocation of both the NAPs and the enzyme [69]. In line with this, the acetylation of a lysine residue (K178) prevents DnaA from binding to ATP and inhibits initiation, whilst the acetylation of another lysine residue (K243) also inhibits initiation but does not affect the ATP/ADP binding affinity of DnaA or the ability of DnaA to bind to the *dnaA* promoter region and to DARS1 [70].

DnaA binds cAMP with a Kd of a similar order to that with which it binds to ATP; indeed, the affinity of DnaA for cAMP is such that most of the cell's DnaA should be bound to cAMP when the latter is present at the physiological concentration of 1 μM [49]. cAMP bound to DnaA is chased by ATP but not by ADP (note that there is only one cAMP binding site on the protein [49]). In vitro, cAMP stimulates DnaA binding to *oriC* and to DnaA sites elsewhere in the chromosome [49]; in vivo, the addition of cAMP to a *cya* mutant (which encodes the adenylate cyclase that catalyses the production of cAMP) increased the level of DnaA [71]. Significantly, despite DnaA's stability in vivo, it has recently been shown to be degraded in vivo in ATP depletion conditions [72], and one possibility is that cAMP helps to both protect DnaA from degradation and regenerate DnaA-ATP from DnaA-ADP (by causing the release of the bound ADP). This raises the question of whether the state of the environment as reflected in a cAMP signal is transduced by the level of DnaA and by the DnaA-ATP: DnaA-ADP ratio into the expression of DnaA-regulated genes and cell cycle timing.

In the case of *Caulobacter crescentus*, the phosphorylation status of CtrA is central to cell cycle progress [73]. The many possible post-translational modifications to DnaA and to other proteins in the initiation and replication hyperstructures, therefore, include phosphorylation, and several other modifications, such as succinylation, methylation, pro-prionylation, malonylation, deamidation of asparagines, and glycosylation (for references see [69]). Another post-translational modification—and one that is largely ignored—is the covalent addition of poly-(R)-3-hydroxybutyrate (PHB) to proteins [74]; one proposed function of such addition to NAPs would be to regulate their interaction with nucleic acids [75]. An important question is, therefore, whether DnaA undergoes modifications like the addition of PHB and, if so, does such modification help the type of hyperstructure into which DnaA assembles?

11. Is DnaA a Controller of Chromosomal Copy Numbers Rather Than a Timer?

Fralick found that the timing of initiation and the number of replicating chromosomes per cell (and the DNA/mass ratio) could be varied independently of one another in a temperature-sensitive *dnaA(ts)* mutant grown at different temperatures. These results were interpreted as DnaA being an essential component of the "replication apparatus" but not itself being the signal that triggers initiation [76,77]. This interpretation would be consistent with the finding that the time of initiation is not advanced by a 50% increase in the concentration of DnaA-ATP, with the authors concluding that although DnaA protein is required for initiation of synchronous and well-timed replication cycles, the accumulation of DnaA-ATP does not control the time of initiation [78]. It should be noted that stopping the transcription of *dnaA* only led to a small increase in cell size, as DnaA was diluted by growth, whilst only disrupting RIDA had a major effect on initiation [34]. Finally, a mathematical model has recently been proposed that combines the titration- and activation-of-DnaA strategies to explain how initiation might be timed at fast and slow growth rates and to give both a precise volume per origin and a constant volume between initiations [79].

12. Does the MMS Operon Play an Important Role in Initiation?

The macromolecular synthesis (MMS) operon is highly conserved [80]; in *E. coli*, it comprises three genes: *rpsU*, which encodes the S21 ribosomal protein, *dnaG*, which encodes the DNA primase involved in the initiation of chromosome replication, and *rpoD*, which encodes the principal sigma subunit of RNA polymerase (sigma70, the "house-keeping" sigma). This operon is subject to a complex pattern of internal and external regulation in which it is possible to regulate each of its three genes independently of the others. In a series of investigations of heterogeneous responses to environmental stresses in *Listeria monocytogenes*, it was found that acid and other stresses primarily selected for *rpsU* variants, in some of which 116 genes were upregulated, mainly those controlled by the alternative stress sigma factor SigB; this leads to the hypothesis (1) that single amino acid substitutions in RpsU enable *L. monocytogenes* to switch between high fitness–low stress resistance and low fitness–high stress resistance and (2) that RpsU interacts with the stressosome, a stress-related hyperstructure responsible for integrating information about multiple environmental stresses and transmitting this as signals [81,82]. How might this relate to hyperstructure dynamics? If the MMS operon exists as a hyperstructure based on coupled transcription–translation, speculative hypotheses that might be entertained include an MMS hyperstructure being physically associated with an initiation/replication hyperstructure or a ribosomal hyperstructure; such association could then supply newly synthesized proteins directly to the appropriate hyperstructure (in the case of the primase, for lagging strand synthesis). Alternatively, the association between the MMS hyperstructure and another hyperstructure could result in the sequestering of newly synthesized proteins leading, in *E. coli*, for example, to a reduction in the level of sigma70 thereby favouring the other sigma factors and the emergence of a stress-adapted phenotype.

13. Does DnaA or a DnaA-Based Initiation Hyperstructure Also Trigger Division?

DNA replication is clearly coupled to cell division, insofar as signalling systems exist to prevent division when DNA has been damaged [83]. These include the SOS system that, when induced by DNA damage, produces the SulA/SfiA protein (along with forty other proteins) to interfere with the action of the key protein in cell division, FtsZ [84]. Using synchronised populations of *E. coli*, it was found that the levels of *ftsZ* mRNA increase at the time of initiation of replication [85,86]. That said, different results have also been obtained [87]. Significantly, the 2 min, or *dcw*, cluster of genes in *E. coli* contains three DnaA boxes upstream of *ftsZ* (within *ftsQA*) that were, however, not found to affect *ftsZ* expression in the conditions tested [86,88]; their role, therefore, remains an intriguing, open question. For example, could these boxes serve to connect physically, via a DnaA polymer, the 2 min transertion stage of the division hyperstructure with the initiation hyperstructure? In *B. subtilis*, DnaA binds the promoter region of *ftsL*, which encodes a key cell division protein [89], whilst, in *C. crescentus*, DnaA binds in vitro to the promoter region of *ftsZ*, which is believed to be part of the DnaA regulon that coordinates the initiation of DNA replication with cell cycle progression [90,91].

In *E. coli*, the putative coupling of DNA replication to cell division is also supported by the fact that several *ts* mutants affected in the initiation or elongation steps of DNA replication stop dividing normally at a non-permissive temperature and form filamentous cells that resemble those formed by a *thy* mutant during thymine starvation [92]. This cessation of division does not, however, necessarily mean that some aspect of the replication of the chromosome (including termination of replication) is responsible for the initiation of cell division. Indeed, after further cultivation of *ts* replication mutants at the non-permissive temperature, division resumes towards the ends of the filamentous cells to produce cells that lack chromosomal DNA; these *anucleate* cells are of almost normal size [92,93]. This production of anucleate cells occurs with *dnaA*, *dnaC*, *dnaG* (*parB*), and *dnaB ts* mutants and requires the absence of the inhibitor of division, SulA/SfiA, and a mutation in *ftsZ* (*sfiB*) [92]. How might this production occur?

It may be significant that the above anucleate cell production also requires cAMP (via either the activity of the wild type *cya* gene or an exogenous supply of cAMP) along with the cAMP receptor protein, CAP [92]. CAP regulates the transcription of over 100 genes in *E. coli*, including those in the *lac* operon. It is therefore conceivable that (1) there are major differences in the structure of the membrane in the presence and absence of cAMP and CAP and (2) these differences could affect transertion (e.g., via Lac permease) and, hence, the membrane domain dynamics that are proposed to time and position division [94,95]. If membrane dynamics do, indeed, underpin the regulation of the cell cycle at a fundamental level, it would make sense for proteins such as DnaA to respond to this fundamental system too, given that these sophisticated proteins presumably evolved some time after protocells had achieved some mastery over replication and division.

It could also be argued that it would have made sense for the earliest protocells to have had the same signalling mechanism leading to both DNA replication and cell division. This is because the RNA and/or DNA in these protocells was probably short and, in the *Ring World* scenario, in the form of a population of ds RNA/DNA rings, each of which catalysed a different reaction [53,96]; hence, the fundamental problem that protocells had to solve was not how to divide after replicating a long stretch of DNA, but rather how to proceed successfully through a complete cell cycle, which is a single decision. Norris has proposed that making this decision requires both intensity-sensing (does a cellular constituent risk becoming limiting for growth?) and quantity-sensing (is there enough material to make viable daughter cells?) [97]. Once a signalling mechanism had been adopted, it would be understandable if modern cells had been constrained to have retained the essence of this mechanism (even if overlain by the complex web of modern macromolecules). It turns out that there is some evidence, based on the relationship between the physical properties of the membrane and the distribution of the nucleoids, consistent with the idea that the initiation of replication and the initiation of division might indeed be triggered at the same time and, if so, logically by the same process (like transertion) [98]. Phospholipids are not just associated with the initiation hyperstructure but also with the division hyperstructure, and in *B. subtilis*, for example, most of the phospholipid synthases are located in the membrane part of the hyperstructure, which is enriched in cardiolipin and phosphatidylethanolamine. Another finding consistent with a close relationship between DNA replication and cell division is that an excess of DnaA (or an effective excess due to a deletion of *datA*) resulted in cell division in the absence of replication to generate anucleate cells [99]. In terms of hyperstructures, one explanation for the dependence of anucleate cell production on cAMP by *dna(ts)* mutants is that the membrane dynamics driving both replication and division hyperstructures are affected by cAMP. For example, cAMP not only affects replication via its binding to DnaA [49] but also affects division [100] via, we propose, the composition and structure of the membrane and cytoplasm. This is because cAMP is central to the induction or repression of many genes, including those in the *lac* operon, which contains a membrane protein, the Lac permease, that has preferences for the physical state and lipid composition of the membrane [101–104] and that affects its bending rigidity [105]. Given that a Lac transertion hyperstructure contains hundreds of macromolecules, the alteration of the membrane and cytoplasm by this cAMP-dependent hyperstructure could well affect division.

14. Do Variations in the Speed of the Elongation Step of DNA Replication Matter?

Oscillations in the speed of the replisome along the *E. coli* chromosome have been reported, and possibly explained as being due to the initiation of new replisomes slowing the progress of existing ones [106]. Temporal oscillations in the speed of the replisome have also been found by others in *E. coli*, *Vibrio cholerae*, and *B. subtilis*, with short pauses at ribosomal genes [107]. These oscillations also showed a time-dependent or bilateral symmetry about the origin, consistent with global variations in the availability of an element essential for replication (see below). Significant variations in the level of ATP between individual *E. coli* cells have been reported [108], whilst complex oscillatory variations in this level occur in individual cells during the cell cycle, with an average maximum of 2.4 mM and minimum of 1.2 mM [109]. Variations in the speed of replication in different places in the chromosome have been proposed to help determine the phenotype [110]. Could such variations be studied at the level of single cells? One technique that might be used is the *CIS* technique (for *Combing and Imaging by Secondary Ion Mass Spectrometry*), which can detect individual DNA fragments labelled in vivo with stable isotopes on the scale of a few hundred base pairs [111,112].

15. Does an Initiation Hyperstructure Sense DNA Supercoiling?

The supercoiling state of chromosomal DNA varies according to the growth phase and to extracellular stresses such as osmotic shock, heat, pH, and antibiotics [113–115]. It can also vary along the chromosome and can form a spatiotemporal gradient running from replication origin to terminus on both arms of the *E. coli* chromosome [116]. DNA gyrase has been proposed to act as a negative regulator of DnaA-dependent replication initiation from *oriC* in *B. subtilis*, since gyrase activity decreases DnaA association with *oriC* and inhibits replication initiation [117]. A deficiency of Topoisomerase I increases negative supercoiling, which results in the formation of transcription-associated RNA-DNA hybrids (R-loops), and DnaA- and *oriC*-independent constitutive stable DNA replication [118]. In other words, the initiation hyperstructure can take more than one form in response to different inputs such as supercoiling and the state of DnaA, thereby acting as a logic gate.

16. Is Ribonucleotide Reductase an Essential Constituent of the Initiation Hyperstructure?

The initiation and elongation steps of chromosome replication are tightly coordinated and mutually dependent in all organisms: the inhibition of initiation results in an increase in elongation rates and vice versa [119–123]. The observation of a negative correlation between initiation and elongation suggests that either directly or indirectly, initiation of DNA replication and elongation of DNA synthesis are interdependent. In eukaryotes, under physiological conditions, a clear negative correlation has been observed between replicon size (length of DNA replicated bidirectionally) and DNA replication fork rates, while inhibiting elongation at replication forks induces the activation of additional replication origins termed "dormant origins" [124]. Could this interrelationship between DNA replication initiation and elongation involve ribonucleoside diphosphate reductase (RNR), which supplies the deoxyribonucleotides (dNTPs) that are essential for replication? Could there be a relationship between the oscillations in ATP during cell growth [109], the oscillations in the speed of the replisome with its pauses at ribosomal genes [107], and the activity of RNR—indeed, could a need to divert ribonucleotides into dNTPs be one explanation for why there is no transcription during the S phase in eukaryotes?

The pool of available dNTPs is critical for successful replication since, with a defective supply, the DNA is likely to be damaged. Decreases in the size of the dNTP pool result in increases in the C period and *vice versa*, consistent with a major role for this pool in replication speed [125–128]. It would make no apparent sense then for initiation of replication to occur without the availability of this pool—and without the ability of RNR to supply dNTPs continuously at the right rate. Localisation of RNR is consistent with this enzyme being part of a replication hyperstructure [129,130] and, importantly, only a very small pool of dNTPs accumulates in the cell, which would allow for no more than one half minute of replication [131,132]. This would suggest the need for ribonucleotide reductase to be present and active at or near the replication forks both at the time of initiation and during elongation.

The question then is whether RNR must be functioning for initiation to occur—and, possibly, functioning in the right place? Put differently, could RNR act as a sensor—or allow the initiation hyperstructure to act as a sensor—in order to couple metabolism and cell growth with replication? If so, could the very activity of RNR determine its presence in the initiation hyperstructure and the ability of this hyperstructure to trigger replication, as proposed for functioning-dependent structures [133]?

The rate of replication fork movement in all organisms depends on the activity of the enzyme RNR, which controls the level and balance of dNTP pool sizes during DNA synthesis (the S phase in eukaryotes and the C period in bacteria). The rate of replication fork movement also depends on a number of other factors, including the lagging strand DNA polymerase DnaE in *B. subtilis* [134] and the replication elongation factor DnaX in *E. coli* [135]. Consistent with an initiation–elongation regulatory circuit, it has been found in *E. coli* that DnaA regulates the *nrdAB* gene, which encodes RNR [136–139]. Low levels/concentrations of DnaA-ATP stimulate *nrdAB* expression (presumably prior to initiation), whereas high levels inhibit *nrdAB* expression (presumably at the time of initiation). Various studies have shown that DnaA-ATP modulates the level of *nrdAB* transcription and RNR activity, such that the active DnaA-ATP form of the protein correlates with both the number of replication forks and dNTP levels [138,139]. Genetic evidence for the existence of such a regulatory circuit has been established with the identification of suppressors of elongation mutants (*dnaX2016*) in *E. coli* that usually map to the *dnaA* gene in both *E. coli* and *B. subtilis* [135,140], while suppressors of the mutant *hda* gene, which overinitiates DNA replication, map to the *nrdAB* gene [141,142]. Suppressors of the *dnaAcos* mutant, which also overinitiates DNA replication, likewise map to the *nrdAB* locus [143,144].

As noted above, initiation in *E. coli* is a complex process involving the formation of a multi-component replication hyperstructure (composed of RNR, DnaA, and possibly SeqA and other factors) that activates initiation at a specific cell mass called the "initiation mass" [129]. The initiation mass is independent, it appears, of cell growth rate. How the cell "senses" the initiation mass, and therefore "knows" when to initiate chromosome duplication, has remained a mystery since the initiation mass concept was first introduced over fifty years ago. It is known, however, that neither DnaA, which controls the frequency of initiation, nor RNR, which controls the rate of DNA synthesis, appears to be the primary determinant of the timing of initiation or of the setting of the initiation mass [78,139] (see above Section 11: **Is DnaA a Controller of Chromosomal Copy Numbers Rather than a Timer?**)

Cells with reduced dNTP levels, however, initiate DNA replication earlier in the cell cycle (immediately after cell division) compared to wild type cells, but at a relatively larger cell size and hence at the same initiation mass [127]. This observation might suggest a link between the rate of elongation and the initiation mass itself, since DNA replication is coupled to cell growth, albeit by an unknown mechanism [134]. This raises an interesting question: does the rate of elongation, which is coupled to cell growth in the mother cell, set the initiation mass in the daughter cells, instead of the initiation mass setting the time of initiation in the daughter cell cycles? If so, then the rate of elongation, rather than the initiation mass itself, might be the decisive parameter that determines the major events driving the bacterial cell cycle. Rephrasing the question more precisely: is the initiation mass a passive consequence of the coupling between replication and growth or is it, as is commonly believed, an active cause of initiation and its timing in the cell cycle?

The independence of the initiation mass from the cell growth rate strongly suggests a metabolic link to the signalling of replication initiation, a link that remains poorly elucidated to date. In eukaryotes, DNA synthesis takes place during the reductive (biosynthetic) phase of the cell cycle and coincides with an abrupt rise in reactive oxygen species (ROS) at the G1 oxidative phase/ S phase transition, which suggests that the cellular metabolic state plays a role in signalling the timing of initiation at a critical cell physiology/mass, at least in eukaryotes [145–147]. As mentioned above (see above Section 6: **Does the Initiation Hyperstructure Contain Glycolytic Enzymes?**), in *B. subtilis*, a number of enzymes involved in metabolism have been shown to be associated with both replication initiation and elongation (specifically, the DnaC helicase, DnaG primase, and DnaE lagging strand polymerase), whilst in *E. coli*, carbon metabolism plays an important role in DNA replication fidelity and correlates with DNA synthesis [148–150].

It is tempting to speculate that levels and balances in metabolite pool sizes play a significant regulatory role in signalling and controlling important cell cycle processes, such as the accumulation of the initiation mass, the timing of the initiation of DNA replication, the rate of chromosome elongation, and cell division. Clearly, these events are tightly coordinated and co-regulated. One way is via post-translational modifications such as the acetylation of DnaA (see above Section 10: **What Modifications Does DnaA Undergo and What Are Their roles?**), and it may be significant that the acetylation of RNR in human cells results in the reduction of the dNTP pool and DNA replication fork stalling [151]. That said, questions concerning the role of metabolism in coordinating and regulating critical cell cycle functions have yet to be fully answered. It would be interesting, for example, to investigate how NADP(H):NAD+, ATP:ADP, and NTP:dNTP pool sizes co-vary during the cell cycle and whether or not they might play a role in determining the initiation mass, and, thus, prove informative in revealing the mysterious mechanism(s) by which the initiation mass appears to coordinate and control the major events of the cell cycle—if, in fact, it does.

17. Miscellaneous Questions

Eberle and collaborators performed a series of experiments that largely entailed shifting a growing culture of a *dnaA(ts)* strain (and sometimes a *dnaC(ts)* strain) to the non-permissive temperature for an hour and then returning that culture to the permissive temperature in the presence or absence of chloramphenicol; in the former case, this resulted in four to five initiation events, as opposed to just one in the latter case [152]. Only ten minutes of inhibition of protein synthesis were needed to produce these extra initiations [153]. Could seeing initiation in terms of hyperstructure dynamics help explain these results? For example, is it possible that the inhibition of protein synthesis, which would disrupt an MMS hyperstructure (or perturb a hyperstructure to which the MMS operon would normally contribute), would, therefore, inhibit an initiation hyperstructure? A complementary possibility is that the drop in temperature resulted in the decondensation of ions from *oriC* and associated proteins within a hyperstructure, leading to the opening of the strands and initiation [97].

As mentioned above (see Section 15 **Does an Initiation Hyperstructure Sense DNA Supercoiling?**), *E. coli* can grow despite the inactivation of *oriC* and *dnaA*, provided cells lack enzymes such as RNase H, which removes RNA-DNA hybrids in the form of R-loops [154]. This is because replication can be initiated at multiple ectopic *oriK* sites (for which a consensus sequence has yet to be defined) elsewhere on the chromosome [155,156]. It has been proposed that this "constitutive stable replication" may be a relic of the replication used by early cells [157]. Assuming that increasing transcription at an *oriK* increases the probability of replication, it is tempting to speculate that such coupling could provide an intensity-sensing mechanism to allow replication of DNA before it becomes limiting for growth [158]. That said, it is difficult to square this simple mechanism with the apparently normal timing of minichromosome replication in conditions in which chromosome replication itself is random (see below).

Eliasson and Nordstrom used an integratively suppressed strain to investigate minichromosome replication [159]. In this strain, the chromosomal *oriC* is inactive and replication occurs at random from a plasmid origin (P1); despite this, the rounds of replication of minichromosomes as seen using density shifts occurred at cell cycle intervals, consistent with a signal for initiation still being generated at the normal time [159]. The authors argued against an artefact due to a lengthening of the eclipse period—a period during which a newly replicated origin is refractory to a second initiation event [160]—but, rather, proposed that the minichromosome replication they observed was not being triggered by the process of chromosome replication; in other words, the system that normally triggers chromosome replication continued working even when chromosome replication was random [159]. This result, therefore, appears to call into question models based on the chromosome being an integral part of the cell cycle clock. Is it possible to explain the result by invoking the operation of a "primitive" physico-chemical clock based, for example, on hyperstructure dynamics? In particular, could the result be explained by the cyclically changing states of metabolic, non-equilibrium hyperstructures [97]? Could it even be based on some sort of long-term cellular memory based on hyperstructures, analogous to the memory of exposure to inducer conferred by the existence of a Lac hyperstructure that, once created by a level of inducer, maintains the capacity to metabolise lactose in the subsequent absence of this high level [161–163]? If such a memory depended on the segregation of hyperstructures with the DNA strands over the generations, it could give a distribution of growth rates and corresponding cell cycle periods in the population [164].

L-forms are bacteria that manage to grow in the absence of a peptidoglycan layer. They can be obtained by different methods and can have different styles of growth [165]. Division still occurs in *E. coli* L-forms even though FtsZ levels are fivefold lower than in the cells from which the L-forms are derived [166]; indeed, division can occur in a *B. subtilis* L-form in the absence of FtsZ [167], which gives a possible insight into the mechanism of division in early cells [168]. What, then, of DNA replication and its relationship to cell division? The high ratio in a *B. subtilis* L-form of the number of genomes (as detected by hybridisation) to the number of colony-forming units was attributed to a weaker coupling between chromosome replication and cell division [169]. A similarly high ratio was found in an L-form of *Listeria monocytogenes*, though it was noted that a third of the L-forms could not form colonies [170]; it was also found in this study that, for large L-form cells, those with a high concentration of DNA divided more frequently than those with a low concentration, which was interpreted as the high density of DNA in the former case contributing to the initiation of membrane perturbations and shape changes [170]. That said, it is important to note that the volume of L-form cells can be much greater than that of their walled parents, in which case the L-forms might contain less DNA per volume unit than parental cells. An *E. coli* L-form revealed a dependence on calcium concentrations in the growth medium, with optimum growth at 32 and 37 °C, in 0.1 or 1.0 mM Ca^{2+}, respectively [171]; this is an intriguing result given the relationship between temperature and ion condensation [172] and the putative role for ion condensation in hyperstructure dynamics and cell cycle regulation [97]. Open questions include whether initiation in L-forms depends on DnaA and *oriC* and where DnaA is located.

18. Discussion

The principles of molecular biology have been to isolate and characterise gene products. The success of this reductionist approach has laid the foundations for the complementary, integrative approach based on physics and physical chemistry that shows how these products interact. This is the approach to the bacterial cell cycle that we have adopted here. Jun and collaborators recently proposed a variant of the initiation-titration model [173] that fully exploits the existence of two forms of DnaA, their interconversion, and the distribution of two types of binding sites on the chromosome that they validated using a physics-based approach [174]. Kleckner and collaborators proposed that, following the completion of "chromosomal and divisome-related events", what they term a "progression control complex"—in other words, a type of hyperstructure—would form [175]; in combination with a mass increase, the changes in this complex would trigger cell division and the release of the terminus regions (for generation 1) along with licensing the subsequent triggering of replication via DnaA, etc. (for generation 2). In this hypothesis, the two events of cell division and nucleoid transition (which leads to initiation of replication) are independent of one another and could be the separate results of a common upstream event. Boye and Nordstrom argued for chromosome replication and cell division having their own, independent, cycles that are coupled by checkpoints to ensure the correct order of events, with replication and division cycles for *E. coli* and replication and mitotic cycles for *Schizosaccharomyces pombe* [176]. The independence of these cycles in bacteria is evidenced when the checkpoints fail: blocking cell division with penicillin does not block chromosome replication, whilst blocking replication in the absence of the SOS system does not (ultimately) block division. In eukaryotic cells, *S. pombe* cells can go through mitosis without a preceding S phase, whilst in meiosis, cells can go through two consecutive reductive cell divisions without intervening DNA replication. Boye, Nordstrom and others propose that these cycles operate in parallel and that they may even be initiated around the same time (for references see [176]). We subscribe to this view but adopt a different approach.

Our approach has been to ask what problems confront systems in general when they must adapt to an environment so as to profit from opportunities for growth and yet survive stresses: this balancing act is "life on the scales", by which we mean that cells are constrained by selective pressures to balance apparently incompatible requirements (e.g., to both grow and not grow) and, hence, to find apparently incompatible solutions (e.g., to invest in both non-equilibrium structures and equilibrium structures) [97]. Successful adaptation requires the selection of regulatory criteria that include sensing when their components risk limiting their growth, sensing when they have enough material for reproducing, sensing when they are becoming too big, avoiding having networks that interfere with one another, and anticipating environmental changes by differentiating. In bacteria, these requirements are met via the cell cycle. To take the case of differentiation, for example, the two daughter cells that result from the cell cycle naturally have different phenotypes unless the species has been selected to prevent this from occurring. This is because bacteria have an abundance of circuits in which locally positive feedback and globally negative feedback are combined; consider, for example, two copies of a gene resulting from replication—one for each future daughter cell—with both copies competing for access to a limited number of RNA polymerases and with the copy being transcribed having a greater chance of continuing to be transcribed (put differently, this is a "rich get richer and the poor get poorer" situation). A similar argument can be made for differentiation in terms of hyperstructures, with each daughter getting a different set [43,69,164]; this leads to the question of whether the generation of daughter cells by the cell cycle actually corresponds to a spandrel [177]).

Once a regulatory system with many interactions between essential components has been constructed, it is well-nigh impossible to replace it completely. With this, and with the above criteria in mind, we have tried to formulate hypotheses for the regulation of the cell cycle that can be grounded in a plausible origins-of-life scenario [53,178,179]. Since this system evolved before the emergence of sophisticated macromolecules, it probably depended on the physical chemistry of the interactions of a host of simple molecules in the form of "composomes" [180], the putative ancestors of hyperstructures. This physical chemistry probably included phase separation, molecular crowding, membrane domain formation, ion condensation, and, in general, the mechanisms responsible for hyperstructure dynamics. In accord with Occam's Razor, we speculate that the regulation of the "cell cycle" of the early cells was a single triggering event.

In the context of a physico-chemical approach based on hyperstructures, we have tried here to frame questions about the actors in the regulation of the cell cycle of modern bacteria. The principal actor in the initiation of chromosome replication is DnaA. The sorts of questions that, therefore, need answering include whether there are chromosomal hyperstructures that depend on DnaA binding to its different sites in the origin, in DARS, in *datA*, and elsewhere; whether there is a membrane hyperstructure that depends on DnaA interacting with lipids; and whether these proposed hyperstructures can form part of a single larger hyperstructure that has a trajectory based on membrane dynamics, DNA supercoiling, crowding, phase separation, etc. (see for example [25]). In this trajectory, the DnaA hyperstructure would initiate not only chromosome replication but also, perhaps, cell division. This hyperstructure might act as a logic gate and take into account: the state of transcription, translation, and replication as interpreted via the putative hyperstructure created by the MMS operon; the state of metabolism as interpreted via the presence of ribonucleotide reductase or via the binding of metabolites to hyperstructure constituents (like that of cAMP to DnaA) or via the putative hyperstructure created by glycolytic enzymes; the state of the chromosome as interpreted via hyperstructures created by supercoiling. The sorts of questions that need answering include "does the DnaA-initiation hyperstructure contain SeqA?", "what is the relationship between strand opening and DnaA binding?", and "what modifications does DnaA undergo and what are their roles?". At a deeper level, questions also include "does DnaA participate in differentiation?" and "does DnaA or a DnaA-based initiation hyperstructure also trigger division?".

At a still deeper level, the fundamental question is whether the initiation of cell cycle events involves a dialogue between separate hyperstructures (e.g., via the exchange of molecules and macromolecules) or whether this initiation involves a single hyperstructure undergoing changes in structure and composition (or both …). Answering this question may require the development of new techniques, for example by using *electro-optic fluorescence* microscopy [181], by combining *Secondary Ion Mass Spectrometry* (which allows 50 nm scale localization of stable isotopes and, hence, cellular activity) [182,183] and *toponomics* [184] (which allows 2 nm scale localization of 100 different proteins) so as to elucidate hyperstructure dynamics, and by revisiting often forgotten papers [152,154,159].

Author Contributions: Conceptualization, M.K., J.H. and V.N.; writing—original draft preparation, M.K., J.H. and V.N.; writing—review and editing, M.K. and V.N. All authors have read and agreed to the published version of the manuscript.

Funding: This research received no external funding.

Acknowledgments: M.K. thanks the Internal Medicine group of Bichat Hospital for providing facilities. V.N. thanks Paul Bourgine for his advice to use a "generous Darwinian fog".

Conflicts of Interest: The authors declare no conflict of interest.

References

1. Yaniv, M.; Kohiyama, M.; Jacob, F.; Gros, F. On the properties of Valyl-sRNA synthetase in various thermosensitive *Escherichia coli* mutants. *C. R. Acad. Hebd. Seances Acad. Sci. D* **1965**, *260*, 6734–6737. [PubMed]
2. Kohiyama, M.; Cousin, D.; Ryter, A.; Jacob, F. Thermosensitive mutants of *Escherichia coli* K12. I. Isolation and rapid characterization. *Ann. L'institut Pasteur* **1966**, *110*, 465–486.
3. Bi, E.F.; Lutkenhaus, J. FtsZ ring structure associated with division in *Escherichia coli*. *Nature* **1991**, *354*, 161–164. [CrossRef] [PubMed]
4. Holmgren, A.; Reichard, P.; Thelander, L. Enzymatic synthesis of deoxyribonucleotides, 8. The effects of ATP and dATP in the CDP reductase system from *E. coli*. *Proc. Natl. Acad. Sci. USA* **1965**, *54*, 830–836. [CrossRef] [PubMed]
5. Jacob, F.; Brenner, S. On the regulation of DNA synthesis in bacteria: The hypothesis of the replicon. *Comptes. Rendus Hebd. Seances L'academie Sci.* **1963**, *256*, 298–300.
6. Kohiyama, M.; Lanfrom, H.; Brenner, S.; Jacob, F. Modifications of indispensable functions in thermosensitive *Escherichia coli* mutants. On a mutation preventing replication of the bacterial chromosome. *Comptes. Rendus Hebd. Seances L'academie Sci.* **1963**, *257*, 1979–1981.
7. Hirota, Y.; Ryter, A.; Jacob, F. Thermosensitive mutants of *E. coli* affected in the processes of DNA synthesis and cellular division. *Cold Spring Harb. Symp. Quant. Biol.* **1968**, *33*, 677–693. [CrossRef] [PubMed]

8. Hirota, Y.; Jacob, F. Production of bacteria without DNA. *C. R. Acad. Hebd. Seances Acad. Sci. D* **1966**, *263*, 1619–1621.
9. Mulder, E.; Woldringh, C.L. Actively replicating nucleoids influence positioning of division sites in *Escherichia coli* filaments forming cells lacking DNA. *J. Bacteriol.* **1989**, *171*, 4303–4314. [CrossRef]
10. Hansen, F.G.; Rasmussen, K.V. Regulation of the DnaA product in *Escherichia coli*. *Mol. Gen. Genet.* **1977**, *155*, 219–225. [CrossRef]
11. Hansen, F.G.; Atlung, T. The DnaA tale. *Front. Microbiol.* **2018**, *9*, 319. [CrossRef] [PubMed]
12. Fuller, R.S.; Kornberg, A. Purified DnaA protein in initiation of replication at the *Escherichia coli* chromosomal origin of replication. *Proc. Natl. Acad. Sci. USA* **1983**, *80*, 5817–5821. [CrossRef] [PubMed]
13. Grimwade, J.E.; Leonard, A.C. Blocking, bending, and binding: Regulation of initiation of chromosome replication during the *Escherichia coli* cell cycle by transcriptional modulators that interact with origin DNA. *Front. Microbiol.* **2021**, *12*, 732270. [CrossRef] [PubMed]
14. Yoshida, R.; Ozaki, S.; Kawakami, H.; Katayama, T. Single-stranded DNA recruitment mechanism in replication origin unwinding by DnaA initiator protein and HU, an evolutionary ubiquitous nucleoid protein. *Nucleic Acids Res.* **2023**, *51*, 6286–6306. [CrossRef] [PubMed]
15. Erzberger, J.P.; Mott, M.L.; Berger, J.M. Structural basis for ATP-dependent DnaA assembly and replication-origin remodeling. *Nat. Struct. Mol. Biol.* **2006**, *13*, 676–683. [CrossRef] [PubMed]
16. Shimizu, M.; Noguchi, Y.; Sakiyama, Y.; Kawakami, H.; Katayama, T.; Takada, S. Near-atomic structural model for bacterial DNA replication initiation complex and its functional insights. *Proc. Natl. Acad. Sci. USA* **2016**, *113*, E8021–E8030. [CrossRef] [PubMed]
17. Luisi, P.L. Defining the transition to life: Self replicating bounded structures and chemical autopoiesis. In *Thinking about Biology; sh studies in the science of complexity;* Stein, W., Ed.; Addison Wesley: Reading, MA, USA, 1993; Volume III, pp. 17–39.
18. Gao, Z.; Zhang, W.; Chang, R.; Zhang, S.; Yang, G.; Zhao, G. Liquid-liquid phase separation: Unraveling the enigma of biomolecular condensates in microbial cells. *Front. Microbiol.* **2021**, *12*, 751880. [CrossRef]
19. Musacchio, A. On the role of phase separation in the biogenesis of membraneless compartments. *EMBO J.* **2022**, *41*, e109952. [CrossRef]
20. Azaldegui, C.A.; Vecchiarelli, A.G.; Biteen, J.S. The emergence of phase separation as an organizing principle in bacteria. *Biophys. J.* **2021**, *120*, 1123–1138. [CrossRef]
21. Lin, S.N.; Wuite, G.J.L.; Dame, R.T. Effect of different crowding agents on the architectural properties of the bacterial nucleoid-associated protein hu. *Int. J. Mol. Sci.* **2020**, *21*, 9553. [CrossRef]
22. Aranovich, A.; Gdalevsky, G.Y.; Cohen-Luria, R.; Fishov, I.; Parola, A.H. Membrane-catalyzed nucleotide exchange on DnaA. Effect of surface molecular crowding. *J. Biol. Chem.* **2006**, *281*, 12526–12534. [CrossRef] [PubMed]
23. Aranovich, A.; Braier-Marcovitz, S.; Ansbacher, E.; Granek, R.; Parola, A.H.; Fishov, I. N-terminal-mediated oligomerization of DnaA drives the occupancy-dependent rejuvenation of the protein on the membrane. *Biosci. Rep.* **2015**, *35*, e00250. [CrossRef] [PubMed]
24. Fuller, R.S.; Kaguni, J.M.; Kornberg, A. Enzymatic replication of the origin of the *Escherichia coli* chromosome. *Proc. Natl. Acad. Sci. USA* **1981**, *78*, 7370–7374. [CrossRef] [PubMed]
25. Woldringh, C.L. The bacterial nucleoid: From electron microscopy to polymer physics-a personal recollection. *Life* **2023**, *13*, 895. [CrossRef] [PubMed]
26. Monterroso, B.; Zorrilla, S.; Sobrinos-Sanguino, M.; Robles-Ramos, M.A.; Lopez-Alvarez, M.; Margolin, W.; Keating, C.D.; Rivas, G. Bacterial FtsZ protein forms phase-separated condensates with its nucleoid-associated inhibitor SlmA. *EMBO Rep.* **2019**, *20*, e45946. [CrossRef] [PubMed]
27. Kitagawa, R.; Ozaki, T.; Moriya, S.; Ogawa, T. Negative control of replication initiation by a novel chromosomal locus exhibiting exceptional affinity for *Escherichia coli* DnaA protein. *Genes Dev.* **1998**, *12*, 3032–3043. [CrossRef]
28. Kasho, K.; Tanaka, H.; Sakai, R.; Katayama, T. Cooperative DnaA binding to the negatively supercoiled *datA* locus stimulates DnaA -ATP hydrolysis. *J. Biol. Chem.* **2017**, *292*, 1251–1266. [CrossRef]
29. Fujimitsu, K.; Senriuchi, T.; Katayama, T. Specific genomic sequences of *E. coli* promote replicational initiation by directly reactivating ADP- DnaA. *Genes Dev.* **2009**, *23*, 1221–1233. [CrossRef]
30. Kasho, K.; Fujimitsu, K.; Matoba, T.; Oshima, T.; Katayama, T. Timely binding of IHF and Fis to *dars2* regulates ATP-DnaA production and replication initiation. *Nucleic Acids Res.* **2014**, *42*, 13134–13149. [CrossRef]
31. Miyoshi, K.; Tatsumoto, Y.; Ozaki, S.; Katayama, T. Negative feedback for *dars2*-Fis complex by ATP-DnaA supports the cell cycle-coordinated regulation for chromosome replication. *Nucleic Acids Res.* **2021**, *49*, 12820–12835. [CrossRef]
32. Frimodt-Moller, J.; Charbon, G.; Krogfelt, K.A.; Lobner-Olesen, A. DNA replication control is linked to genomic positioning of control regions in *Escherichia coli*. *PLoS Genet.* **2016**, *12*, e1006286. [CrossRef] [PubMed]
33. Kato, J.; Katayama, T. Hda, a novel DnaA-related protein, regulates the replication cycle in *Escherichia coli*. *EMBO J.* **2001**, *20*, 4253–4262. [CrossRef] [PubMed]
34. Knoppel, A.; Brostrom, O.; Gras, K.; Elf, J.; Fange, D. Regulatory elements coordinating initiation of chromosome replication to the *Escherichia coli* cell cycle. *Proc. Natl. Acad. Sci. USA* **2023**, *120*, e2213795120. [CrossRef] [PubMed]
35. Kepes, F. Periodic transcriptional organization of the *E. coli* genome. *J. Mol. Biol.* **2004**, *340*, 957–964. [CrossRef] [PubMed]
36. Yung, B.Y.; Kornberg, A. Membrane attachment activates DnaA protein, the initiation protein of chromosome replication in *Escherichia coli*. *Proc. Natl. Acad. Sci. USA* **1988**, *85*, 7202–7205. [CrossRef] [PubMed]

37. Castuma, C.E.; Crooke, E.; Kornberg, A. Fluid membranes with acidic domains activate DnaA, the initiator protein of replication in *Escherichia coli*. *J. Biol. Chem.* **1993**, *268*, 24665–24668. [CrossRef]

38. Xia, W.; Dowhan, W. In vivo evidence for the involvement of anionic phospholipids in initiation of DNA replication in *Escherichia coli*. *Proc. Natl. Acad. Sci. USA* **1995**, *92*, 783–787. [CrossRef] [PubMed]

39. Newman, G.; Crooke, E. DnaA, the initiator of *Escherichia coli* chromosomal replication, is located at the cell membrane. *J. Bacteriol.* **2000**, *182*, 2604–2610. [CrossRef]

40. Regev, T.; Myers, N.; Zarivach, R.; Fishov, I. Association of the chromosome replication initiator DnaA with the *Escherichia coli* inner membrane in vivo: Quantity and mode of binding. *PLoS ONE* **2012**, *7*, e36441. [CrossRef]

41. Hou, Y.; Kumar, P.; Aggarwal, M.; Sarkari, F.; Wolcott, K.M.; Chattoraj, D.K.; Crooke, E.; Saxena, R. The linker domain of the initiator DnaA contributes to its ATP binding and membrane association in *E. coli* chromosomal replication. *Sci. Adv.* **2022**, *8*, eabq6657. [CrossRef]

42. Zheng, W.; Li, Z.; Skarstad, K.; Crooke, E. Mutations in DnaA protein suppress the growth arrest of acidic phospholipid-deficient *Escherichia coli* cells. *EMBO J.* **2001**, *20*, 1164–1172. [CrossRef] [PubMed]

43. Norris, V.; Madsen, M.S. Autocatalytic gene expression occurs via transertion and membrane domain formation and underlies differentiation in bacteria: A model. *J. Mol. Biol.* **1995**, *253*, 739–748. [CrossRef] [PubMed]

44. Woldringh, C.L. The role of co-transcriptional translation and protein translocation (transertion) in bacterial chromosome segregation. *Mol. Microbiol.* **2002**, *45*, 17–29. [CrossRef]

45. Fishov, I.; Namboodiri, S. A nonstop thrill ride from genes to the assembly of the T3SS injectisome. *Nat. Commun.* **2023**, *14*, 1973. [CrossRef] [PubMed]

46. Fernandez-Coll, L.; Maciag-Dorszynska, M.; Tailor, K.; Vadia, S.; Levin, P.A.; Szalewska-Palasz, A.; Cashel, M. The absence of (p)ppGpp renders initiation of *Escherichia coli* chromosomal DNA synthesis independent of growth rates. *mBio* **2020**, *11*, e03223-19. [CrossRef] [PubMed]

47. Chiaramello, A.E.; Zyskind, J.W. Coupling of DNA replication to growth rate in *Escherichia coli*: A possible role for guanosine tetraphosphate. *J. Bacteriol.* **1990**, *172*, 2013–2019. [CrossRef] [PubMed]

48. Gonzalez, D.; Collier, J. Effects of (p)ppGpp on the progression of the cell cycle of *Caulobacter crescentus*. *J. Bacteriol.* **2014**, *196*, 2514–2525. [CrossRef]

49. Hughes, P.; Landoulsi, A.; Kohiyama, M. A novel role for cAMP in the control of the activity of the *E. coli* chromosome replication initiator protein, DnaA. *Cell* **1988**, *55*, 343–350. [CrossRef]

50. Tymecka-Mulik, J.; Boss, L.; Maciag-Dorszynska, M.; Matias Rodrigues, J.F.; Gaffke, L.; Wosinski, A.; Cech, G.M.; Szalewska-Palasz, A.; Wegrzyn, G.; Glinkowska, M. Suppression of the *Escherichia coli dnaA46* mutation by changes in the activities of the pyruvate-acetate node links DNA replication regulation to central carbon metabolism. *PLoS ONE* **2017**, *12*, e0176050. [CrossRef]

51. Horemans, S.; Pitoulias, M.; Holland, A.; Pateau, F.; Lechaplais, C.; Ekaterina, D.; Perret, A.; Soultanas, P.; Janniere, L. Pyruvate kinase, a metabolic sensor powering glycolysis, drives the metabolic control of DNA replication. *BMC Biol.* **2022**, *20*, 87. [CrossRef]

52. Holland, A.; Pitoulias, M.; Soultanas, P.; Janniere, L. The replicative DnaE polymerase of *Bacillus subtilis* recruits the glycolytic pyruvate kinase (PykA) when bound to primed DNA templates. *Life* **2023**, *13*, 965. [CrossRef] [PubMed]

53. Norris, V.; Demongeot, J. The ring world: Eversion of small double-stranded polynucleotide circlets at the origin of DNA double helix, RNA polymerization, triplet code, twenty amino acids, and strand asymmetry. *Int. J. Mol. Sci.* **2022**, *23*, 12915. [CrossRef] [PubMed]

54. Ogden, G.B.; Pratt, M.J.; Schaechter, M. The replicative origin of the *E. coli* chromosome binds to cell membranes only when hemimethylated. *Cell* **1988**, *54*, 127–135. [CrossRef]

55. Shakibai, N.; Ishidate, K.; Reshetnyak, E.; Gunji, S.; Kohiyama, M.; Rothfield, L. High-affinity binding of hemimethylated *oriC* by *Escherichia coli* membranes is mediated by a multiprotein system that includes SeqA and a newly identified factor, SeqB. *Proc. Natl. Acad. Sci. USA* **1998**, *95*, 11117–11121. [CrossRef] [PubMed]

56. d'Alencon, E.; Taghbalout, A.; Kern, R.; Kohiyama, M. Replication cycle dependent association of SeqA to the outer membrane fraction of *E. coli*. *Biochimie* **1999**, *81*, 841–846. [CrossRef] [PubMed]

57. Landoulsi, A.; Malki, A.; Kern, R.; Kohiyama, M.; Hughes, P. The *E. coli* cell surface specifically prevents the initiation of DNA replication at *oriC* on hemimethylated DNA templates. *Cell* **1990**, *63*, 1053–1060. [CrossRef] [PubMed]

58. Hiraga, S.; Ichinose, C.; Niki, H.; Yamazoe, M. Cell cycle-dependent duplication and bidirectional migration of SeqA-associated DNA-protein complexes in *E. coli*. *Mol. Cell* **1998**, *1*, 381–387. [CrossRef]

59. Helgesen, E.; Saetre, F.; Skarstad, K. Topoisomerase IV tracks behind the replication fork and the SeqA complex during DNA replication in *Escherichia coli*. *Sci. Rep.* **2021**, *11*, 474. [CrossRef]

60. Norris, V.; Fralick, J.; Danchin, A. A SeqA hyperstructure and its interactions direct the replication and sequestration of DNA. *Mol. Microbiol.* **2000**, *37*, 696–702. [CrossRef]

61. Sekimizu, K.; Bramhill, D.; Kornberg, A. ATP activates DnaA protein in initiating replication of plasmids bearing the origin of the *E. coli* chromosome. *Cell* **1987**, *50*, 259–265. [CrossRef]

62. Messer, W.; Egan, B.; Gille, H.; Holz, A.; Schaefer, C.; Woelker, B. The complex of *oriC* DNA with the DnaA initiator protein. *Res. Microbiol.* **1991**, *142*, 119–125. [CrossRef] [PubMed]

63. Nievera, C.; Torgue, J.J.; Grimwade, J.E.; Leonard, A.C. SeqA blocking of DnaA-*oriC* interactions ensures staged assembly of the *E. coli* pre-rc. *Mol. Cell* **2006**, *24*, 581–592. [CrossRef]

64. Sakiyama, Y.; Kasho, K.; Noguchi, Y.; Kawakami, H.; Katayama, T. Regulatory dynamics in the ternary DnaA complex for initiation of chromosomal replication in *Escherichia coli*. *Nucleic Acids Res.* **2017**, *45*, 12354–12373. [CrossRef] [PubMed]

65. Zorman, S.; Seitz, H.; Sclavi, B.; Strick, T.R. Topological characterization of the DnaA-*oriC* complex using single-molecule nanomanipuation. *Nucleic Acids Res.* **2012**, *40*, 7375–7383. [CrossRef] [PubMed]

66. Landoulsi, A.; Kohiyama, M. DnaA protein dependent denaturation of negative supercoiled *oriC* DNA minicircles. *Biochimie* **2001**, *83*, 33–39. [CrossRef] [PubMed]

67. Filutowicz, M. Requirement of DNA gyrase for the initiation of chromosome replication in *Escherichia coli* k-12. *Mol. Gen. Genet.* **1980**, *177*, 301–309. [CrossRef] [PubMed]

68. Hupp, T.R.; Kaguni, J.M. Activation of mutant forms of DnaA protein of *Escherichia coli* by DnaK and GrpeE proteins occurs prior to DNA replication. *J. Biol. Chem.* **1993**, *268*, 13143–13150. [CrossRef] [PubMed]

69. Norris, V.; Kayser, C.; Muskhelishvili, G.; Konto-Ghiorghi, Y. The roles of nucleoid-associated proteins and topoisomerases in chromosome structure, strand segregation and the generation of phenotypic heterogeneity in bacteria. *FEMS Microbiol. Rev.* **2022**. [CrossRef]

70. Li, S.; Zhang, Q.; Xu, Z.; Yao, Y.F. Acetylation of lysine 243 inhibits the oric binding ability of DnaA in *Escherichia coli*. *Front. Microbiol.* **2017**, *8*, 699. [CrossRef]

71. Landoulsi, A.; Kohiyama, M. Initiation of DNA replication in delta *cya* mutants of *Escherichia coli* K12. *Biochimie* **1999**, *81*, 827–834. [CrossRef]

72. Charbon, G.; Mendoza-Chamizo, B.; Campion, C.; Li, X.; Jensen, P.R.; Frimodt-Moller, J.; Lobner-Olesen, A. Energy starvation induces a cell cycle arrest in *Escherichia coli* by triggering degradation of the DnaA initiator protein. *Front. Mol. Biosci.* **2021**, *8*, 629953. [CrossRef] [PubMed]

73. Coppine, J.; Kaczmarczyk, A.; Petit, K.; Brochier, T.; Jenal, U.; Hallez, R. Regulation of bacterial cell cycle progression by redundant phosphatases. *J. Bacteriol.* **2020**, *202*, e00345-20. [CrossRef] [PubMed]

74. Seebach, D. No life on this planet without PHB. *Helv. Chim. Acta* **2023**, *106*, e202200205. [CrossRef]

75. Reusch, R.N.; Shabalin, O.; Crumbaugh, A.; Wagner, R.; Schroder, O.; Wurm, R. Posttranslational modification of *E. coli* histone-like protein H-NS and bovine histones by short-chain poly-(r)-3-hydroxybutyrate (cPHB). *FEBS Lett.* **2002**, *527*, 319–322. [CrossRef] [PubMed]

76. Fralick, J.A. Studies on the regulation of initiation of chromosome replication in *Escherichia coli*. *J. Mol. Biol.* **1978**, *122*, 271–286. [CrossRef]

77. Fralick, J.A. Is DnaA the 'pace-maker' of chromosome replication? An old paper revisited. *Mol. Microbiol.* **1999**, *31*, 1011–1012. [CrossRef] [PubMed]

78. Flatten, I.; Fossum-Raunehaug, S.; Taipale, R.; Martinsen, S.; Skarstad, K. The DnaA protein is not the limiting factor for initiation of replication in *Escherichia coli*. *PLoS Genet.* **2015**, *11*, e1005276. [CrossRef]

79. Berger, M.; Wolde, P.R.T. Robust replication initiation from coupled homeostatic mechanisms. *Nat. Commun.* **2022**, *13*, 6556. [CrossRef]

80. Versalovic, J.; Koeuth, T.; Britton, R.; Geszvain, K.; Lupski, J.R. Conservation and evolution of the *rpsU-dnaG-rpoD* macromolecular synthesis operon in bacteria. *Mol. Microbiol.* **1993**, *8*, 343–355. [CrossRef]

81. Metselaar, K.I.; den Besten, H.M.; Boekhorst, J.; van Hijum, S.A.; Zwietering, M.H.; Abee, T. Diversity of acid stress resistant variants of *Listeria monocytogenes* and the potential role of ribosomal protein s21 encoded by *rpsU*. *Front. Microbiol.* **2015**, *6*, 422. [CrossRef]

82. Koomen, J.; Huijboom, L.; Ma, X.; Tempelaars, M.H.; Boeren, S.; Zwietering, M.H.; den Besten, H.M.W.; Abee, T. Amino acid substitutions in ribosomal protein RpsU enable switching between high fitness and multiple-stress resistance in *Listeria monocytogenes*. *Int. J. Food Microbiol.* **2021**, *351*, 109269. [CrossRef] [PubMed]

83. Huisman, O.; D'Ari, R.; George, J. Inducible *sfi* dependent division inhibition in *Escherichia coli*. *Mol. Gen. Genet.* **1980**, *177*, 629–636. [CrossRef] [PubMed]

84. Jones, C.; Holland, I.B. Role of the SulB (FtsZ) protein in division inhibition during the SOS response in *Escherichia coli*: FtsZ stabilizes the inhibitor SulA in maxicells. *Proc. Natl. Acad. Sci. USA* **1985**, *82*, 6045–6049. [CrossRef] [PubMed]

85. Robin, A.; Joseleau-Petit, D.; D'Ari, R. Transcription of the *ftsZ* gene and cell division in *Escherichia coli*. *J. Bacteriol.* **1990**, *172*, 1392–1399. [CrossRef]

86. Garrido, T.; Sanchez, M.; Palacios, P.; Aldea, M.; Vicente, M. Transcription of *ftsZ* oscillates during the cell cycle of *Escherichia coli*. *EMBO J.* **1993**, *12*, 3957–3965. [CrossRef] [PubMed]

87. Zhou, P.; Helmstetter, C.E. Relationship between *ftsZ* gene expression and chromosome replication in *Escherichia coli*. *J. Bacteriol.* **1994**, *176*, 6100–6106. [CrossRef] [PubMed]

88. Masters, M.; Paterson, T.; Popplewell, A.G.; Owen-Hughes, T.; Pringle, J.H.; Begg, K.J. The effect of DnaA protein levels and the rate of initiation at *oriC* on transcription originating in the *ftsQ* and *ftsA* genes: In vivo experiments. *Mol. Gen. Genet.* **1989**, *216*, 475–483. [CrossRef]

89. Goranov, A.I.; Katz, L.; Breier, A.M.; Burge, C.B.; Grossman, A.D. A transcriptional response to replication status mediated by the conserved bacterial replication protein DnaA. *Proc. Natl. Acad. Sci. USA* **2005**, *102*, 12932–12937. [CrossRef]

90. Hottes, A.K.; Shapiro, L.; McAdams, H.H. DnaA coordinates replication initiation and cell cycle transcription in *Caulobacter crescentus*. *Mol. Microbiol.* **2005**, *58*, 1340–1353. [CrossRef]

91. Fernandez-Fernandez, C.; Gonzalez, D.; Collier, J. Regulation of the activity of the dual-function DnaA protein in *Caulobacter crescentus*. *PLoS ONE* **2011**, *6*, e26028. [CrossRef]

92. Jaffe, A.; D'Ari, R.; Norris, V. SOS-independent coupling between DNA replication and cell division in *Escherichia coli*. *J. Bacteriol.* **1986**, *165*, 66–71. [CrossRef] [PubMed]

93. Jaffe, A.; D'Ari, R. Regulation of chromosome segregation in *Escherichia coli*. *Ann. Inst. Pasteur. Microbiol.* **1985**, *136A*, 159–164. [PubMed]

94. Norris, V.; Woldringh, C.; Mileykovskaya, E. A hypothesis to explain division site selection in *Escherichia coli* by combining nucleoid occlusion and Min. *FEBS Lett.* **2004**, *561*, 3–10. [CrossRef]

95. Matsumoto, K.; Hara, H.; Fishov, I.; Mileykovskaya, E.; Norris, V. The membrane: Transertion as an organizing principle in membrane heterogeneity. *Front. Microbiol.* **2015**, *6*, 572. [CrossRef] [PubMed]

96. Demongeot, J.; Moreira, A. A possible circular RNA at the origin of life. *J. Theor. Biol.* **2007**, *249*, 314–324. [CrossRef]

97. Norris, V.; Amar, P. Chromosome replication in *Escherichia coli*: Life on the scales. *Life* **2012**, *2*, 286–312. [CrossRef] [PubMed]

98. Fishov, I.; Woldringh, C.L. Visualization of membrane domains in *Escherichia coli*. *Mol. Microbiol.* **1999**, *32*, 1166–1172. [CrossRef] [PubMed]

99. Morigen, M.; Flatten, I.; Skarstad, K. The *Escherichia coli datA* site promotes proper regulation of cell division. *Microbiology* **2014**, *160*, 703–710. [CrossRef]

100. D'Ari, R.; Jaffe, A.; Bouloc, P.; Robin, A. Cyclic AMP and cell division in *Escherichia coli*. *J. Bacteriol.* **1988**, *170*, 65–70. [CrossRef]

101. Picas, L.; Montero, M.T.; Morros, A.; Vazquez-Ibar, J.L.; Hernandez-Borrell, J. Evidence of phosphatidylethanolamine and phosphatidylglycerol presence at the annular region of lactose permease of *Escherichia coli*. *Biochim. Biophys. Acta* **2010**, *1798*, 291–296. [CrossRef]

102. Picas, L.; Carretero-Genevrier, A.; Montero, M.T.; Vazquez-Ibar, J.L.; Seantier, B.; Milhiet, P.E.; Hernandez-Borrell, J. Preferential insertion of lactose permease in phospholipid domains: Afm observations. *Biochim. Biophys. Acta* **2010**, *1798*, 1014–1019. [CrossRef] [PubMed]

103. Bogdanov, M.; Dowhan, W. Lipid-dependent generation of dual topology for a membrane protein. *J. Biol. Chem.* **2012**, *287*, 37939–37948. [CrossRef] [PubMed]

104. Suarez-Germa, C.; Hernandez-Borrell, J.; Prieto, M.; Loura, L.M. Modeling FRET to investigate the selectivity of lactose permease of *Escherichia coli* for lipids. *Mol. Membr. Biol.* **2014**, *31*, 120–130. [CrossRef] [PubMed]

105. Lopez Mora, N.; Findlay, H.E.; Brooks, N.J.; Purushothaman, S.; Ces, O.; Booth, P.J. The membrane transporter lactose permease increases lipid bilayer bending rigidity. *Biophys. J.* **2021**, *120*, 3787–3794. [CrossRef]

106. Bhat, D.; Hauf, S.; Plessy, C.; Yokobayashi, Y.; Pigolotti, S. Speed variations of bacterial replisomes. *eLife* **2022**, *11*, e75884. [CrossRef] [PubMed]

107. Huang, D.; Johnson, A.E.; Sim, B.S.; Lo, T.W.; Merrikh, H.; Wiggins, P.A. The in vivo measurement of replication fork velocity and pausing by lag-time analysis. *Nat. Commun.* **2023**, *14*, 1762. [CrossRef] [PubMed]

108. Yaginuma, H.; Kawai, S.; Tabata, K.V.; Tomiyama, K.; Kakizuka, A.; Komatsuzaki, T.; Noji, H.; Imamura, H. Diversity in ATP concentrations in a single bacterial cell population revealed by quantitative single-cell imaging. *Sci. Rep.* **2014**, *4*, 6522. [CrossRef]

109. Lin, W.H.; Jacobs-Wagner, C. Connecting single-cell ATP dynamics to overflow metabolism, cell growth, and the cell cycle in *Escherichia coli*. *Curr. Biol.* **2022**, *32*, 3911–3924. [CrossRef]

110. Norris, V.; Koch, I.; Amar, P.; Kepes, F.; Janniere, L. Hypothesis: Local variations in the speed of individual DNA replication forks determine the phenotype of daughter cells. *Med. Res. Arch.* **2017**, *5*, 1–18.

111. Cabin-Flaman, A.; Monnier, A.F.; Coffinier, Y.; Audinot, J.N.; Gibouin, D.; Wirtz, T.; Boukherroub, R.; Migeon, H.N.; Bensimon, A.; Janniere, L.; et al. Combed single DNA molecules imaged by secondary ion mass spectrometry. *Anal. Chem.* **2011**, *83*, 6940–6947. [CrossRef]

112. Cabin-Flaman, A.; Monnier, A.F.; Coffinier, Y.; Audinot, J.N.; Gibouin, D.; Wirtz, T.; Boukherroub, R.; Migeon, H.N.; Bensimon, A.; Janniere, L.; et al. Combining combing and secondary ion mass spectrometry to study DNA on chips using (13)C and (15)N labeling. *F1000Research* **2016**, *5*, 1437. [CrossRef] [PubMed]

113. Dorman, C.J. Flexible response: DNA supercoiling, transcription and bacterial adaptation to environmental stress. *Trends Microbiol.* **1996**, *4*, 214–216. [CrossRef] [PubMed]

114. Blot, N.; Mavathur, R.; Geertz, M.; Travers, A.; Muskhelishvili, G. Homeostatic regulation of supercoiling sensitivity coordinates transcription of the bacterial genome. *EMBO Rep.* **2006**, *7*, 710–715. [CrossRef] [PubMed]

115. Lal, A.; Dhar, A.; Trostel, A.; Kouzine, F.; Seshasayee, A.S.; Adhya, S. Genome scale patterns of supercoiling in a bacterial chromosome. *Nat. Commun.* **2016**, *7*, 11055. [CrossRef] [PubMed]

116. Muskhelishvili, G.; Forquet, R.; Reverchon, S.; Meyer, S.; Nasser, W. Coherent domains of transcription coordinate gene expression during bacterial growth and adaptation. *Microorganisms* **2019**, *7*, 694. [CrossRef] [PubMed]

117. Samadpour, A.N.; Merrikh, H. DNA gyrase activity regulates DnaA-dependent replication initiation in *Bacillus subtilis*. *Mol. Microbiol.* **2018**, *108*, 115–127. [CrossRef] [PubMed]

118. Leela, J.K.; Raghunathan, N.; Gowrishankar, J. Topoisomerase I essentiality, DnaA-independent chromosomal replication, and transcription-replication conflict in *Escherichia coli*. *J. Bacteriol.* **2021**, *203*, e0019521. [CrossRef]

119. Hand, R. Regulation of DNA replication on subchromosomal units of mammalian cells. *J. Cell Biol.* **1975**, *64*, 89–97. [CrossRef]

120. Conti, C.; Sacca, B.; Herrick, J.; Lalou, C.; Pommier, Y.; Bensimon, A. Replication fork velocities at adjacent replication origins are coordinately modified during DNA replication in human cells. *Mol. Biol. Cell* **2007**, *18*, 3059–3067. [CrossRef]

121. Petermann, E.; Woodcock, M.; Helleday, T. Chk1 promotes replication fork progression by controlling replication initiation. *Proc. Natl. Acad. Sci. USA* **2010**, *107*, 16090–16095. [CrossRef]

122. Guzmán, E.; Salguero, I.; Mata Martín, C.; López-Acedo, E.; Guarino Almeida, E.; Sánchez-Romero, M.; Norris, V.; Jiménez-Sánchez, A. Relationship between fork progression and initiation of chromosome replication in *E. coli*. In *DNA Replication*; Seligmann, H., Ed.; InTechOpen: Rijeka, Croatia, 2011; pp. 203–220.

123. Menolfi, D.; Lee, B.J.; Zhang, H.; Jiang, W.; Bowen, N.E.; Wang, Y.; Zhao, J.; Holmes, A.; Gershik, S.; Rabadan, R.; et al. ATR kinase supports normal proliferation in the early S phase by preventing replication resource exhaustion. *Nat. Commun.* **2023**, *14*, 3618. [CrossRef] [PubMed]

124. Blow, J.J.; Ge, X.Q.; Jackson, D.A. How dormant origins promote complete genome replication. *Trends Biochem. Sci.* **2011**, *36*, 405–414. [CrossRef] [PubMed]

125. Zaritsky, A.; Pritchard, R.H. Changes in cell size and shape associated with changes in the replication time of the chromosome of *Escherichia coli*. *J. Bacteriol.* **1973**, *114*, 824–837. [CrossRef] [PubMed]

126. Churchward, G.; Bremer, H. Determination of deoxyribonucleic acid replication time in exponentially growing *Escherichia coli* b/r. *J. Bacteriol.* **1977**, *130*, 1206–1213. [CrossRef] [PubMed]

127. Odsbu, I.; Morigen; Skarstad, K. A reduction in ribonucleotide reductase activity slows down the chromosome replication fork but does not change its localization. *PLoS ONE* **2009**, *4*, e7617. [CrossRef] [PubMed]

128. Zhu, M.; Dai, X.; Guo, W.; Ge, Z.; Yang, M.; Wang, H.; Wang, Y.P. Manipulating the bacterial cell cycle and cell size by titrating the expression of ribonucleotide reductase. *mBio* **2017**, *8*, e01741-17. [CrossRef] [PubMed]

129. Guzman, E.C.; Caballero, J.L.; Jimenez-Sanchez, A. Ribonucleoside diphosphate reductase is a component of the replication hyperstructure in *Escherichia coli*. *Mol. Microbiol.* **2002**, *43*, 487–495. [CrossRef]

130. Sanchez-Romero, M.A.; Molina, F.; Jimenez-Sanchez, A. Organization of ribonucleoside diphosphate reductase during multifork chromosome replication in *Escherichia coli*. *Microbiology* **2011**, *157*, 2220–2225. [CrossRef]

131. Werner, R. Nature of DNA precursors. *Nat. New Biol.* **1971**, *233*, 99–103. [CrossRef]

132. Pato, M.L. Alterations of deoxyribonucleoside triphosphate pools in *Escherichia coli*: Effects on deoxyribonucleic acid replication and evidence for compartmentation. *J. Bacteriol.* **1979**, *140*, 518–524. [CrossRef]

133. Thellier, M.; Legent, G.; Amar, P.; Norris, V.; Ripoll, C. Steady-state kinetic behaviour of functioning-dependent structures. *FEBS J.* **2006**, *273*, 4287–4299. [CrossRef] [PubMed]

134. Nouri, H.; Monnier, A.F.; Fossum-Raunehaug, S.; Maciag-Dorszynska, M.; Cabin-Flaman, A.; Kepes, F.; Wegrzyn, G.; Szalewska-Palasz, A.; Norris, V.; Skarstad, K.; et al. Multiple links connect central carbon metabolism to DNA replication initiation and elongation in *Bacillus subtilis*. *DNA Res.* **2018**, *25*, 641–653. [CrossRef] [PubMed]

135. Skovgaard, O.; Lobner-Olesen, A. Reduced initiation frequency from *oriC* restores viability of a temperature-sensitive *Escherichia coli* replisome mutant. *Microbiology* **2005**, *151*, 963–973. [CrossRef] [PubMed]

136. Fuchs, J.A.; Karlstrom, H.O. Mapping of *nrdA* and *nrdB* in *Escherichia coli* k-12. *J. Bacteriol.* **1976**, *128*, 810–814. [CrossRef] [PubMed]

137. Augustin, L.B.; Jacobson, B.A.; Fuchs, J.A. *Escherichia coli* Fis and DnaA proteins bind specifically to the *nrd* promoter region and affect expression of an *nrd-lac* fusion. *J. Bacteriol.* **1994**, *176*, 378–387. [CrossRef] [PubMed]

138. Gon, S.; Camara, J.E.; Klungsoyr, H.K.; Crooke, E.; Skarstad, K.; Beckwith, J. A novel regulatory mechanism couples deoxyribonucleotide synthesis and DNA replication in *Escherichia coli*. *EMBO J.* **2006**, *25*, 1137–1147. [CrossRef] [PubMed]

139. Olliver, A.; Saggioro, C.; Herrick, J.; Sclavi, B. DnaA-ATP acts as a molecular switch to control levels of ribonucleotide reductase expression in *Escherichia coli*. *Mol. Microbiol.* **2010**, *76*, 1555–1571. [CrossRef]

140. Janniere, L.; Canceill, D.; Suski, C.; Kanga, S.; Dalmais, B.; Lestini, R.; Monnier, A.F.; Chapuis, J.; Bolotin, A.; Titok, M.; et al. Genetic evidence for a link between glycolysis and DNA replication. *PLoS ONE* **2007**, *2*, e447. [CrossRef]

141. Fujimitsu, K.; Su'etsugu, M.; Yamaguchi, Y.; Mazda, K.; Fu, N.; Kawakami, H.; Katayama, T. Modes of overinitiation, DnaA gene expression, and inhibition of cell division in a novel cold-sensitive I mutant of *Escherichia coli*. *J. Bacteriol.* **2008**, *190*, 5368–5381. [CrossRef]

142. Babu, V.M.P.; Itsko, M.; Baxter, J.C.; Schaaper, R.M.; Sutton, M.D. Insufficient levels of the *nrdAB*-encoded ribonucleotide reductase underlie the severe growth defect of the delta *hda E. coli* strain. *Mol. Microbiol.* **2017**, *104*, 377–399. [CrossRef]

143. Charbon, G.; Campion, C.; Chan, S.H.; Bjorn, L.; Weimann, A.; da Silva, L.C.; Jensen, P.R.; Lobner-Olesen, A. Re-wiring of energy metabolism promotes viability during hyperreplication stress in *E. coli*. *PLoS Genet.* **2017**, *13*, e1006590. [CrossRef] [PubMed]

144. Charbon, G.; Riber, L.; Lobner-Olesen, A. Countermeasures to survive excessive chromosome replication in *Escherichia coli*. *Curr. Genet.* **2018**, *64*, 71–79. [CrossRef] [PubMed]

145. Havens, C.G.; Ho, A.; Yoshioka, N.; Dowdy, S.F. Regulation of late G1/S phase transition and APC Cdh1 by reactive oxygen species. *Mol. Cell. Biol.* **2006**, *26*, 4701–4711. [CrossRef] [PubMed]

146. Burhans, W.C.; Heintz, N.H. The cell cycle is a redox cycle: Linking phase-specific targets to cell fate. *Free Radic. Biol. Med.* **2009**, *47*, 1282–1293. [CrossRef] [PubMed]

147. Kirova, D.G.; Judasova, K.; Vorhauser, J.; Zerjatke, T.; Leung, J.K.; Glauche, I.; Mansfeld, J. A ROS-dependent mechanism promotes CDK2 phosphorylation to drive progression through S phase. *Dev. Cell* **2022**, *57*, 1712–1727. [CrossRef] [PubMed]
148. Maciag, M.; Nowicki, D.; Janniere, L.; Szalewska-Palasz, A.; Wegrzyn, G. Genetic response to metabolic fluctuations: Correlation between central carbon metabolism and DNA replication in *Escherichia coli*. *Microb. Cell Fact.* **2011**, *10*, 19. [CrossRef] [PubMed]
149. Maciag, M.; Nowicki, D.; Szalewska-Palasz, A.; Wegrzyn, G. Central carbon metabolism influences fidelity of DNA replication in *Escherichia coli*. *Mutat. Res.* **2012**, *731*, 99–106. [CrossRef]
150. Krause, K.; Maciag-Dorszynska, M.; Wosinski, A.; Gaffke, L.; Morcinek-Orlowska, J.; Rintz, E.; Bielanska, P.; Szalewska-Palasz, A.; Muskhelishvili, G.; Wegrzyn, G. The role of metabolites in the link between DNA replication and central carbon metabolism in *Escherichia coli*. *Genes* **2020**, *11*, 447. [CrossRef]
151. Chen, G.; Luo, Y.; Warncke, K.; Sun, Y.; Yu, D.S.; Fu, H.; Behera, M.; Ramalingam, S.S.; Doetsch, P.W.; Duong, D.M.; et al. Acetylation regulates ribonucleotide reductase activity and cancer cell growth. *Nat. Commun.* **2019**, *10*, 3213. [CrossRef]
152. Eberle, H.; Forrest, N.; Hrynyszyn, J.; Van Knapp, J. Regulation of DNA synthesis and capacity for initiation in DNA temperature sensitive mutants of *Escherichia coli* i. Reinitiation and chain elongation. *Mol. Gen. Genet.* **1982**, *186*, 57–65. [CrossRef]
153. Eberle, H.; Forrest, N. Regulation of DNA synthesis and capacity for initiation in DNA temperature sensitive mutants of *Escherichia coli*. Ii. Requirements for acquisition and expression of initiation capacity. *Mol. Gen. Genet.* **1982**, *186*, 66–70. [CrossRef] [PubMed]
154. Kogoma, T.; Subia, N.L.; von Meyenburg, K. Function of ribonuclease H in initiation of DNA replication in *Escherichia coli* K-12. *Mol. Gen. Genet.* **1985**, *200*, 103–109. [CrossRef] [PubMed]
155. Maduike, N.Z.; Tehranchi, A.K.; Wang, J.D.; Kreuzer, K.N. Replication of the *Escherichia coli* chromosome in RNase HI-deficient cells: Multiple initiation regions and fork dynamics. *Mol. Microbiol.* **2014**, *91*, 39–56. [CrossRef] [PubMed]
156. Veetil, R.T.; Malhotra, N.; Dubey, A.; Seshasayee, A.S.N. Laboratory evolution experiments help identify a predominant region of constitutive stable DNA replication initiation. *mSphere* **2020**, *5*, e00939-19. [CrossRef] [PubMed]
157. von Meyenburg, K.; Boye, E.; Skarstad, K.; Koppes, L.; Kogoma, T. Mode of initiation of constitutive stable DNA replication in RNase H-defective mutants of *Escherichia coli* K-12. *J. Bacteriol.* **1987**, *169*, 2650–2658. [CrossRef] [PubMed]
158. Norris, V. Hypothesis: Transcriptional sensing and membrane-domain formation initiate chromosome replication in *Escherichia coli*. *Mol. Microbiol.* **1995**, *15*, 985–987. [CrossRef] [PubMed]
159. Eliasson, A.; Nordstrom, K. Replication of minichromosomes in a host in which chromosome replication is random. *Mol. Microbiol.* **1997**, *23*, 1215–1220. [CrossRef] [PubMed]
160. Zaritsky, A.; Rabinovitch, A.; Liu, C.; Woldringh, C.L. Does the eclipse limit bacterial nucleoid complexity and cell width? *Synth. Syst. Biotechnol.* **2017**, *2*, 267–275. [CrossRef]
161. Novick, A.; Weiner, M. Enzyme induction as an all-or-none phenomenon. *Proc. Natl. Acad. Sci. USA* **1957**, *43*, 553–566. [CrossRef]
162. Cohn, M.; Horibata, K. Inhibition by glucose of the induced synthesis of the beta-galactoside-enzyme system of *Escherichia coli*. Analysis of maintenance. *J. Bacteriol.* **1959**, *78*, 601–612. [CrossRef]
163. Laurent, M.; Charvin, G.; Guespin-Michel, J. Bistability and hysteresis in epigenetic regulation of the lactose operon. Since Delbruck, a long series of ignored models. *Cell. Mol. Biol.* **2005**, *51*, 583–594. [PubMed]
164. Norris, V.; Ripoll, C. Generation of bacterial diversity by segregation of DNA strands. *Front. Microbiol.* **2021**, *12*, 550856. [CrossRef] [PubMed]
165. Glover, W.A.; Yang, Y.; Zhang, Y. Insights into the molecular basis of L-form formation and survival in *Escherichia coli*. *PLoS ONE* **2009**, *4*, e7316. [CrossRef] [PubMed]
166. Onoda, T.; Enokizono, J.; Kaya, H.; Oshima, A.; Freestone, P.; Norris, V. Effects of calcium and calcium chelators on growth and morphology of *Escherichia coli* L-form NC-7. *J. Bacteriol.* **2000**, *182*, 1419–1422. [CrossRef] [PubMed]
167. Leaver, M.; Dominguez-Cuevas, P.; Coxhead, J.M.; Daniel, R.A.; Errington, J. Life without a wall or division machine in *Bacillus subtilis*. *Nature* **2009**, *457*, 849–853. [CrossRef]
168. Mercier, R.; Kawai, Y.; Errington, J. Excess membrane synthesis drives a primitive mode of cell proliferation. *Cell* **2013**, *152*, 997–1007. [CrossRef] [PubMed]
169. Waterhouse, R.N.; Allan, E.J.; Amijee, F.; Undril, V.J.; Glover, L.A. An investigation of enumeration and DNA partitioning in *Bacillus subtilis* L-form bacteria. *J. Appl. Bacteriol.* **1994**, *77*, 497–503. [CrossRef]
170. Studer, P.; Staubli, T.; Wieser, N.; Wolf, P.; Schuppler, M.; Loessner, M.J. Proliferation of *Listeria monocytogenes* L-form cells by formation of internal and external vesicles. *Nat. Commun.* **2016**, *7*, 13631. [CrossRef]
171. Onoda, T.; Oshima, A. Effects of Ca2+ and a protonophore on growth of an *Escherichia coli* L-form. *J. Gen. Microbiol.* **1988**, *134*, 3071–3077. [CrossRef]
172. Ripoll, C.; Norris, V.; Thellier, M. Ion condensation and signal transduction. *BioEssays* **2004**, *26*, 549–557. [CrossRef]
173. Hansen, F.G.; Christensen, B.B.; Atlung, T. The initiator titration model: Computer simulation of chromosome and minichromosome control. *Res. Microbiol.* **1991**, *142*, 161–167. [CrossRef] [PubMed]
174. Fu, H.; Xiao, F.; Jun, S. Replication initiation in bacteria: Precision control based on protein counting. *bioRxiv* **2023**. [CrossRef]
175. Kleckner, N.E.; Chatzi, K.; White, M.A.; Fisher, J.K.; Stouf, M. Coordination of growth, chromosome replication/segregation, and cell division in *E. coli*. *Front. Microbiol.* **2018**, *9*, 1469. [CrossRef] [PubMed]
176. Boye, E.; Nordstrom, K. Coupling the cell cycle to cell growth. *EMBO Rep.* **2003**, *4*, 757–760. [CrossRef] [PubMed]
177. Amir, A. Is cell size a spandrel? *eLife* **2017**, *6*, e22186. [CrossRef] [PubMed]

178. Hunding, A.; Kepes, F.; Lancet, D.; Minsky, A.; Norris, V.; Raine, D.; Sriram, K.; Root-Bernstein, R. Compositional complementarity and prebiotic ecology in the origin of life. *Bioessays* **2006**, *28*, 399–412. [CrossRef] [PubMed]
179. Hassenkam, T.; Deamer, D. Visualizing RNA polymers produced by hot wet-dry cycling. *Sci. Rep.* **2022**, *12*, 10098. [CrossRef] [PubMed]
180. Segre, D.; Ben-Eli, D.; Lancet, D. Compositional genomes: Prebiotic information transfer in mutually catalytic noncovalent assemblies. *Proc. Natl. Acad. Sci. USA* **2000**, *97*, 4112–4117. [CrossRef]
181. Bowman, A.J.; Huang, C.; Schnitzer, M.J.; Kasevich, M.A. Wide-field fluorescence lifetime imaging of neuron spiking and subthreshold activity in vivo. *Science* **2023**, *380*, 1270–1275. [CrossRef]
182. Lechene, C.; Hillion, F.; McMahon, G.; Benson, D.; Kleinfeld, A.M.; Kampf, J.P.; Distel, D.; Luyten, Y.; Bonventre, J.; Hentschel, D.; et al. High-resolution quantitative imaging of mammalian and bacterial cells using stable isotope mass spectrometry. *J. Biol.* **2006**, *5*, 20. [CrossRef]
183. Gangwe Nana, G.Y.; Ripoll, C.; Cabin-Flaman, A.; Gibouin, D.; Delaune, A.; Janniere, L.; Grancher, G.; Chagny, G.; Loutelier-Bourhis, C.; Lentzen, E.; et al. Division-based, growth rate diversity in bacteria. *Front. Microbiol.* **2018**, *9*, 849. [CrossRef]
184. Schubert, W.; Gieseler, A.; Krusche, A.; Serocka, P.; Hillert, R. Next-generation biomarkers based on 100-parameter functional super-resolution microscopy TIS. *N. Biotechnol.* **2012**, *29*, 599–610. [CrossRef]

Opinion

The Quantification of Bacterial Cell Size: Discrepancies Arise from Varied Quantification Methods

Qian'andong Cao [1,2], Wenqi Huang [1,2], Zheng Zhang [1,2], Pan Chu [1,2], Ting Wei [1,2], Hai Zheng [1,2,*] and Chenli Liu [1,2,*]

[1] Shenzhen Institute of Synthetic Biology, Shenzhen Institutes of Advanced Technology, Chinese Academy of Sciences, Shenzhen 518055, China
[2] University of Chinese Academy of Sciences, Beijing 100049, China
* Correspondence: hai.zheng@siat.ac.cn (H.Z.); cl.liu@siat.ac.cn (C.L.)

Abstract: The robust regulation of the cell cycle is critical for the survival and proliferation of bacteria. To gain a comprehensive understanding of the mechanisms regulating the bacterial cell cycle, it is essential to accurately quantify cell-cycle-related parameters and to uncover quantitative relationships. In this paper, we demonstrate that the quantification of cell size parameters using microscopic images can be influenced by software and by the parameter settings used. Remarkably, even if the consistent use of a particular software and specific parameter settings is maintained throughout a study, the type of software and the parameter settings can significantly impact the validation of quantitative relationships, such as the constant-initiation-mass hypothesis. Given these inherent characteristics of microscopic image-based quantification methods, it is recommended that conclusions be cross-validated using independent methods, especially when the conclusions are associated with cell size parameters that were obtained under different conditions. To this end, we presented a flexible workflow for simultaneously quantifying multiple bacterial cell-cycle-related parameters using microscope-independent methods.

Keywords: bacterial cell cycle; microscopic images; cell size; initiation mass

Citation: Cao, Q.; Huang, W.; Zhang, Z.; Chu, P.; Wei, T.; Zheng, H.; Liu, C. The Quantification of Bacterial Cell Size: Discrepancies Arise from Varied Quantification Methods. *Life* **2023**, *13*, 1246. https://doi.org/10.3390/life13061246

Academic Editors: Vic Norris, Itzhak Fishov and Arieh Zaritsky

Received: 22 April 2023
Revised: 21 May 2023
Accepted: 21 May 2023
Published: 24 May 2023

1. Introduction

Growth and division are fundamental needs of all cells. During its cell cycle, a cell needs to coordinate its growth with genome replication and cell division to achieve faithful self-replication under various conditions. In eukaryotes, the cell cycle is divided into four ordered phases, G1, S, G2, and M2. Multiple checkpoints exist to control the order and timing of cell-cycle transitions through protein phosphorylation [1]. However, in bacteria, no obvious checkpoint has been identified. During rapid growth, many bacteria can initiate new rounds of DNA replication before the completion of the previous round, resulting in overlapping cell cycles [2]. How bacteria achieve cell-cycle control to coordinate cell growth with genome replication and cell division has been the subject of frequent investigations. These investigations into bacterial cell-cycle regulation are not only helpful in controlling bacterial growth for industrial production; they may also be instructive for building a synthetic cell from the bottom up.

When considering the developmental history of bacterial physiology, the significant progress in our understanding of the bacterial cell cycle is often attributed to the improvement of relevant experimental methods and concepts [3], leading to the identification of new quantitative relations and inspiring new models. For example, by developing rigorously quantitative experimental methods and focusing on steady-state growth instead of the "obligatory life cycle" of the bacteria, Maaløe and Kjeldgaard were able to ensure high reproducibility of their experiments. Based on such reproducible quantitative data, they discovered the SMK growth law in 1958, i.e., that the population-averaged cell mass scales exponentially with the growth rate [4]. In addition, the baby machine invented by

Charles E. Helmstetter [5,6] facilitated the synchronization of bacterial cell populations and enabled temporal analysis of the bacterial cell cycle [7]. By combining the baby machine with radioactive pulse labeling, Helmstetter carefully quantified the DNA synthesis rates of *E. coli* under various conditions [8,9]. These measurements provided a quantitative basis for Helmstetter and Stephen Cooper to establish the CH model, which quantitively states constant C and D periods of 40 and 20 min, respectively, for *E. coli* cells, with a doubling time of less than 60 min [2]. The C period refers to the period between the initiation and the corresponding termination of bacterial chromosome replication, while the D period refers to the interval between DNA replication termination and corresponding cell division. Subsequently, Donachie integrated the SMK growth law with the CH model and proposed the constant-initiation-mass hypothesis [10]. This hypothesis states that the replication of the chromosome is initiated when the ratio of cellular mass to the number of chromosome origins reaches a growth-rate-independent constant, termed the initiation mass (m_i), and the corresponding cell division always follows DNA replication initiation by the $C + D$ period. As this hypothesis provides a simple interpretation of how bacterial cells coordinate cell growth, DNA replication, and cell division, it has significantly impacted studies on the bacterial cell cycle. However, in repeated investigations into this hypothesis over the past few decades, both confirmation [11–17] and contradictions [18–23] have emerged.

While empirical observations of bulk populations have contributed to the establishment of several quantitative relationships among bacterial cell cycle parameters, the population-averaged cell behavior masks variation among individuals and does not reflect the typical behavior of single cells. Recent advances in microfluidics [24–27], high-throughput imaging [28,29], and automated image analysis [30–32] have enlivened the study of single-cell bacterial physiology [33] and provided novel opportunities to explore problems that are challenging at the population level, such as cell-size homeostasis. Through the dynamic tracking of numerous cells with single-cell resolution, the universal strategy for bacterial cell-size maintenance known as the "division adder correlation" has been discovered [34–36]. Furthermore, the combination of single-molecule fluorescent labeling and single-cell tracking has significantly facilitated the investigation of chromosome organization [37–40], replisome dynamics [22,41–43], and stochasticity or noise in the bacterial cell cycle [44–46]. By fluorescently labeling the relevant molecules of different cell cycle events, the cell cycle progression in individual bacterial cells can be monitored. These long-term observations can generate a large amount of single-cell quantitative data that aid in identifying correlations between different cell-cycle events and in uncovering quantitative laws of cell-cycle control [14,17,47–49]. As an example, in 2019, Si et al. employed the fluorescently labeled replisome protein to visualize replication cycles and investigated both the division adder and initiation adder under various perturbations [47]. More recently, Govers et al. quantified and analyzed broad phenotypes of the fluorescently labeled *E. coli* and numerous gene deletion derivatives in various media using microscopic images, then identified four new quantitative relations that were related to nucleoid segregation and different steps of cell division [17].

As part of the advancements in single-cell related techniques, many types of software tools have been developed to facilitate the high-throughput and automated extraction of cell-cycle-related parameters of bacterial cells from microscopic images. These software tools have been widely adopted in many studies. However, little attention has been paid to the impact of using different software, or the same software with different parameter settings, on the results and on the relevant conclusions. This paper demonstrates that discrepancies exist when analyzing identical datasets with different software or with the same software with different parameter settings. Importantly, these discrepancies can lead to different conclusions when validating quantitative relations, even if the consistent use of a particular software and specific parameter settings is maintained throughout a study. Therefore, it is recommended that conclusions be cross-validated using microscope-independent methods, and a flexible workflow is presented for this purpose.

2. Quantification Methods Based on Microscopic Images

Due to the presence of the diffraction limit and the small size of bacterial cells, accurately determining the actual boundary of bacterial cells in microscopic images can be challenging. A variety of high-throughput software has been developed to automatically obtain the properties of bacterial cells [32]. Generally, the pipeline for automatic cell-size evaluation includes image brightness correction, cell segmentation, and morphology extraction [30]. The cell segmentation is pivotal for high-quality cell-size characterization. Current bacterial cell segmentation algorithms broadly fall into two categories: classical computer vision and machine-learning-based algorithms. The former requires manual optimization of tunable parameters through the visual inspection of segmentation results, as was carried out by MicrobeJ [50], Oufti [51], BacStalk [52], CellProfiler [53], and Cell-Shape [54]. Machine-learning based algorithms rely on training with labeled ground-true data and their performance is largely dependent on the quality and size of the training dataset. Among such algorithms, deep neural networks (DNNs) have emerged as superior tools for cell segmentation [55,56]. As several excellent studies have comprehensively introduced or compared these algorithms/software tools [31,32,56–59], we will not conduct a quantitative evaluation of their segmentation quality here. Instead, we focus on discussing the influence of software and parameter settings on quantitative outcomes when the segmentation results are satisfactory.

To demonstrate this, the following experimental and analytical procedures were implemented. First, for reliable quantification of cell size, it was necessary to establish a steady-state growth status of the cells. Otherwise, significant variations in the results of characterizing cell-cycle-related parameters may have occurred when samples were taken at different time points [60–62]. Therefore, we captured phase-contrast images of *E. coli* K12 substr. NCM3722 grown in four different media. In brief, the steady-state growth was established by serial dilution, as previously described [23]. The cells were immobilized using a 1% agarose pad (prepared with 0.9% NaCl (w/v)) when OD_{600} reached ~0.2. The immobilized cells were imaged within 5 min at room temperature (RT), using an inverted microscope (IX-83, Olympus, Tokyo, Japan) equipped with a 100× oil objective (Olympus), an automated xy-stage (ASI, MS2000), and a sCMOS camera (Prime BSI, photometrics). Three types of software, MicrobeJ, Oufti, and BacStalk, were selected to process these images. Various parameter settings were achieved by adjusting the *auto-threshold offset* of MicrobeJ and the *cellwidth* and *meshwidth* of Oufti. For BacStalk, we used its default setting. The satisfactory segmentation performance was verified through visual inspection (Figure 1a) and the outlier were excluded by manual correction or filtered according to intensity and cell area. Cell size parameters, including cell length, cell width, and cell area, were obtained directly from the software output. In addition, we developed customized image-processing scripts based on deep-learning algorithms. The processing pipeline can be summarized in four steps: first, segmenting individual cells using U-Net [63,64]; second, determining edge details using Otsu's thresholding; third, calculating the midlines of cells through interpolation; and last, calculating size parameters including length, width, and area. Except for Oufti, the cell volume (V) was calculated by the software based on cell length (L) and width (W) and the formula $V = \frac{4}{3}\pi\left(\frac{W}{2}\right)^3 + \pi\left(\frac{W}{2}\right)^2(L-W)$, assuming that *E. coli* is a cylinder with hemispherical polar caps. All of these size parameters, which represent the population-averaged values for more than 4500 cells in each growth condition, are listed in Table 1.

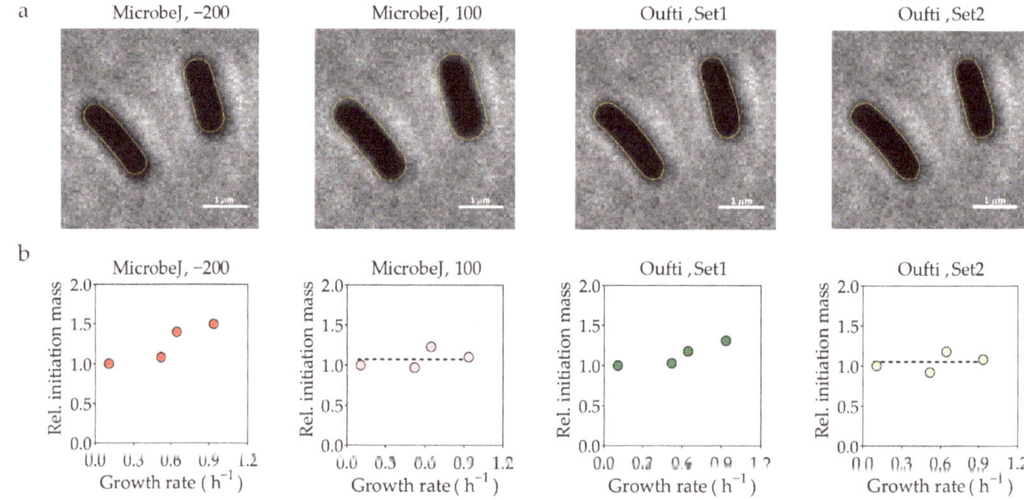

Figure 1. Different software and different parameter settings can yield divergent conclusions. (**a**) The detected contours of cells grown in MOPS medium with glutamine as the sole carbon with different parameters and software are shown by yellow lines. With MicrobeJ, we took *auto-threshold offset* = −200 and *auto-threshold offset* = 100 as an example. For Oufti Set1, we set *cellwidth* of cells grown in MOPS + glutamine, MOPS + alanine, MOPS + glycerol, and MOPS + glucose at 8, 9, 9, and 12, and for Oufti set2, we set *cellwidth* at 9, 9, 10, 11. The values of *meshwidth* were set to the corresponding *cellwidth* plus 2. The additional parameters are documented in Supplementary Materials Table S1 and were maintained consistently for both Set1 and Set2. (**b**) The relative initiation mass obtained by different software and different parameter settings. The relative initiation mass was calculated based on cell volume data obtained with different software and different parameter settings, and the population-averaged *oriC* number (see Section 3). The horizontal dashed line represents the average relative initiation mass for cells in four growth conditions.

Table 1. Cell features quantified with different software or parameters.

Features [1]	Medium [2]	MicrobeJ −200	MicrobeJ 100	Oufti Set1	Oufti Set2	BacStalk	Custom Scripts [3]
Length (μm)	Glutamine	1.85 ± 0.44	2.16 ± 0.47	2.03 ± 0.47	2.04 ± 0.47	2.25 ± 0.48	2.11 ± 0.54
	Alanine	2.22 ± 0.59	2.51 ± 0.67	2.38 ± 0.56	2.38 ± 0.56	2.63 ± 0.66	2.35 ± 0.45
	Glycerol	2.61 ± 0.67	2.94 ± 0.72	2.77 ± 0.65	2.78 ± 0.64	3.04 ± 0.69	2.79 ± 0.88
	Glucose	2.73 ± 0.65	2.97 ± 0.68	2.81 ± 0.67	2.80 ± 0.66	3.11 ± 0.67	2.88 ± 0.60
Width [4] (μm)	Glutamine	0.55 ± 0.06	0.75 ± 0.06	0.52 ± 0.03	0.55 ± 0.03	0.79 ± 0.06	0.55 ± 0.05
	Alanine	0.61 ± 0.06	0.81 ± 0.06	0.57 ± 0.02	0.57 ± 0.02	0.83 ± 0.06	0.60 ± 0.05
	Glycerol	0.67 ± 0.08	0.88 ± 0.08	0.60 ± 0.03	0.64 ± 0.03	0.90 ± 0.07	0.65 ± 0.05
	Glucose	0.79 ± 0.08	0.96 ± 0.08	0.72 ± 0.04	0.69 ± 0.03	1.00 ± 0.07	0.74 ± 0.06
Area (μm²)	Glutamine	0.94 ± 0.26	1.51 ± 0.37	1.04 ± 0.25	1.10 ± 0.27	1.17 ± 0.26	1.02 ± 0.22
	Alanine	1.28 ± 0.36	1.90 ± 0.54	1.35 ± 0.33	1.35 ± 0.33	1.47 ± 0.36	1.25 ± 0.24
	Glycerol	1.66 ± 0.48	2.46 ± 0.66	1.64 ± 0.42	1.74 ± 0.43	1.86 ± 0.44	1.60 ± 0.30
	Glucose	2.03 ± 0.52	2.70 ± 0.67	2.00 ± 0.51	1.92 ± 0.49	2.13 ± 0.47	1.90 ± 0.39

Table 1. *Cont.*

Features [1]	Medium [2]	MicrobeJ −200	MicrobeJ 100	Oufti Set1	Oufti Set2	BacStalk	Custom Scripts [3]
Volume [5] (μm^3)	Glutamine	0.39 ± 0.13	0.85 ± 0.24	0.44 ± 0.11	0.49 ± 0.13	0.96 ± 0.25	0.46 ± 0.13
	Alanine	0.59 ± 0.19	1.15 ± 0.37	0.63 ± 0.16	0.63 ± 0.16	1.26 ± 0.36	0.59 ± 0.12
	Glycerol	0.85 ± 0.31	1.62 ± 0.51	0.80 ± 0.22	0.90 ± 0.23	1.74 ± 0.48	0.84 ± 0.26
	Glucose	1.21 ± 0.37	1.92 ± 0.56	1.18 ± 0.31	1.09 ± 0.29	2.18 ± 0.57	1.11 ± 0.25

[1] The length, width, area, volume presented in the table correspond to the average value of these parameters of the cells in the population. [2] Four media refers to MOPS media with glucose, glycerol, alanine, or glutamine as the sole carbon source, with corresponding growth rates of $0.93 \pm 0.06, 0.65 \pm 0.02, 0.52 \pm 0.01, 0.11 \pm 0.02\ h^{-1}$, respectively. [3] The custom scripts have been made openly available and can be accessed through the link provided in Supplementary Materials Table S1. [4] MicrobeJ provides the mean width of the cell as the cell width. BacStalk employs the maximum width of the cell body and our custom scripts used the fitted mean cell width. In Oufti, we defined cell width by the mean width of cell mesh, as Oufti does not directly provide cell width. [5] For Oufti, we used the cell volume directly provided by the software instead of that calculated by the length and width.

It is apparent that the absolute values of cell-size parameters, such as cell length, cell width, area, and volume, are affected by the software and the parameter settings. Therefore, investigators should use consistent criteria, including the same software and parameter settings, to process microscopic images in a study. However, we questioned whether this alone was sufficient to produce conclusive results.

To this end, we considered the validation of the constant-initiation-mass hypothesis as an example here. This hypothesis proposes that the initiation mass, which refers to the cellular mass per *oriC* at the time of replication initiation, remains constant at different growth rates. To validate this hypothesis, investigators should assess the initiation mass of wild-type cells cultivated in diverse growth media exhibiting varying growth rates. The assessment of initiation mass can be carried out by employing time-lapse images of cells that have been fluorescently labeled to indicate replication initiation events, or by utilizing snapshot images in combination with techniques that facilitate the quantification of the population-averaged *oriC* number. Here, we employed the latter method, since we already had the cell volume data in four different growth media. The population-averaged *oriC* numbers were quantified by analyzing the DAPI-stained samples of run-out experiments with flow cytometry, as described previously [23].

The conclusion regarding the validation of the constant-initiation-mass hypothesis is affected by the software and parameter settings utilized for the analysis of the microscopic images. We calculated the initiation mass (m_i) based on the widely used equation [65], $m_i = \frac{\overline{V}}{\overline{o}} \times \frac{1}{ln2}$, where $\overline{V}$ and $\overline{o}$ are the population-averaged cell volume and the *oriC* number, respectively. As shown in Table 1, the absolute value of the cell volume, i.e., $\overline{V}$, is largely affected by the software and parameter settings. Thus, the absolute value of m_i is also subject to these effects. More importantly, as the relative cell volume across different growth conditions is also influenced by the software and parameter settings, the relative initiation mass across the four media exhibited different trends depending on the software and parameters employed. Upon calculating the relative initiation mass based on cell volume data obtained via MicrobeJ, with an auto-threshold offset set to −200, we found a gradual increase in relative initiation mass as growth rates increased from 0.1 to 0.9 h^{-1}, with cells grown in MOPS + glucose exhibiting a notable ~50% increase, compared to cells in MOPS + glutamine. This implied that the initiation mass was not constant, but growth-rate-dependent (Figure 1b, left panel). However, divergent conclusions may emerge when adopting an *auto-threshold offset* setting of 100. In that case, the calculated relative initiation mass varied by only ~10% between cells grown in MOPS + glucose and MOPS + glutamine (Figure 1b, second panel). When using Oufti for image processing, adjustments to cell width and mesh width parameters yielded similar discrepancies. The relative initiation mass displayed a growth-rate-dependent or growth-rate-independent pattern, depending on whether parameter Set1 or Set2 was used, respectively, for image processing (Figure 1b, third and fourth panel). Since the Set1 and Set2 in Oufti can produce almost the same cell length, the observed change in trends was mainly attributed to the variation in cell width. These findings suggest that the constancy of initiation mass can be influenced by software and parameter settings. Given that the time-lapse imaging approaches for quantifying the initiation mass also need to calculate the cell volume based on microscopic images, the effect of software and parameter settings are expected to be same.

Collectively, these results strongly imply that when drawing conclusions that rely heavily on comparing the sizes of cells cultured under different growth conditions, using the same software and consistent parameter settings for image analysis is not sufficient to produce conclusive results, and additional efforts are required to enhance the credibility of the conclusion. One possible solution is to calibrate the parameter settings of the specific software using standard nanoparticles with a known diameter that is comparable to the cell width. However, it is noteworthy that variations in the optical properties of bacterial cells and nanoparticles can still give rise to inconsistencies. Therefore, we recommend cross-validation of the conclusions, whenever possible, using techniques that are not dependent on microscopic images.

3. Quantification Methods Not Reliant on Microscopic Images

This section provides an overview of microscope-independent techniques that are capable of measuring the cell size, the cellular *oriC* number, and the initiation mass. By integrating these methods, we introduced a flexible workflow for concurrently quantifying these parameters (Figure 2). This workflow can be utilized either individually or for the corroboration of findings obtained from microscopic images.

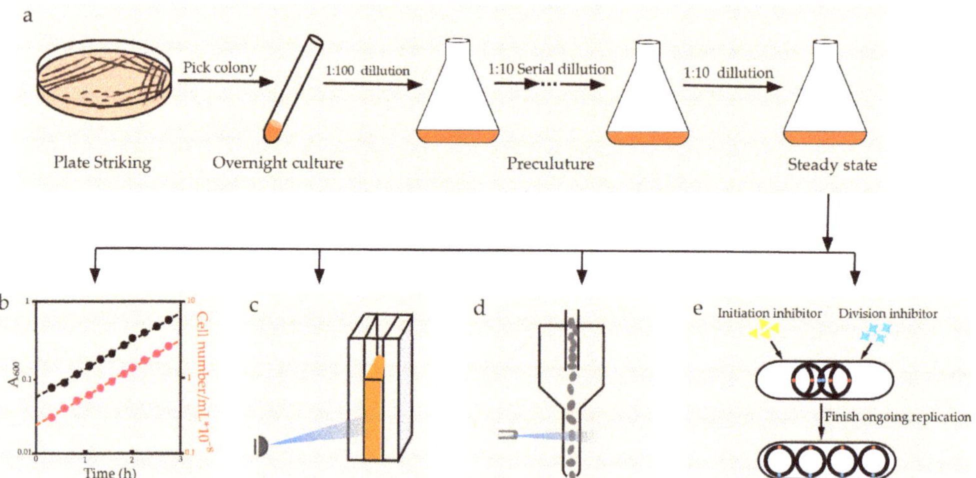

Figure 2. A workflow for simultaneous quantification of multiple parameters at the population level. (**a**) Establishment of the steady-state growth by serial dilution. (**b**) Verification of the steady-state growth by monitoring the growth rate of total biomass and cell number. The cell number (red lines) and cell mass (black lines) growth curves can form two parallel lines in semi-log plots if the steady-state growth has been achieved. (**c**) OD measurement. (**d**) Cell counting by flow cytometry. (**e**) Quantification of averaged cellular *oriC* number by runout experiments.

To establish a steady-state growth, bacteria should be maintained in exponential growth for at least 10 generations via serial dilution (Figure 2a), and the growth rates of the total biomass and the cell numbers should be monitored to verify the steady-state growth (Figure 2b). Once the steady-state growth is established and verified, samples for quantifying cell-cycle-related parameters can be taken at any time point, as the average cell composition per cell is expected to be constant [66].

The population-averaged cellular mass ($\overline{m}$), which is closely related to the cell volume, can be characterized by dividing the total biomass by the total cell numbers of the population. Dry weight and OD are two common metrics for quantifying bacterial biomass. Compared with the dry weight measurement, determining the OD of liquid cultures with a spectrophotometer was more convenient (Figure 2c). Plate counting and flow cytometry (Figure 2d) are two methods for the absolute enumeration of bacterial cells. The plate counting often requires serial dilution of the cell culture to ensure a countable range of colony numbers (25–250 colony forming units, or CFU, bacteria on a standard petri dish) [67]. Compared with plate counts, flow cytometry requires sophisticated hardware, but it is a faster and more accurate technique for measuring cell densities. For an apparatus with a controllable sample flow rate, e.g., CytoFLEX (Beckman Coulter), the cell concentration of samples can be conveniently measured with appropriate dilution and staining methods [23]. If the sample flow rate of the flow cytometer is unknown, suspensions of microspheres with standard densities must be used as references [68]. It is worth noting that, although the forward scatter (FSC) determined by flow cytometry can also reflect the relative size of the cell, it is difficult to compare the FSC obtained from different instruments, even when using the same parameter settings [23].

The population-averaged cellular *oriC* number ($\overline{o}$) can be determined through a run-out experiment and, by combining this with $\overline{m}$, the initiation mass (m_i) can also be obtained. To carry out a run-out experiment, cephalexin and rifamycin were added to the cell suspension to inhibit cell division and DNA replication initiation, respectively. Cells were then allowed to grow under the same culture conditions for 2–3 mass doubling time to complete the on-going replication [23]. Consequently, the number of fully replicated chromosomes after run-out was equal to the number of *oriC* at the time of the addition of the compounds (Figure 2e). After fixation with 70% ethanol and staining with appropriate DNA dye (such as DAPI), the samples of the run-out experiments could be analyzed with flow cytometry or a fluorescence microscope. It should be noted that, rifamycin-resistant replication initiation will invalidate the run-out experiment in several genetic backgrounds [69,70], and the mechanism for this resistance is still unclear. According to derivation process presented in [65], once we have $\overline{o}$ and $\overline{m}$ for cells in a steady-state growth status, m_i can be calculated using the following equation: $m_i = \frac{\overline{m}}{\overline{o}} \times \frac{1}{ln2}$.

Consider the aforementioned validation of the constant-initiation-mass hypothesis as an example. Both $\overline{m}$ and $\overline{o}$ are found to be positively correlated with the growth rates. Specifically, when compared with cells grown in MOPS + glutamine, the $\overline{m}$ and $\overline{o}$ of cells grown in MOPS + glucose increased 2-fold and 1-fold, respectively (Figure 3a,b). As a result, rather than remaining constant, the initiation mass (m_i) was growth-rate-dependent. It increased continuously as the growth rates increased, exhibiting a ~50% increase in cells grown in MOPS + glucose compared to those cultivated in MOPS + glutamine (Figure 3c), which was comparable to the result obtained based on microscopic images using MicrobeJ with the auto-threshold offset set to -200. Therefore, it can be concluded that the initiation mass of *E. coli* K12 substr. NCM3722 cells is not constant, but dependent on the growth rate when grown within the range of 0.1 to 0.9 h^{-1}.

Two points should be noted at the end of this section. First, as the optimal approach may vary depending on the individual characteristics of each case, in this study, we did not intend to endorse the utilization of any specific software or parameter setting for accurately quantifying cell volume based on microscopic images. Second, the presented microscope-independent workflow was derived from our daily practice and may be limited, to some extent, by our scope of knowledge or the available instruments. Other well-established methods that have not been mentioned in the current workflow can certainly be adopted for cross-validation purposes or used directly to address certain scientific questions.

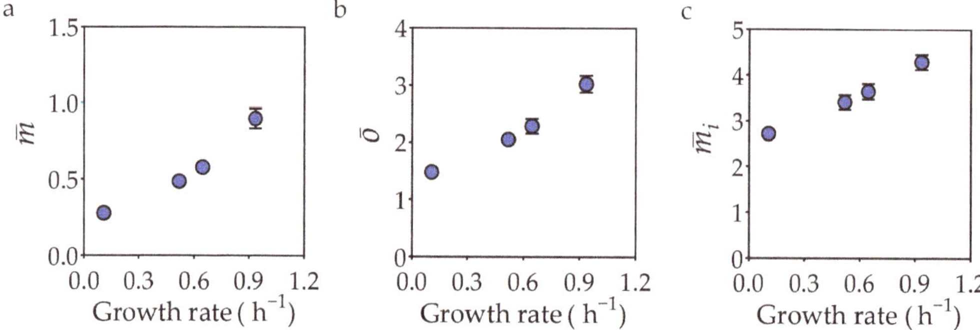

Figure 3. Cross-validation of the growth-rate-dependence of the initiation mass by the microscope-independent method. Growth rate dependence of the population-averaged cellular mass (**a**), population averaged cellular *oriC* number (**b**), and initiation mass (**c**). The $\bar{m}$ and $\bar{o}$ are characterized as described previously [23]. The units for $\bar{m}$ and m_i are OD_{600} mL per 10^9 cells and 10^{-10} OD_{600} mL, respectively. Error bar represents the standard error of biological replicates.

4. Conclusions

Microbiology is primarily an experimental science, and the use of different experimental methods may sometimes generate conflicting conclusions for the same scientific question. Therefore, researchers should understand the pros and cons of the various methods to effectively design experiments and to evaluate results. In this paper, we found that the application of different software and different parameters for image analysis can yield variations in both the absolute and relative sizes of cells grown in various conditions, leading to divergent conclusions. Therefore, it is of paramount importance to employ microscope-independent approaches to cross-validate the conclusions drawn solely from image analysis, especially when it comes to a quantitative comparison of cell volume across various perturbations.

Supplementary Materials: The following supporting information can be downloaded at: https://www.mdpi.com/article/10.3390/life13061246/s1, Table S1: Output of the image processing software and corresponding parameter settings.

Author Contributions: Q.C.: investigation, formal analysis, visualization, writing—original draft preparation, and writing—review and editing; W.H.: investigation, writing—original draft preparation, and writing—review and editing; Z.Z., P.C. and T.W.: software, writing—editing; H.Z.: conceptualization, formal analysis, writing—original draft preparation, and writing—review and editing; C.L.: conceptualization, supervision, and writing—review and editing. All authors have read and agreed to the published version of the manuscript.

Funding: This research was funded by the National Key R&D Program of China (2018YFA0902701, 2018YFA0903600), the National Natural Science Foundation of China (32025022, 32230062, 32170042), the Strategic Priority Research Program of the Chinese Academy of Sciences (XDB0480000), a Joint NSFC-ISF Research Grant (32061143021), and the Youth Innovation Promotion Association CAS (number 2022369).

Data Availability Statement: Data associated with this manuscript are available in the Supplementary Materials.

Acknowledgments: We are grateful to Terence Hwa for providing the *E. coli* NCM strain and to Taku A. Tokuyasu and Nan Luo for reviewing the manuscript. We thank numerous colleagues for discussions.

Conflicts of Interest: The authors declare no conflict of interest.

References

1. Elledge, S.J. Cell cycle checkpoints: Preventing an identity crisis. *Science* **1996**, *274*, 1664–1672. [CrossRef] [PubMed]
2. Cooper, S.; Helmstetter, C.E. Chromosome replication and the division cycle of *Escherichia coli* B/r. *J. Mol. Biol.* **1968**, *31*, 519–540. [CrossRef] [PubMed]
3. Jun, S.; Si, F.; Pugatch, R.; Scott, M. Fundamental principles in bacterial physiology-history, recent progress, and the future with focus on cell size control: A review. *Rep. Prog. Phys.* **2018**, *81*, 056601. [CrossRef] [PubMed]
4. Schaechter, M.; Maaløe, O.; Kjeldgaard, N.O. Dependency on medium and temperature of cell size and chemical composition during balanced growth of *Salmonella typhimurium*. *Microbiology* **1958**, *19*, 592–606. [CrossRef]
5. Helmstetter, C.E.; Cummings, D.J. Bacterial Synchronization by Selection of Cells at Division. *Proc. Natl. Acad. Sci. USA* **1963**, *50*, 767–774. [CrossRef]
6. Helmstetter, C.E.; Eenhuis, C.; Theisen, P.; Grimwade, J.; Leonard, A.C. Improved bacterial baby machine: Application to *Escherichia coli* K-12. *J. Bacteriol.* **1992**, *174*, 3445–3449. [CrossRef]
7. Bates, D.; Epstein, J.; Boye, E.; Fahrner, K.; Berg, H.; Kleckner, N. The *Escherichia coli* baby cell column: A novel cell synchronization method provides new insight into the bacterial cell cycle. *Mol. Microbiol.* **2005**, *57*, 380–391. [CrossRef]
8. Helmstetter, C.E. Rate of DNA synthesis during the division cycle of *Escherichia coli* B/r. *J. Mol. Biol.* **1967**, *24*, 417–427. [CrossRef]
9. Helmstetter, C.E.; Cooper, S. DNA synthesis during the division cycle of rapidly growing *Escherichia coli* B/r. *J. Mol. Biol.* **1968**, *31*, 507–518. [CrossRef]
10. Donachie, W.D. Relationship between cell size and time of initiation of DNA replication. *Nature* **1968**, *219*, 1077–1079. [CrossRef]
11. Koppes, L.; Meyer, M.; Oonk, H.; De Jong, M.; Nanninga, N. Correlation between size and age at different events in the cell division cycle of *Escherichia coli*. *J. Bacteriol.* **1980**, *143*, 1241–1252. [CrossRef] [PubMed]
12. Boye, E.; Stokke, T.; Kleckner, N.; Skarstad, K. Coordinating DNA replication initiation with cell growth: Differential roles for DnaA and SeqA proteins. *Proc. Natl. Acad. Sci. USA* **1996**, *93*, 12206–12211. [CrossRef] [PubMed]
13. Hill, N.S.; Kadoya, R.; Chattoraj, D.K.; Levin, P.A. Cell size and the initiation of DNA replication in bacteria. *PLoS Genet.* **2012**, *8*, e1002549. [CrossRef] [PubMed]
14. Wallden, M.; Fange, D.; Lundius, E.G.; Baltekin, Ö.; Elf, J. The synchronization of replication and division cycles in individual *E. coli* cells. *Cell* **2016**, *166*, 729–739. [CrossRef] [PubMed]
15. Zheng, H.; Ho, P.-Y.; Jiang, M.; Tang, B.; Liu, W.; Li, D.; Yu, X.; Kleckner, N.E.; Amir, A.; Liu, C. Interrogating the *Escherichia coli* cell cycle by cell dimension perturbations. *Proc. Natl. Acad. Sci. USA* **2016**, *113*, 15000–15005. [CrossRef] [PubMed]
16. Si, F.; Li, D.; Cox, S.E.; Sauls, J.T.; Azizi, O.; Sou, C.; Schwartz, A.B.; Erickstad, M.J.; Jun, Y.; Li, X. Invariance of initiation mass and predictability of cell size in *Escherichia coli*. *Curr. Biol.* **2017**, *27*, 1278–1287. [CrossRef]
17. Govers, S.K.; Campos, M.; Tyagi, B.; Laloux, G.; Jacobs-Wagner, C. Apparent simplicity and emergent robustness in bacterial cell cycle control. *bioRxiv* **2023**. [CrossRef]
18. Helmstetter, C.E. Initiation of chromosome replication in *Escherichia coli*: II. Analysis of the control mechanism. *J. Mol. Biol.* **1974**, *84*, 21–36. [CrossRef]
19. Churchward, G.; Estiva, E.; Bremer, H. Growth rate-dependent control of chromosome replication initiation in *Escherichia coli*. *J. Bacteriol.* **1981**, *145*, 1232–1238. [CrossRef]
20. Zaritsky, A.; Zabrovitz, S. DNA synthesis in *Escherichia coli* during a nutritional shift-up. *Mol. Gen. Genet.* **1981**, *181*, 564–566. [CrossRef]
21. Wold, S.; Skarstad, K.; Steen, H.; Stokke, T.; Boye, E. The initiation mass for DNA replication in *Escherichia coli* K-12 is dependent on growth rate. *EMBO J.* **1994**, *13*, 2097–2102. [CrossRef] [PubMed]
22. Bates, D.W.; Kleckner, N. Chromosome and replisome dynamics in *E. coli*: Loss of sister cohesion triggers global chromosome movement and mediates chromosome segregation. *Cell* **2005**, *121*, 899–911. [CrossRef]
23. Zheng, H.; Bai, Y.; Jiang, M.; Tokuyasu, T.A.; Huang, X.; Zhong, F.; Wu, Y.; Fu, X.; Kleckner, N.; Hwa, T.; et al. General quantitative relations linking cell growth and the cell cycle in *Escherichia coli*. *Nat. Microbiol.* **2020**, *5*, 995–1001. [CrossRef] [PubMed]
24. Yin, H.; Marshall, D. Microfluidics for single cell analysis. *Curr. Opin. Biotechnol.* **2012**, *23*, 110–119. [CrossRef] [PubMed]
25. Grünberger, A.; Wiechert, W.; Kohlheyer, D. Single-cell microfluidics: Opportunity for bioprocess development. *Curr. Opin. Biotechnol.* **2014**, *29*, 15–23. [CrossRef]
26. Duncombe, T.A.; Tentori, A.M.; Herr, A.E. Microfluidics: Reframing biological enquiry. *Nat. Rev. Mol. Cell Biol.* **2015**, *16*, 554–567. [CrossRef]
27. Potvin-Trottier, L.; Luro, S.; Paulsson, J. Microfluidics and single-cell microscopy to study stochastic processes in bacteria. *Curr. Opin. Microbiol.* **2018**, *43*, 186–192. [CrossRef]
28. Gahlmann, A.; Moerner, W.E. Exploring bacterial cell biology with single-molecule tracking and super-resolution imaging. *Nat. Rev. Microbiol.* **2014**, *12*, 9–22. [CrossRef]
29. Kuwada, N.J.; Traxler, B.; Wiggins, P.A. High-throughput cell-cycle imaging opens new doors for discovery. *Curr. Genet.* **2015**, *61*, 513–516. [CrossRef]
30. Caicedo, J.C.; Cooper, S.J.; Heigwer, F.; Warchal, S.J.; Qiu, P.; Molnar, C.; Vasilevich, A.; Barry, J.D.; Bansal, H.S.; Kraus, O.Z. Data-analysis strategies for image-based cell profiling. *Nat. Methods* **2017**, *14*, 849–863. [CrossRef]
31. Smith, K.; Piccinini, F.; Balassa, T.; Koos, K.; Danka, T.; Azizpour, H.; Horvath, P. Phenotypic Image Analysis Software Tools for Exploring and Understanding Big Image Data from Cell-Based Assays. *Cell Syst.* **2018**, *6*, 636–653. [CrossRef] [PubMed]

32. Jeckel, H.; Drescher, K. Advances and opportunities in image analysis of bacterial cells and communities. *FEMS Microbiol. Rev.* **2021**, *45*, fuaa062. [CrossRef] [PubMed]

33. Taheri-Araghi, S.; Brown, S.D.; Sauls, J.T.; McIntosh, D.B.; Jun, S. Single-Cell Physiology. *Annu. Rev. Biophys.* **2015**, *44*, 123–142. [CrossRef] [PubMed]

34. Campos, M.; Surovtsev, I.V.; Kato, S.; Paintdakhi, A.; Beltran, B.; Ebmeier, S.E.; Jacobswagner, C. A Constant Size Extension Drives Bacterial Cell Size Homeostasis. *Cell* **2014**, *159*, 1433–1446. [CrossRef]

35. Taheriaraghi, S.; Bradde, S.; Sauls, J.T.; Hill, N.S.; Levin, P.A.; Paulsson, J.; Vergassola, M.; Jun, S. Cell-Size Control and Homeostasis in Bacteria. *Curr. Biol.* **2015**, *25*, 385–391. [CrossRef]

36. Jun, S.; Taheri-Araghi, S. Cell-size maintenance: Universal strategy revealed. *Trends Microbiol.* **2015**, *23*, 4–6. [CrossRef]

37. Lau, I.; Filipe, S.R.; Soballe, B.; Okstad, O.; Barre, F.; Sherratt, D.J. Spatial and temporal organization of replicating *Escherichia coli* chromosomes. *Mol. Microbiol.* **2004**, *49*, 731–743. [CrossRef]

38. Wang, X.; Possoz, C.; Sherratt, D.J. Dancing around the divisome: Asymmetric chromosome segregation in *Escherichia coli*. *Genes Dev.* **2005**, *19*, 2367–2377. [CrossRef]

39. Joshi, M.C.; Bourniquel, A.; Fisher, J.K.; Ho, B.T.; Magnan, D.; Kleckner, N.; Bates, D. *Escherichia coli* sister chromosome separation includes an abrupt global transition with concomitant release of late-splitting intersister snaps. *Proc. Natl. Acad. Sci. USA* **2011**, *108*, 2765–2770. [CrossRef]

40. Santi, I.; Mckinney, J.D. Chromosome Organization and Replisome Dynamics in *Mycobacterium smegmatis*. *Mbio* **2015**, *6*, e01999-14. [CrossRef]

41. Reyes-Lamothe, R.; Sherratt, D.J.; Leake, M.C. Stoichiometry and architecture of active DNA replication machinery in *Escherichia coli*. *Science* **2010**, *328*, 498–501. [CrossRef] [PubMed]

42. Trojanowski, D.; Ginda, K.; Pioro, M.; Holowka, J.; Skut, P.; Jakimowicz, D.; Zakrzewskaczerwinska, J. Choreography of the Mycobacterium Replication Machinery during the Cell Cycle. *mBio* **2015**, *6*, e02125-14. [CrossRef] [PubMed]

43. Mangiameli, S.M.; Veit, B.T.; Merrikh, H.; Wiggins, P.A. The Replisomes Remain Spatially Proximal throughout the Cell Cycle in Bacteria. *PLoS Genet.* **2017**, *13*, e1006582. [CrossRef] [PubMed]

44. Iyer-Biswas, S.; Wright, C.S.; Henry, J.T.; Lo, K.; Burov, S.; Lin, Y.; Crooks, G.E.; Crosson, S.; Dinner, A.R.; Scherer, N.F. Scaling laws governing stochastic growth and division of single bacterial cells. *Proc. Natl. Acad. Sci. USA* **2014**, *111*, 15912–15917. [CrossRef]

45. Tanouchi, Y.; Pai, A.; Park, H.; Huang, S.; Stamatov, R.; Buchler, N.E.; You, L. A noisy linear map underlies oscillations in cell size and gene expression in bacteria. *Nature* **2015**, *523*, 357–360. [CrossRef]

46. Adiciptaningrum, A.; Osella, M.; Moolman, M.C.; Cosentino Lagomarsino, M.; Tans, S.J. Stochasticity and homeostasis in the *E. coli* replication and division cycle. *Sci. Rep.* **2015**, *5*, 18261. [CrossRef]

47. Si, F.; Treut, G.L.; Sauls, J.T.; Vadia, S.; Levin, P.A.; Jun, S. Mechanistic Origin of Cell-Size Control and Homeostasis in Bacteria. *Curr. Biol.* **2019**, *29*, 1760. [CrossRef]

48. Witz, G.; van Nimwegen, E.; Julou, T. Initiation of chromosome replication controls both division and replication cycles in *E. coli* through a double-adder mechanism. *Elife* **2019**, *8*, e48063. [CrossRef]

49. Colin, A.; Micali, G.; Faure, L.; Cosentino Lagomarsino, M.; van Teeffelen, S. Two different cell-cycle processes determine the timing of cell division in *Escherichia coli*. *Elife* **2021**, *10*, e67495. [CrossRef]

50. Ducret, A.; Quardokus, E.M.; Brun, Y.V. MicrobeJ, a tool for high throughput bacterial cell detection and quantitative analysis. *Nat Microbiol* **2016**, *1*, 16077. [CrossRef]

51. Paintdakhi, A.; Parry, B.; Campos, M.; Irnov, I.; Elf, J.; Surovtsev, I.; Jacobs-Wagner, C. Oufti: An integrated software package for high-accuracy, high-throughput quantitative microscopy analysis. *Mol. Microbiol.* **2016**, *99*, 767–777. [CrossRef] [PubMed]

52. Hartmann, R.; van Teeseling, M.C.F.; Thanbichler, M.; Drescher, K. BacStalk: A comprehensive and interactive image analysis software tool for bacterial cell biology. *Mol. Microbiol.* **2020**, *114*, 140–150. [CrossRef]

53. McQuin, C.; Goodman, A.; Chernyshev, V.; Kamentsky, L.; Cimini, B.A.; Karhohs, K.W.; Doan, M.; Ding, L.; Rafelski, S.M.; Thirstrup, D.; et al. CellProfiler 3.0: Next-generation image processing for biology. *PLoS Biol.* **2018**, *16*, e2005970. [CrossRef] [PubMed]

54. Goñi-Moreno, Á.; Kim, J.; de Lorenzo, V. CellShape: A user-friendly image analysis tool for quantitative visualization of bacterial cell factories inside. *Biotechnol. J.* **2017**, *12*, 1600323. [CrossRef]

55. Shal, K.; Choudhry, M.S. Evolution of Deep Learning Algorithms for MRI-Based Brain Tumor Image Segmentation. *Crit. Rev. Biomed. Eng.* **2021**, *49*, 77–94. [CrossRef] [PubMed]

56. Cutler, K.J.; Stringer, C.; Lo, T.W.; Rappez, L.; Stroustrup, N.; Brook Peterson, S.; Wiggins, P.A.; Mougous, J.D. Omnipose: A high-precision morphology-independent solution for bacterial cell segmentation. *Nat. Methods* **2022**, *19*, 1438–1448. [CrossRef] [PubMed]

57. Seo, H.; Badiei Khuzani, M.; Vasudevan, V.; Huang, C.; Ren, H.; Xiao, R.; Jia, X.; Xing, L. Machine learning techniques for biomedical image segmentation: An overview of technical aspects and introduction to state-of-art applications. *Med. Phys.* **2020**, *47*, e148–e167. [CrossRef] [PubMed]

58. Malhotra, P.; Gupta, S.; Koundal, D.; Zaguia, A.; Enbeyle, W. Deep Neural Networks for Medical Image Segmentation. *J. Health Eng.* **2022**, *2022*, 9580991. [CrossRef]

59. Spahn, C.; Gómez-de-Mariscal, E.; Laine, R.F.; Pereira, P.M.; von Chamier, L.; Conduit, M.; Pinho, M.G.; Jacquemet, G.; Holden, S.; Heilemann, M.; et al. DeepBacs for multi-task bacterial image analysis using open-source deep learning approaches. *Commun. Biol.* **2022**, *5*, 688. [CrossRef]

60. Fishov, I.; Zaritsky, A.; Grover, N.B. On microbial states of growth. *Mol. Microbiol.* **1995**, *15*, 789–794. [CrossRef]

61. Neidhardt, F.C. Bacterial Growth: Constant Obsession with dN/dt. *J. Bacteriol.* **1999**, *181*, 7405–7408. [CrossRef] [PubMed]

62. Neidhardt, F.C. Apples, oranges and unknown fruit. *Nat. Rev. Microbiol.* **2006**, *4*, 876. [CrossRef] [PubMed]

63. Falk, T.; Mai, D.; Bensch, R.; Çiçek, Ö.; Abdulkadir, A.; Marrakchi, Y.; Böhm, A.; Deubner, J.; Jäckel, Z.; Seiwald, K.; et al. U-Net: Deep learning for cell counting, detection, and morphometry. *Nat. Methods* **2019**, *16*, 67–70. [CrossRef] [PubMed]

64. Lugagne, J.B.; Lin, H.; Dunlop, M.J. DeLTA: Automated cell segmentation, tracking, and lineage reconstruction using deep learning. *PLoS Comput. Biol.* **2020**, *16*, e1007673. [CrossRef]

65. Bremer, H.; Churchward, G.; Young, R. Relation between growth and replication in bacteria. *J. Theor. Biol.* **1979**, *81*, 533–545. [CrossRef]

66. Cooper, S. 1-Bacterial Growth. In *Bacterial Growth and Division*; Cooper, S., Ed.; Academic Press: San Diego, CA, USA, 1991; pp. 7–17.

67. Sutton, S. Accuracy of Plate Counts. *J. Valid. Technol.* **2011**, *17*, 42–46.

68. Ou, F.; Mcgoverin, C.; Swift, S.; Vanholsbeeck, F. Absolute bacterial cell enumeration using flow cytometry. *J. Appl. Microbiol.* **2017**, *123*, 464–477. [CrossRef]

69. Von Freiesleben, U.; Rasmussen, K.V.; Atlung, T.; Hansen, F.G. Rifampicin-resistant initiation of chromosome replication from *oriC* in *ihf* mutants. *Mol. Microbiol.* **2000**, *37*, 1087–1093. [CrossRef]

70. Morigen; Molina, F.; Skarstad, K. Deletion of the *datA* site does not affect once-per-cell-cycle timing but induces rifampin-resistant replication. *J. Bacteriol.* **2005**, *187*, 3913–3920. [CrossRef]

Article

The Replicative DnaE Polymerase of *Bacillus subtilis* Recruits the Glycolytic Pyruvate Kinase (PykA) When Bound to Primed DNA Templates

Alexandria Holland [1], Matthaios Pitoulias [1], Panos Soultanas [1,*] and Laurent Janniere [2,*]

1   Biodiscovery Institute, School of Chemistry, University of Nottingham, Nottingham NG7 2RD, UK
2   Génomique Métabolique, Genoscope, Institut François Jacob, CEA, CNRS,
    Université Evry, Université Paris-Saclay, 91057 Evry, CEDEX, France
*   Correspondence: panos.soultanas@nottingham.ac.uk (P.S.); laurent.janniere@univ-evry.fr (L.J.)

Abstract: The glycolytic enzyme PykA has been reported to drive the metabolic control of replication through a mechanism involving PykA moonlighting functions on the essential DnaE polymerase, the DnaC helicase and regulatory determinants of PykA catalytic activity in *Bacillus subtilis*. The mutants of this control suffer from critical replication and cell cycle defects, showing that the metabolic control of replication plays important functions in the overall rate of replication. Using biochemical approaches, we demonstrate here that PykA interacts with DnaE for modulating its activity when the replication enzyme is bound to a primed DNA template. This interaction is mediated by the CAT domain of PykA and possibly allosterically regulated by its PEPut domain, which also operates as a potent regulator of PykA catalytic activity. Furthermore, using fluorescence microscopy we show that the CAT and PEPut domains are important for the spatial localization of origins and replication forks, independently of their function in PykA catalytic activity. Collectively, our data suggest that the metabolic control of replication depends on the recruitment of PykA by DnaE at sites of DNA synthesis. This recruitment is likely highly dynamic, as DnaE is frequently recruited to and released from replication machineries to extend the several thousand RNA primers generated from replication initiation to termination. This implies that PykA and DnaE continuously associate and dissociate at replication machineries for ensuring a highly dynamic coordination of the replication rate with metabolism.

Keywords: microbiology; metabolism; replication; PykA; DnaE; moonlighting activity

Citation: Holland, A.; Pitoulias, M.; Soultanas, P.; Janniere, L. The Replicative DnaE Polymerase of *Bacillus subtilis* Recruits the Glycolytic Pyruvate Kinase (PykA) When Bound to Primed DNA Templates. *Life* **2023**, *13*, 965. https://doi.org/10.3390/life13040965

Academic Editors: Lluís Ribas de Pouplana and Ron Elber

Received: 28 February 2023
Revised: 21 March 2023
Accepted: 5 April 2023
Published: 7 April 2023

1. Introduction

Pyruvate kinase is the enzyme responsible for the final stage of glycolysis, catalyzing the formation of pyruvate and ATP from phosphoenolpyruvate (PEP) and ADP. Mammals contain four isoforms of pyruvate kinase, whereas most bacteria contain a single isoform. The mammalian isoform PKM2 and some bacterial isoform PykA exhibit non-metabolic roles in addition to their metabolic functions. Mammalian pyruvate kinase has been shown to regulate transcription by phosphorylating transcription factors or binding to transcription factors to enhance gene expression levels [1,2]. Jannière et al. first observed in *Bacillus subtilis* the suppression of temperature-sensitive mutations in genes coding for some replication proteins by the dysfunction of certain genes coding for glycolytic enzymes [3]. Particularly, mutants in *pgm*, *pgk*, *eno*, and *pykA* genes reversed the phenotypes of mutations in genes coding for DnaE (the lagging strand DNA polymerase), DnaC (the replicative helicase—the homologue of *Escherichia coli* DnaB), and DnaG (the primase), strongly suggesting the existence of a genetic connection between glycolysis and DNA replication. Subsequently, PykA was shown to exhibit an important regulatory role linking metabolism to replication [4], and recently, PykA has been shown to typify a novel family of cross-species replication regulators that drive the metabolic control of replication through a mechanism involving regulatory determinants of PykA catalytic activity [5]. Surprisingly, the disruption of this regulatory control causes dramatic replication and cell cycle defects, showing that the metabolic control of replication is important for the overall rate of DNA synthesis [5]. As failures in replication control increase the risk of replication errors and double-stranded DNA breaks, dysregulation of the metabolic control of replication may pave the way for genetic instability and carcinogenesis [6–9].

The mammalian PykA isoform PKM2 is a homotetramer that is affected by allosteric regulators, such as fructose 1,6-biphosphate (FBP) and serine, which promote tetramer assembly, while cysteine promotes tetramer dissociation [10–12]. The structure of the PykA tetramer from many organisms has been solved, highlighting the conservation of their global architecture and active site [13]. Almost all pyruvate kinases are homotropically activated by the substrate PEP and regulated by allosteric heterotropic effectors [13]. The allosteric regulation mechanism involves conformational changes between neighboring subunits around the tetramer, resulting in different structural states. In *Bacillus subtilis*, the heterotrophic effectors of PykA are AMP and ribose 5-phosphate [14], which appear to stabilize the active conformer [15]. Each monomeric subunit of the *Geobacillus stearothermophilus* PykA protein, which is 72.4% identical to *B. subtilis* PykA, comprises a catalytic domain (CAT, residues 1–473), containing the substrate and effector binding sites and an additional C-terminal domain with a structure resembling the PEP utilizer (PEPut, residues 474–586) domain of pyruvate phosphate dikinase and other metabolic enzymes (Figure 1) [16,17]. In these enzymes, PEPut is phosphorylated at a conserved TSH motif (residues 536–539) at the expense of PEP and ATP to drive sugar imports and catalytic or regulatory activities [18–28]. The PEPut domain is not required for the catalytic activity of PykA [16]. Instead, it has a regulatory role on the enzyme activity via its interaction with CAT and the phosphorylation status of its TSH motif [5]. PEPut interacts with CAT via a hydrogen bond between E209 and L536, assisting the heterotrophic effectors in stabilizing the active R-state conformation [15,29]. The CAT and PEPut domains were shown to exert different effects on DNA replication. CAT is a stimulator of elongation and PEPut is a nutrient dependent inhibitor of initiation [5]. These activities depend on substrates binding to CAT, the CAT-PEPut interaction, and the phosphorylation status of the TSH motif. The direct functional effects of PykA on the activities of the DnaE and DnaC replication proteins were demonstrated with in vitro primer extension assays and helicase assays, respectively [5]. Despite the functional evidence of an interaction between PykA and the replication proteins DnaE and DnaC, there is no direct physical evidence of a protein–protein interaction. Here, we demonstrate a direct protein–protein interaction between PykA and DnaE, and further show that this interaction is mediated by CAT, depends on the binding of DnaE to primed DNA templates, and that PEPut somehow regulates the moonlighting function of PykA on DnaE activity. Overall, this study suggests that the metabolic control of replication depends on a dynamic recruitment of PykA at sites of DNA synthesis by DnaE and that the impact of this recruitment on DNA synthesis may be allosterically regulated by PEPut.

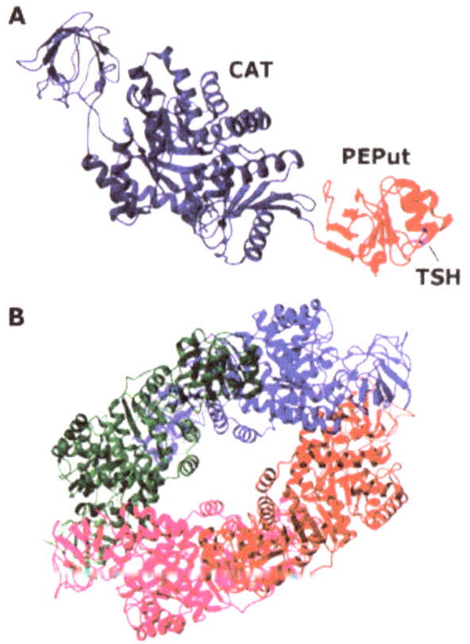

Figure 1. Structure of the *Geobacillus stearothermophilus* PykA protein. (**A**) Structure of the monomer (PDB2E28; 72.48% sequence identity to the *B. subtilis* PykA). The CAT and PEPut domains are colored blue and red, respectively, and the location of the TSH motif is shown. (**B**) Structure of the functional tetramer (PDB2E28). The four monomers are colored differently.

2. Materials and Methods

2.1. Cloning of the pykA Gene and Production and Purification of the Protein

2.1.1. Cloning of *pykA*

A DNA fragment of 1755 bp carrying the *B. subtilis pykA* gene was amplified from genomic DNA using the PyKAF (5′-TACTTCCAATCCAATGCAAGAAAAACTAAAATTG-TTTGTACCATCG-3′) and PykAR (5′-TTATCCACTTCCAATGTTATTAAAGAACGCTCG-CACG-3′) forward and reverse primers, respectively, in colony PCR reactions using Q5 high-fidelity DNA polymerase according to supplier instructions. Typically, a *B. subtilis* single colony was suspended in 20 mL LB and grown at 37 °C until the optical density reached 0.4–0.8 at 600 nm. Thereafter, colony PCR reactions were carried out with genomic DNA (10 μL) acting as the template at a 10-fold dilution. The PCR reactions were carried out in a volume of 50 μL with one unit of Q5 high-fidelity DNA polymerase, 0.5 μM PykAF and PykAR, and 0.25 mM dNTPs in 1XQ5 polymerase buffer. PCR products were cleaned up with the Clean-up kit (Thermo Scientific, Waltham, MA, USA), resolved by agarose electrophoresis, and gel extracted using the GeneJET Gel Extraction kit (Thermo Scientific) and cloned into the p2CT plasmid (gift by James Berger) (Supplementary Figure S1) using ligation-independent cloning to construct the p2CT-*pykA* production vector which produces a terminally His-MBP-tagged PykA protein.

2.1.2. Expression of *pykA*

For heterologous expression of the *B. subtilis pykA*, the p2CT-BsuPykA production vector was introduced into competent Rosetta (DE3) *E. coli* cells. Single colonies were used to inoculate two 600 mL 2xYT cultures, supplemented with 60 μL carbenicillin (50 mg/mL) in 2 L conical flasks. The flasks were incubated at 37 °C, with shaking (180 rpm) until the optical density reached 0.6–0.8 at 600 nm. Expression of *pykA* was induced by the addition of 0.5 mM IPTG and after further 3 h of growth, the cells were harvested by centrifugation at 3000× *g* for 15 min. Cells were suspended in 30 mL of buffer A (500 mM NaCl, 50 mM Tris-HCl pH 7.5, and 20 mM imidazole), supplemented with 1 mM PMSF and 50 μL protease inhibitor cocktail (Fischer, Waltham, MA, USA), and sonicated using a SANYO Soniprep 150 set at 15 amplitude microns 4 times for 1 min with 1 min intervals on ice in between. Then, benzonase (20 μL) was added to the cell lysate, which was further clarified at $40,000 \times g$ for 40 min. The clarified soluble crude extract was isolated and filtered through a 0.22 μm filter.

2.1.3. Purification of PykA

The PykA protein was purified from the filtered crude extract using a combination of IMAC (Immobilized Metal Affinity Chromatography) and gel filtration. First, the filtered crude extract was loaded onto a 5 mL HisTrap HP column (GE Healthcare, Chicago, IL, USA) equilibrated in buffer A. The column was washed thoroughly with buffer A and the PykA protein was eluted using gradient elution with buffer B (500 mM NaCl, 50 mM Tris-HCl pH 7.5, and 1 M imidazole). The eluted PykA protein was collected and quantified spectrophotometrically (extinction coefficient 76,780 M^{-1} cm^{-1}). TEV protease was added at 1:20 (TEV protease:PykA) molar ratio while dialyzing the protein solution overnight in dialysis buffer (500 mM NaCl and 50 mM Tris-HCl pH 7.5) at 4 °C in order to remove the His-MBP tag. The untagged PykA protein was then loaded back onto a 5 mL HisTrap HP column equilibrated in buffer A and the flow-through containing the untagged PykA was collected. Finally, the PykA protein solution was concentrated to 5–7 mL using a vivaspin 10 kDa cut-off filter. EDTA was added to 1 mM and the PykA preparation was then loaded onto a HiLoad 26/60 Superdex 200 Prep Grade gel filtration column (GE Healthcare) equilibrated in buffer C (500 mM NaCl, 50 mM Tris -HCl pH 7.5, and 1 mM EDTA). Fractions containing the PykA protein were pooled, and the purified PykA protein was quantified spectrophotometrically (extinction coefficient 8940 M^{-1} cm^{-1}), aliquoted, and stored in -80 °C.

2.1.4. PykA Activity Assay

The activity of purified PykA was assayed coupling the PykA catalyzed reaction (conversion of phosphoenolpyruvate to pyruvate) to the conversion of pyruvate into lactate catalyzed by LDH (Lactate Dehydrogenase) in the presence of NADH at 25 °C. The oxidation of NADH to NAD, which is directly proportional to the activity of PykA, was followed spectrophotometrically at 340 nm. The LDH-catalyzed reaction was first optimized to ensure that it did not become a limiting factor when measuring the activity of PykA. Then, PykA catalyzed reactions were carried out in 96-well plates using the reaction master mix 10 mM Tri-HCl pH 7.5, 10 mM $MgCl_2$, 50 mM KCl, 0.5 mM NADH, 2 mM ADP, 9.375×10^{-4} mg/mL LDH, and 5.7 μg/mL PykA or CAT, at 25 °C. Using the GraphPad Prism 9 software, data were fit to an allosteric sigmoidal curve with reaction rates on the Y-axis and PEP concentration on the X-axis.

2.2. Cloning of a DNA Fragment Coding for the PEPut Domain of PykA and Production and Purification of the Peptide

2.2.1. Fragment Cloning

The DNA fragment coding for the PEPut domain, with the preceding 10 amino acids, was isolated by PCR using genomic DNA and the pepF (5′-TACTTCCAATCCAATGCAGC-ACAAAATGCAAAAGAAGCT-3′) and pepR (5′-TTATCCACTTCCAATGTTATTAAAG-AACGCTCGCACG-3′) primers. Typically, a *B. subtilis* single colony was suspended in 20 mL of LB and grown at 37 °C until the optical density reached 0.4–0.8 at 600 nm. Thereafter, PCR reactions were carried out with 10 μL of undiluted, 10× and 100× diluted samples in 1XQ5 polymerase buffer, 0.25 mM dNTPs, 0.5 μM primers (pepF and pepR), and Q5 polymerase (1 unit) in a total volume of 50 μL, using the following program: 1Xcycle 98 °C 5 min, 30Xcycles 98 °C 1 min, Ta (optimal annealing temperature) 30 s, 72 °C 30 s/kb, and 1Xcycle 72 °C 5 min. The resulting PEPut DNA fragment was cloned into the p2CT plasmid using ligation-independent cloning to produce the p2CT-PEPut production vector, in a process similar to that described above for the *pykA* (see also Supplementary Figure S1).

The resulting p2CT-PEPut vector produced an N-terminally His-MBP-tagged PEPut protein. The His-MBP tag was removable by proteolysis using the TEV protease.

2.2.2. Production and Purification of PEPut

Production and purification of the PEPut were carried out as described for the full-length PykA protein but the last gel filtration step during purification was omitted. Quantification of the final untagged PEPut (MW 9582.8 Da) was carried out spectrophotometrically using the extinction coefficients 69,330 M^{-1} cm^{-1} (for the His-MBP tagged PEPut) and 1490 M^{-1} cm^{-1} (for the untagged PEPut after TEV protease treatment).

2.3. Cloning of DNA Fragments Coding for CAT or Mutated PykA, Expression, and Purification of the Proteins

Template DNA of p2CT-BsuPykA plasmid containing the gene for wild-type PykA was used to generate the *pykA* mutants using site directed mutagenesis (NEB Q5 site directed mutagenesis kit) and the CAT construct. The primers listed in Table 1 were purchased from Sigma. Gene overexpression and purification of all mutated proteins and the CAT domain were carried out as described for the wild-type PykA protein.

Table 1. A list of the oligonucleotides used to create the PykA mutations and the CAT construct. The mutated sequences are shown in bold small caps.

PykA R32A Fwd	GAACGTGGCT**gcA**TTAAACTTTTC
PykA R32A Rev	ATTCCTGACTCCATTAATTTC
PykA G245A D246A Fwd	GGTTGCACGC**gcagca**TTAGGTGTGG
PykA G245A D246A Rev	ATTAAGCCGTCAGACACTTC
PykA TSH T537D Fwd	AGGCGGTTTG**gat**AGCCATGCTG
PykA TSH T537D Rev	TCTTCTGTAATAAGAGCAGAC
PykA TSH S538D Fwd	CGGTTTGACT**gat**CATGCTGCGG
PykA TSH S538D Rev	CCTTCTTCTGTAATAAGAGC
PykA TSH H539D Fwd	TTTGACTAGC**gat**GCTGCGGTAG
PykA TSH H539D Rev	CCGCCTTCTTCTGTAATAAG
FWD PykA_TSH DDD	CGGTTTGGAT**gatg**ATGCTGCGGTAG
REV PykA_TSH DDD	CCTTCTTCTGTAATAAGAGC
PykA CAT fwd	TAATAACATTGGAAGTGGATAAC
PykA CAT rev	GCCGACAGTATGAACCTTC

2.4. Production and Purification of the B. subtilis DnaE Polymerase

The *B. subtilis dnaE* gene was cloned and the encoded protein was produced and purified, as described previously [30,31].

2.5. Polymerase Assays

Time-course polymerase assays were carried out by monitoring the DnaE primer extension activity using a 5′-end fluorescently labeled (Cy5; Sigma) 15 mer synthetic oligonucleotide (5′-AAGGGGGTGTGTGTG-3′) annealed onto a 110 mer oligonucleotide (5′-CACACACACACACACACACACACACACACACACACACACACACACACACACACA-CACACCCCCTTTAAAAAAAAAAAAAAAAAAGCCAAAAGCAGTGCCAAGCTTGCAT-GCC-3′ in 50 mM Tris-HCL pH7.5, 50 mM NaCl, 10 mM $MgCl_2$, 1 mM DTT, 1 mM dNTPs, and DnaE (4 nM) in the presence or absence of PykA, CAT, and PEPut (the concentrations of these proteins are specified in the relevant figure legends). The DNA products of the reactions were resolved with denaturing urea PAGE (15% gels, prepared and run in 1 XTBE). Visualization and image capturing were carried out using Licor Odyssey Fc Imaging System.

2.6. Surface Plasmon Resonance (SPR)

Initial SPR experiments were carried out in the absence of DNA, with only purified DnaE and PykA proteins. SPR experiments were carried out with a four-channel sensor C1 chip and a Biacore T200. The details of DnaE immobilization on the sensor chip surface, using EDC/NHS {1-ethyl-3(-dimethylaminopropyl)carboiimide hydrochloride/*N*-hydroxysuccinimide} chemistry, were as described elsewhere [26]. Subsequent experiments with DNA were carried out using a primed probe. Probes were constructed by annealing 30 mer (5′-AGGGGGTGTGTGTGTGTGTGTGTGTGTGTG-3′) to the 5′-end biotinylated 110 mer (5′-CACACACACACACACACACACACACACACACACACACACACACAC-ACACACACACACACACCCCCTTTAAAAAAAAAAAAAAAAAGCCAAAAGCAGTG-CCAAGCTTGCATGCC-3′). The annealing reaction comprised of the biotinylated 110 mer oligonucleotide and the 30 mer oligonucleotide template in slight molar excess (1.2:1), incubated at 95 °C for 2 min, and subsequently allowed to slowly cool to room temperature. Experiments were carried out at the Research Complex at Harwell in Oxford using a Biacore T200 at 200 µL/min with running buffer (25 mM NaCl, 25 mM $MgCl_2$, 150 mM Tris-HCl pH 7.5, and 1 mM DTT) at 25 °C. 5′ biotinylated DNA probe was immobilized to a streptavidin-coated chip surface (Biacore SA chip) to yield an increase of approximately 100 response units (RUs) per flow cell. When required to remove proteins that were bound to immobilized DNA, flow cells were regenerated by washing with 0.5% *w*/*v* SDS.

2.7. B. subtilis Strains and Growth Conditions

B. subtilis strains are listed in Supplementary Table S1. They were constructed by transforming competent cells with genomic DNA. The genotypes of constructed strains were checked by phenotypic analyses, endonuclease restriction, PCR analysis, and/or Sanger sequencing (Eurofins Genomics, Germany). Routinely, *B. subtilis* cells were grown at 37 °C in LB plus malate (0.4% *w*/*v*) with or without antibiotics at the following concentrations: spectinomycin (Sp, 60 µg/mL); tetracycline (Tet, 7.5 µg/mL). Microscopy studies were carried out with cells grown at 37 °C in the MC medium [5].

2.8. Microscopy

Cells were first grown overnight at 30 °C in MC supplemented with antibiotics. Upon saturation, cultures were diluted 1000-fold in the same medium without antibiotics and growth at 37 °C was carefully monitored using spectrophotometry. At OD_{600nm} = 0.15, 1.5 mL of cells were centrifuged at room temperature (11,000 rpm for 1 min), resuspended in 10 µL MC medium, and immediately mounted onto a glass slide covered with a 1.0% *w*/*v* agarose pad in MC. A 0.17 mm glass coverslip was then placed on top of the agarose pad. Microscopy was carried out using an epifluorescence microscope (Zeiss, Axio Observer.Z1) with a 100× magnification oil-immersion objective (Plan-APOCHROMAT Ph3) and a CMOS camera (Orca Flash 4.0 LT Hamamatsu). Digital images were acquired and analyzed using the Zen 2.6 (blue edition) software.

3. Results

3.1. The CAT Domain Alone Elicits a Functional Effect on the DnaE Polymerase Activity and PEPut Modulates This Effect and PykA Catalytic Activity

PykA stimulates the polymerase activity of DnaE in primer extension in vitro assays, implying that a physical interaction between the two proteins mediates this functional effect [5]. In an effort to first identify whether an individual domain, either CAT or PEPut, of PykA mediates this interaction or whether a full-length PykA protein is essential for this interaction, we overproduced and purified the CAT and PEPut domains separately (Figure 2A). We then carried out primer extension in vitro assays using a primed probe constructed from synthetic oligonucleotides to establish whether any one of the individual domains of PykA are capable of exerting a functional effect on the DnaE polymerase activity (Figure 2B). Data confirmed the stimulation of DnaE polymerase activity by PykA, and showed that CAT significantly inhibits this activity whereas PEPut alone does not appear to affect it (Figure 2C,D). This result suggests that CAT mediates the physical and functional interaction with DnaE and that PEPut modulates allosterically the effect of CAT on DnaE.

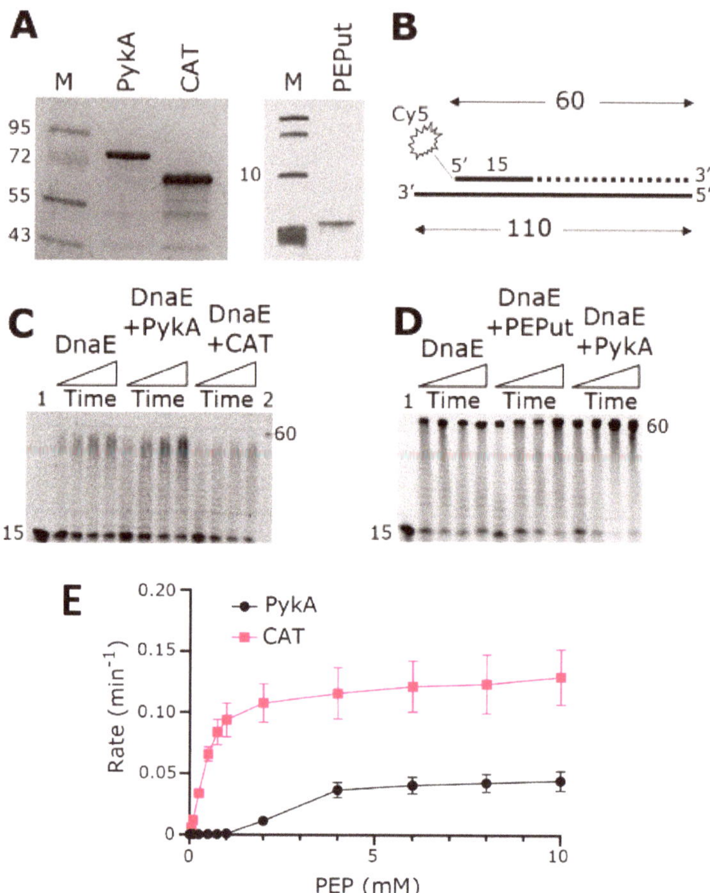

Figure 2. Replication and metabolic activities of purified PykA, CAT, and PEPut proteins. (**A**) SDS-PAGE gels showing purified PykA, CAT, and PEPut. Molecular weight markers in lanes M are shown in kDa units. (**B**) The radioactively labeled synthetic oligonucleotide probe used in primer extension assays is shown schematically. The 5′-end fluorescently labeled 15 mer oligonucleotide is annealed onto a 110 mer oligonucleotide. DnaE extends the 15 mer oligonucleotide to form a 60 mer product, as shown schematically by the dotted line. (**C**) Primer extension assays showing the biochemical effects of PykA and CAT on the DnaE polymerase activity. Time-course reactions (30, 60, 90, and 120 s, represented schematically from left to right by the rectangular triangles) were carried out with 4 nM DnaE and 16 nM PykA monomers (4 nM PykA tetramers) or 16 nM CAT, as described in the Section 2.5. Lanes 1 and 2 show the 15 mer and 60 mer oligonucleotide markers, respectively. (**D**) Primer extension assays showing the biochemical effects of PykA and PEPut on the DnaE polymerase activity. Time-course reactions (30, 60, 90, and 120 s, represented schematically from left to right by the rectangular triangles) were carried out with 4 nM DnaE and 16 nM PykA monomers (4 nM PykA tetramer) or 16 nM PEPut, as described in the Section 2.5. Lane 1 shows the position of the 15 mer oligonucleotide marker while the position of the 60 mer oligonucleotide is indicated by the side of the gel. (**E**) Activity assays of the purified full-length PykA and CAT. The assays were carried out as described in Section 2.1.4, PykA activity assay. All assays were carried out in triplicates and the mean reaction rates were plotted against PEP concentration with the error bars showing the standard error.

It was previously shown that PEPut has no metabolic activity per se but modulates the catalytic activity of PykA [5,16]. To confirm this, we compared the PykA and CAT catalytic activities using a standard biochemical assay (see Material and Methods for details). The results showed that CAT alone, when not linked to PEPut, is significantly more active than PykA (Figure 2E), confirming the negative allosteric control of PEPut on PykA catalytic activity.

3.2. The CAT Domain Mediates the Physical Interaction with DnaE in a Complex with a Primed Template and PEPut Reduces the Strength of This Interaction

Although a functional interaction between the PykA and DnaE proteins has been demonstrated, there is still no direct evidence of a physical interaction between the two proteins. To address this issue, we used SPR (Figure 3A). Our initial investigations with just purified DnaE and PykA proteins by immobilizing DnaE on the surface and then adding PykA did not reveal an interaction despite significant efforts (Supplementary Figure S2). One important component that was missing from our set up was DNA; hence, we constructed a primed probe composed of a 30 mer annealed to a 110 mer synthetic oligonucleotide. Next, the probe biotinylated at the 5′-end of the long oligonucleotide was immobilized on the streptavidin-coated surface of a chip (Figure 3A). The initial addition of DnaE at increasing concentrations (0, 10, 35, 87.5, 175, 350, and 700 nM) produced characteristic binding curves indicating a strong binding of DnaE to the probe (Supplementary Figure S3A). The maximum RU values against the DnaE concentration were fitted to a non-linear regression one site-specific binding curve using GraphPad Prism 9, revealed that DnaE binds to the probe with a K_d = 146.5 nM (R2 = 0.9587) (Supplementary Figure S3B). The subsequent experiments adding DnaE (35 nM) with increasing concentrations of PykA (0, 10, 25, 50, 100, 250, 500, and 1000 nM) produced characteristic binding curves, indicating an interaction between DnaE and PykA (Figure 3B).The maximum RU values of each curve against the (PykA) fitted to a non-linear regression as above, which revealed that the DnaE–PykA interaction has a K_d = 14.57 nM (R2 = 0.8515), indicating a strong and stable interaction. Similar SPR experiments with the CAT and PEPut domains revealed a strong interaction with CAT (K_d = 7.33 nM, R^2 = 0.9672) (Figure 3C), but no interaction with PEPut (Figure 3D). Control SPR experiments confirmed that PykA, CAT, and PEPut do not interact with the DNA probe (Supplementary Figure S4). Thus, we conclude from these data that the physical interaction of PykA with DnaE is mediated by CAT and that, because of our inability to detect an interaction between the two proteins in the absence of DNA, this interaction is DNA-dependent. The two-fold higher affinity of CAT (K_d = 7.33 nM) over PykA (K_d = 14.57 nM) for the DnaE-DNA complex further highlights the importance of PEPut in allosterically regulating PykA replication functions.

3.3. The TSH Motif of PEPut Impacts the Catalytic Activity of PykA but Not Its Effect on DnaE Activity

To acquired insights concerning the involvement of PEPut in PykA replication and metabolic functions, PykA proteins mutated in the TSH motif of PEPut were purified. This mutagenesis study was motivated by previous data showing that TSH mutations have different and contrasted effects on replication initiation and PykA catalytic activities [5], and that this motif is phosphorylated in other metabolic enzymes to ensure catalytic and regulatory functions (see above). In the purified mutant proteins, the three residues of the TSH motif were replaced individually or collectively by a D for mimicking phosphorylation. We also purified two proteins that were mutated in the catalytic site of PykA as controls. These mutations (R32A and G245AD246A) stimulate replication elongation in vivo and strongly reduce PykA activity [5]. The results showed that all mutant proteins stimulated the DnaE polymerase activity just like the wild-type PykA protein, suggesting that none of the tested mutations disrupt the PykA-DnaE interaction, its effect on DnaE activity, and its modulation by PEPut (Figure 4). In contrast, the catalytic activity of PykA is differently affected by the mutations. While the PEPut TSH > DDD and TSH > TDH mutations increase PykA activities, the TSH > DSH and TSH > TSD mutations decrease it or have no effect, respectively. Mutations in the catalytic site either reduced PykA activity (G245AD246A) or resulted in a catalytically dead protein (R32A). Hence, PykA catalytic activity is strongly modulated by PEPut, with T and S phosphorylation having opposite effects (inhibition and stimulation, respectively). Moreover, while the CAT and PykATSH > <u>DDD</u> proteins exhibit similar, very strong, catalytic activities, they oppositely impact DnaE activity (inhibition and stimulation, respectively) (compare the top panels of Figure 4A with Figure 2C,E). This shows that, whereas the TSH > <u>DDD</u> mutation mimics PEPut deletion in the metabolic assay, it does not do so in the replication assay. This suggests that the TSH > <u>DDD</u> mutation abrogates the metabolic but not the replication function of PEPut. Overall, these results highlight the complex and somewhat distinct functions played by the PEPut domain in modulating the effect of PykA on DnaE activity and PykA catalytic activity.

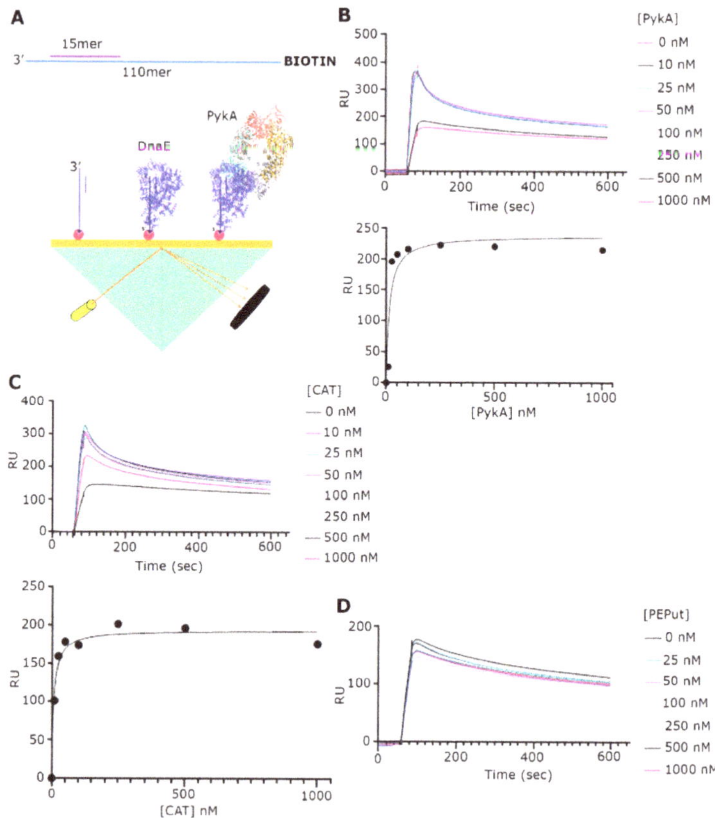

Figure 3. SPR analysis of the PykA-DnaE interaction. (**A**) Schematic diagrams explaining the primed DNA probe that was immobilized on the chip surface and the general SPR set up. (**B**) SPR sensograms with DnaE (35 nM) and increasing concentrations of full-length PykA, as indicated. Underneath the sensograms, the graph shows the fitting of the data to a one site-specific binding curve, using GraphPrism 9. (**C**) SPR sensograms with DnaE (35 nM) and increasing concentrations of CAT, as indicated. Underneath the sensograms, the graph shows the fitting of the data to a one site-specific binding curve, using GraphPrism 9. (**D**) SPR sensograms with DnaE (35 nM) and increasing concentrations of PEPut, as indicated, showing no interaction between DnaE and PEPut.

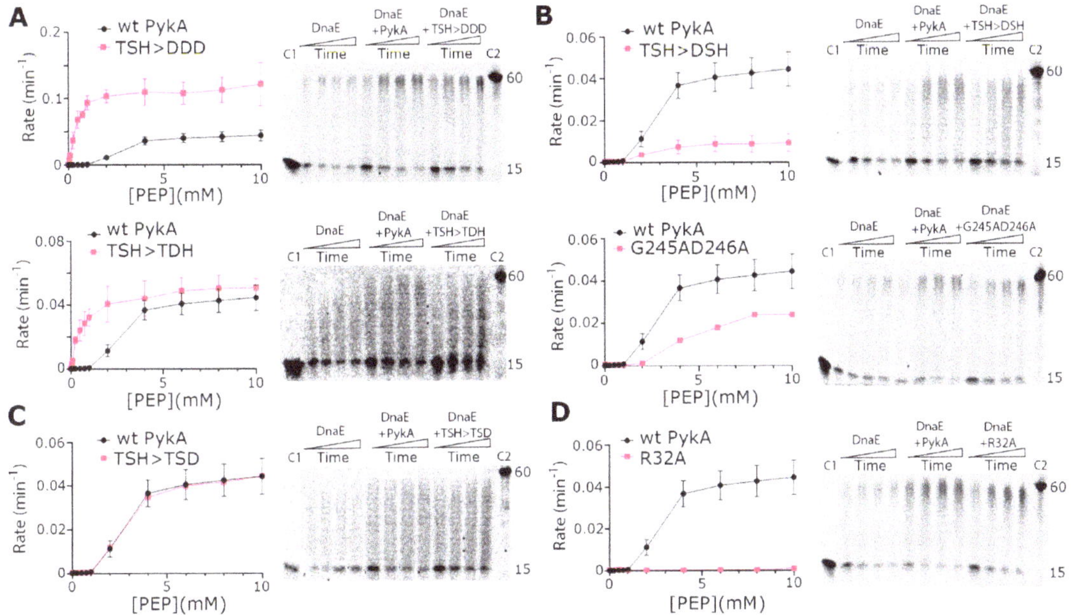

Figure 4. Biochemical characterization of PykA mutant proteins. PykA activity assays were carried out as described in 'Materials and Methods' and different mutations were found to increase (panel **A**), decrease (panel **B**), have no effect (panel **C**), or abolish (panel **D**) the PykA activity. All PykA activity assays were performed in triplicates and the mean rates were plotted against PEP concentration using GraphPad Prism 9, with the error bars showing the standard error. All PykA mutants were tested for the ability to stimulate the primer extension activity of DnaE and were all found able to stimulate DnaE, just like the wild-type PykA did. Primer extension assays were carried out as described in Section 2.5 with 4 nM DnaE and 16 nM PykA monomers (4 nM tetramers) in time courses 30, 60, 90, and 120 s, which are represented schematically from left to right by the rectangular triangles. C1 and C2 represent the annealed 15 mer primer and the maximum length 60 mer product of the primer extension reaction, respectively.

3.4. PEPut and Cat Mutations Impact Origin and Replisome Localization

The finding that PykA physically interacts with DnaE suggests that this metabolic enzyme can be recruited to the orisome and replisome to influence replication initiation, and elongation. Unfortunately, and despite significant efforts, we were unable to clearly localize PykA-fluorescent protein fusions by epi-fluorescence microscopy [5]. In an attempt to find indirect evidence of PykA recruitment at orisomes and replisomes, we analyzed *oriCs* and replication fork distribution in live wild-type and *pykA* cells. The visualization of *oriCs* and replication forks was carried out in cells encoding the Spo0J-GFP or DnaX-GFP protein fusions, respectively [32], and the foci distribution was analyzed in wild-type and representative CAT and PEPut mutants. Cells were grown in the MC gluconeogenic medium (malate + casein pancreatic hydrolysate), which was previously extensively used to analyze PykA metabolic and replication functions [5]. Most of the mutants (14/18) exhibited an average number of foci per cell and a ratio of cells with four over two foci similar to the wild-type strain (Figure 5A). However, these parameters significantly increased for *oriCs* foci (Spo0J-GFP context) in the Cat (PykA$_{E209A}$) and PEPut (PykA$_{TSH>AAA}$) mutants and notably decreased for replication forks foci (DnaX-GFP context) in the Cat (PykA$_{JP}$) and PEPut (PykA$_{TSH>AAA}$) mutants. Moreover, changes in foci positioning along the long cell axis were found in one mutant (in *pykA$_{TSH>AAA}$* cells with four DnaX-GFP foci, Figure 5B).

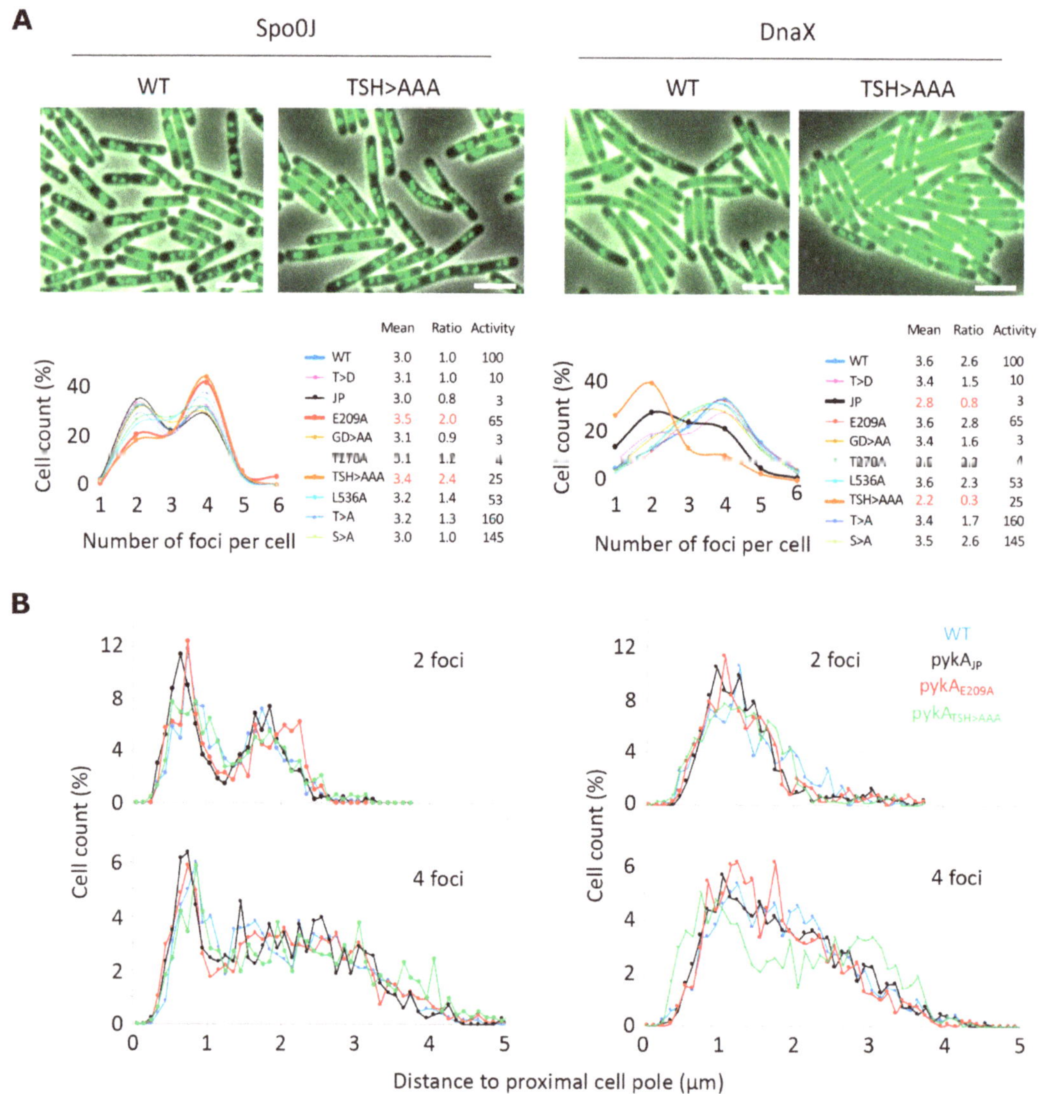

Figure 5. Pattern of Spo0J-GFP and DnaX-GFP foci in wild-type cells and *pykA* mutants. Cells encoding the GFP fusion proteins were grown in MC and foci distribution was analyzed by microscopy at $OD_{650\,nm} = 0.15$ in populations of about 2500 bacteria. (**A**) Proportion of cells with 1 to 6 foci. Top row: representative overlays of the phase contrast and GFP images of wild-type (WT) and $pykA_{TSH>AAA}$ (TSH > AAA) cells. Scale bar: 4 µm. Bottom row: Distribution of cells with n Spo0J-GFP or DnaX-GFP foci. Mean: Mean number of foci per cell. Ratio: Number of cells with 4 over 2 foci. Activity: Pyruvate kinase activity (%) [5]. Mutant proteins are described in the Supplementary Table S1. (**B**) Distance pole-foci along the long cell axis (using as reference the pole closest to the first foci). Note that the mean cell size of $pykA_{E209A}$, $pykA_{T537A}$, $pykA_{T537D}$, and $pykA_{TSH>AAA}$ cells varies by <5% compared to the wild-type strain (4.1 µm; 400 cells analyzed by strain; Spo0J context).

Theoretically, changes in foci patterns may be caused by changes in cell division. In *B. subtilis*, this process depends on pyruvate and the UDP-glucose concentration [33,34].

Given that PykA produces pyruvate from PEP in the last reaction of glycolysis and that UDP-glucose is a derivative of the first metabolite of glycolysis (glucose-6P), one can hypothesize that PykA mutations change foci patterns by altering pyruvate and/or UDP-glucose concentration and hence, cell division. However, the metabolome analysis argues against this hypothesis [5] as well as the lack of covariation between PykA activity and foci patterns (Figure 5A, inserted table). Strikingly, we found wild-type numbers of foci per cell and wild-type ratios of cells with four over two foci in strains encoding a PykA activity ranging from 3 to 165% of the parental strain, while these parameters were altered in mutants with a PykA activity of 3, 25, and 65%. Moreover, no significant cell size changes (<5%) were found in representative *pykA* mutants compared to the wild-type strain (Figure 5). Collectively, our results suggest that CAT and PEPut are important for the spatial localization of origins and replisomes, independent of their function in PykA catalytic activity. We suggest that these results provide indirect evidence for PykA recruitment at orisomes and replisomes.

4. Discussion

Despite our exquisite knowledge of the molecular mechanisms that underpin replication and cellular metabolism, our understanding on how these two of the most fundamental functions of life regulate each other's activities is limited. Recently, PykA has emerged as a new family of cross-species replication control regulators that drives the metabolic control of replication through a mechanism involving regulatory determinants of PykA catalytic activity, namely, CAT-substrate interactions, CAT-PEPut interaction, and PEPut TSH phosphorylation [5]. The CAT domain of PykA was shown to stimulate the replication fork speed whereas the PEPut domain of PykA was shown to act as a nutrient-dependent inhibitor of replication initiation [5]. Purified PykA forms an active, stable tetramer that stimulates the polymerase activity of DnaE, suggesting a functional interaction, but no direct protein–protein interaction between DnaE and PykA has been demonstrated to date [5]. Here, we show that PykA physically interacts with DnaE via a CAT–DnaE interaction when DnaE is bound to a primed DNA template and that PEPut modulates this interaction, the effect of PykA on DnaE activity, and the PykA catalytic activity.

The SPR data presented in this paper uncover a physical protein–protein interaction between DnaE and PykA. This interaction occurs when the polymerase is bound to a primed template and is mediated by the CAT domain of PykA. In contrast, the purified PEPut domain has no effect on DnaE activity and does not show any signs of a physical interaction with the DnaE. Interestingly, PykA and CAT have distinct effects on DnaE: while PykA stimulates DnaE activity CAT inhibits it, and the interaction of the DnaE-DNA complex with PykA is weaker than with CAT (K_d = 14.57 versus 7.33 nM, respectively). As the two proteins differ by the presence/absence of PEPut, and as PEPut interacts with CAT through hydrogen bonding [15,29], the distinct effects of PykA and CAT on DnaE may result from some allosteric changes induced by PEPut. If so, this may assign a regulatory function of PEPut on the replication moonlighting activity of PykA. Unfortunately, we were unable to identify the residues that were important for this regulatory function, as the replication phenotypes of purified PykA mutants (two were mutated in CAT and four in the TSH motif of PEPut) were indistinguishable from that of the native protein. This is in contrast to the in vivo data, which showed that the CAT mutations used here stimulate the rate of replication forks while those in the TSH motif abrogate or leave intact the PEPut-driven inhibitory activity of initiation [5]. Clearly, the simplified assay used here does not recapitulate the in vivo situation and needs to be significantly improved to approach the complexity of live cells.

A parallel study was carried out to further characterize the factors regulating the PykA catalytic activity. On its own, the CAT domain exhibits a significantly increased activity compared to the full-length PykA protein. As most of the bacterial PykA proteins lack PEPut [15], and as the related PEPut-containing the *Geobacillus stearothermophilus* protein forms an active tetramer with and without its PEPut domain [29], our results suggest that the difference in CAT and PykA activities results from a PEPut-mediated allosteric regulation of PykA activity, as proposed previously [5,15]. This notion is supported by our data, showing that the PykA catalytic activity is either increased (TSH > T<u>D</u>H and TSH > <u>DDD</u>), impaired (TSH > <u>D</u>SH), or left intact (TSH > TS<u>D</u>) in the mutants of the TSH motif of PEPut. Although these results are in good agreement with our previous study [5], some apparent contradictions were found: the TSH > T<u>D</u>H and TSH > <u>DDD</u> mutants were found to be highly active here but they were poorly active in the previous study. We envision that these differences result from differences in the experimental set up (standard assay versus crude extract assay), indicating that the standard assay needs to be complemented with additional factors to fully recapitulate the complexity of the allosteric regulation of PykA activity by PEPut.

The detection of a strong interaction between PykA and DnaE when bound to a primed DNA template is an important discovery toward the molecular understanding of the metabolic control of replication. Indeed, this result indicates that PykA is recruited by DnaE at sites of DNA synthesis and that the thousands of cytosolic PykA molecules cannot trap the tens of free (not DNA-bound) DnaE proteins far away from replication sites. The PykA recruitment at sites of DNA synthesis is likely highly dynamic, as DnaE is frequently recruited to and released from replication machineries to extend RNA primers during replication initiation and lagging strand synthesis [30,31,35–37]. This implies that the PykA-DnaE interactions are incessantly dissociated and reconstituted, allowing new PykA molecules to be recruited at replication machineries to ensure a highly dynamic modulation of the replication rate with metabolism. Obviously, this hypothesis needs to be substantiated by additional data, including direct evidence of PykA recruitment at origins and replication forks in live cells. Interestingly, the microscopy data presented here provide indirect support for this recruitment, showing that the CAT and PEPut domains are important for a proper spatial localization of origins and replication forks, independent of their functions in PykA catalytic activity.

Our inability to detect an interaction between the PEPut domain and DnaE raises the question of how PEPut modulates replication initiation. Several hypotheses can be put forward to address this issue. First, it can be considered that our assay, which monitors DNA synthesis and not DNA initiation, does not contain structures such as supercoiled DNA and melted double-strand DNA needed for PykA recruitment via a DnaE-PEPut interaction. Alternatively, it may be that the PEPut domain interacts with some other components of the *B. subtilis* orisome (e.g., DnaA, DnaC, etc. . . .) via an as yet unidentified physical interaction. It may also be that only the CAT domain physically interacts with DnaE (and perhaps other replication proteins) with the PEPut domain controlling the replication moonlighting functions of CAT on initiation or elongation. This control could depend on the growth conditions. As previously shown, suggesting that the PEPut initiation function is turned on in neoglucogenic media and off in glycolytic media [5].

A long history of investigations suggest that the metabolism-replication links described here and in our previous work [5] are more the tip of an iceberg than the exception. In *B. subtilis*, comprehensive studies uncovered a toolbox of intermingled metabolism-replication links that temporalizes replication in the cell cycle in a nutritional-dependent manner. This toolbox comprises on one side, reactions ensuring the 3C part of glycolysis (including PykA) and the downstream pyruvate metabolism, and, on the other side, the replication enzymes DnaC, DnaG, and DnaE [3,4,38]. Similar results were found in *E. coli* [39–41] and possibly *Caulobacter crescentus* [42]. In *E. coli*, evidence for the modulation of initiation by metabolites (acetyl-CoA and cAMP) was also provided [43,44]. In eukaryotes, the timing of the origin firing depends on an increase in acetyl-CoA, which promotes histone acetylation [45]. This increase is geared by a metabolic cycle in yeast and by nuclear forms of the ATP-citrate lyase and PDH complexes in mammalian cells [46–48]. Moreover, the glyceraldehyde-3-phosphate dehydrogenase (GAPDH) and the lactate dehydrogenase (LDH) enter the nucleus to induce histone H2B production and promote S-phase progression [49,50]. Another metabolic enzyme, the phosphoglycerate kinase PGK, interacts with the protein kinase CDC7 in the nucleus to stimulate replication initiation by enhancing the CDC7-mediated activation of the replicative MCM helicase [51]. Additionally, the impaired expression of genes of the central carbon metabolism delays the entry of human fibroblasts into the S phase or decreases the number of cells in this phase [52–54]. Finally, PGK, GAPDH, and LDH modulate the activity of the three eukaryotic replicative polymerases (Polα, Polε, and Polδ) in vitro [55–57]. Collectively, these results suggest that DNA replication is under a metabolic control geared by determinants of the central carbon metabolism throughout the evolution tree. Since Schaechter's seminal studies in *Salmonella typhimurium* in 1958, it is now well-established that this control compartmentalizes DNA synthesis in the cell cycle in a wide range of nutritional conditions in bacteria, while it directs the entry and progression of the S phase in the reduction phase of an oscillating redox metabolic cycle in eukaryote cells [58–61]. By extrapolating knowledge on PykA moonlighting activities in DNA replication, we proposed that the metabolic control of replication from bacteria to eukaryote is orchestrated by a mechanism in which metabolites and proteins of central carbon metabolism that signal and sense nutritional stimuli for regulating cellular metabolism, ensure moonlighting activities to convey this metabolic information to the replication machinery for coordinating replication to metabolism. As the metabolic control of replication plays an important role in the overall replication rate [5], and as failures in this control cause chromosomal lesions (double-strand DNA breaks and nucleotide misincorporation) increasing the risk of genetic diseases, such as cancer [6–9], we propose that metabolic changes underpinning the Warburg effect and disrupting the metabolic control of replication may form an additional, intrinsic root cause of genetic instability and cancer initiation.

5. Conclusions

Our report provides novel molecular insights into the metabolic control of DNA replication in *B. subtilis*. It suggests that (i) this control depends on a direct physical interaction between the CAT domain of PykA and the DNA polymerase DnaE, which is essential for replication initiation and lagging strand synthesis; (ii) this interaction occurs at primed sites, allowing PykA recruitment within the orisome and replisome; and (iii) the PEPut domain of PykA regulates, at least in part, the moonlighting replication functions of PykA and is also a key regulator of PykA catalytic activity, thereby connecting replication to metabolism.

Supplementary Materials: The following supporting information can be downloaded at: https://www.mdpi.com/article/10.3390/life13040965/s1, Figure S1: The plasmid map of the p2CT-*pykA* expression vector. Figure S2: SPR data showing no interaction between DnaE (immobilized on the chip surface) and PykA. Figure S3: SPR sensograms of increasing concentrations of DnaE; Figure S4: Control SPR experiments with increasing concentrations of PykA, CAT and PEPut; Table S1: Strains.

Author Contributions: Conceptualization, P.S. and L.J.; formal analysis, P.S. and L.J.; funding acquisition, P.S. and L.J.; investigation, A.H., M.P. and L.J.; methodology, P.S. and L.J.; supervision, P.S.; writing—original draft, P.S. and L.J.; writing—review and editing, P.S. and L.J. All authors have read and agreed to the published version of the manuscript.

Funding: This work was supported by a Biotechnology Biological Sciences Research Council (BBSRC) grant: BB/R013357/1 to P.S., a University of Nottingham sub-contract RIS1165589 to L.J., and by recurring research funds from CNRS (Centre National de la Recherche Scientifique) and UEVE (Univesité d'Evry Val d'Essonne) to L.J. M.P. was partially supported by a University of Nottingham Vice Chancellor's Excellence PhD award.

Institutional Review Board Statement: Not applicable.

Informed Consent Statement: Not applicable.

Data Availability Statement: All data are contained within the article.

Acknowledgments: We thank James Berger and Lyle A. Simmons for the gift of plasmid p2CT and fluorescent *B. subtilis* strains (BTS8 and LAS26), respectively, Jemma Harris for her help with the SPR in the Research Complex at Harwell, Natasha Preston and Francesca Slack for their help with the PykA mutant activity assays, and Marina Koutsidou for her help with some of the DnaE polymerase assays.

Conflicts of Interest: The authors declare no conflict of interest. The funders had no role in the design of the study; in the collection, analyses or interpretation of data; in the writing of the manuscript; or in the decision to publish the results.

References

1. He, C.-L.; Bian, Y.-Y.; Xue, Y.; Liu, Z.-X.; Zhou, K.-Q.; Yao, C.-F.; Lin, Y.; Zou, H.-F.; Luo, F.-X.; Qu, Y.-Y.; et al. Pyruvate Kinase M2 Activates MTORC1 by Phosphorylating AKT1S1. *Sci. Rep.* **2016**, *6*, 21524. [CrossRef] [PubMed]
2. Chen, X.; Chen, S.; Yu, D. Protein Kinase Function of Pyruvate Kinase M2 and Cancer. *Cancer Cell Int.* **2020**, *20*, 523. [CrossRef] [PubMed]
3. Jannière, L.; Canceill, D.; Suski, C.; Kanga, S.; Dalmais, B.; Lestini, R.; Monnier, A.-F.; Chapuis, J.; Bolotin, A.; Titok, M.; et al. Genetic Evidence for a Link between Glycolysis and DNA Replication. *PLoS ONE* **2007**, *2*, e447. [CrossRef] [PubMed]
4. Nouri, H.; Monnier, A.-F.; Fossum-Raunehaug, S.; Maciąg-Dorszyńska, M.; Cabin-Flaman, A.; Képès, F.; Węgrzyn, G.; Szalewska-Pałasz, A.; Norris, V.; Skarstad, K.; et al. Multiple Links Connect Central Carbon Metabolism to DNA Replication Initiation and Elongation in *Bacillus subtilis*. *DNA Res.* **2018**, *25*, 641–653. [CrossRef]
5. Horemans, S.; Pitoulias, M.; Holland, A.; Pateau, E.; Lechaplais, C.; Ekaterina, D.; Perret, A.; Soultanas, P.; Janniere, L. Pyruvate Kinase, a Metabolic Sensor Powering Glycolysis, Drives the Metabolic Control of DNA Replication. *BMC Biol.* **2022**, *20*, 87. [CrossRef]
6. Mathews, C.K. Deoxyribonucleotide Metabolism, Mutagenesis and Cancer. *Nat. Rev. Cancer* **2015**, *15*, 528–539. [CrossRef]
7. Tubbs, A.; Nussenzweig, A. Endogenous DNA Damage as a Source of Genomic Instability in Cancer. *Cell* **2017**, *168*, 644–656. [CrossRef]
8. Rodgers, K.; McVey, M. Error-Prone Repair of DNA Double-Strand Breaks: Error-prone repair of DNA DSBS. *J. Cell. Physiol.* **2016**, *231*, 15–24. [CrossRef]
9. Saxena, S.; Zou, L. Hallmarks of DNA Replication Stress. *Mol. Cell* **2022**, *82*, 2298–2314. [CrossRef]
10. Yang, J.; Liu, H.; Liu, X.; Gu, C.; Luo, R.; Chen, H.-F. Synergistic Allosteric Mechanism of Fructose-1,6-Bisphosphate and Serine for Pyruvate Kinase M2 via Dynamics Fluctuation Network Analysis. *J. Chem. Inf. Model.* **2016**, *56*, 1184–1192. [CrossRef]
11. Chaneton, B.; Hillmann, P.; Zheng, L.; Martin, A.C.L.; Maddocks, O.D.K.; Chokkathukalam, A.; Coyle, J.E.; Jankevics, A.; Holding, F.P.; Vousden, K.H.; et al. Serine Is a Natural Ligand and Allosteric Activator of Pyruvate Kinase M2. *Nature* **2012**, *491*, 458–462. [CrossRef]
12. Nakatsu, D.; Horiuchi, Y.; Kano, F.; Noguchi, Y.; Sugawara, T.; Takamoto, I.; Kubota, N.; Kadowaki, T.; Murata, M. L-Cysteine Reversibly Inhibits Glucose-Induced Biphasic Insulin Secretion and ATP Production by Inactivating PKM2. *Proc. Natl. Acad. Sci. USA* **2015**, *112*, E1067–E1076. [CrossRef]
13. Schormann, N.; Hayden, K.L.; Lee, P.; Banerjee, S.; Chattopadhyay, D. An Overview of Structure, Function, and Regulation of Pyruvate Kinases. *Protein Sci.* **2019**, *28*, 1771–1784. [CrossRef]
14. Sakai, H.; Suzuki, K.; Imahori, K. Purification and Properties of Pyruvate Kinase from *Bacillus stearothermophilus*. *J. Biochem.* **1986**, *99*, 1157–1167. [CrossRef]
15. Morgan, H.P.; Zhong, W.; McNae, I.W.; Michels, P.A.M.; Fothergill-Gilmore, L.A.; Walkinshaw, M.D. Structures of Pyruvate Kinases Display Evolutionarily Divergent Allosteric Strategies. *R. Soc. Open Sci.* **2014**, *1*, 140120. [CrossRef]
16. Sakai, H. Possible Structure and Function of the Extra C-Terminal Sequence of Pyruvate Kinase from *Bacillus stearothermophilus*. *J. Biochem.* **2004**, *136*, 471–476. [CrossRef]
17. Nguyen, C.C.; Saier, M.H. Phylogenetic Analysis of the Putative Phosphorylation Domain in the Pyruvate Kinase of *Bacillus stearothermophilus*. *Res. Microbiol.* **1995**, *146*, 713–719. [CrossRef]
18. Alpert, C.A.; Frank, R.; Stueber, K.; Deutscher, J.; Hengstenberg, W. Phosphoenolpyruvate-Dependent Protein Kinase Enzyme I of *Streptococcus faecalis*: Purification and Properties of the Enzyme and Characterization of Its Active Center. *Biochemistry* **1985**, *24*, 959–964. [CrossRef]
19. Teplyakov, A.; Lim, K.; Zhu, P.-P.; Kapadia, G.; Chen, C.C.H.; Schwartz, J.; Howard, A.; Reddy, P.T.; Peterkofsky, A.; Herzberg, O. Structure of Phosphorylated Enzyme I, the Phosphoenolpyruvate:Sugar Phosphotransferase System Sugar Translocation Signal Protein. *Proc. Natl. Acad. Sci. USA* **2006**, *103*, 16218–16223. [CrossRef]
20. Goss, N.H.; Evans, C.T.; Wood, H.G. Pyruvate Phosphate Dikinase: Sequence of the Histidyl Peptide, the Pyrophosphoryl and Phosphoryl Carrier. *Biochemistry* **1980**, *19*, 5805–5809. [CrossRef]
21. Herzberg, O.; Chen, C.C.; Kapadia, G.; McGuire, M.; Carroll, L.J.; Noh, S.J.; Dunaway-Mariano, D. Swiveling-Domain Mechanism for Enzymatic Phosphotransfer between Remote Reaction Sites. *Proc. Natl. Acad. Sci. USA* **1996**, *93*, 2652–2657. [CrossRef] [PubMed]
22. Tolentino, R.; Chastain, C.; Burnell, J. Identification of the Amino Acid Involved in the Regulation of Bacterial Pyruvate, Orthophosphate Dikinase and Phosphoenolpyruvate Synthetase. *Adv. Biol. Chem.* **2013**, *3*, 12–21. [CrossRef]

23. Burnell, J.N.; Chastain, C.J. Cloning and Expression of Maize-Leaf Pyruvate, Pi Dikinase Regulatory Protein Gene. *Biochem. Biophys. Res. Commun.* **2006**, *345*, 675–680. [CrossRef] [PubMed]

24. Burnell, J.N. Cloning and Characterization of Escherichia Coli DUF299: A Bifunctional ADP-Dependent Kinase—Pi-Dependent Pyrophosphorylase from Bacteria. *BMC Biochem.* **2010**, *11*, 1. [CrossRef]

25. Burnell, J.N.; Hatch, M.D. Regulation of C4 Photosynthesis: Identification of a Catalytically Important Histidine Residue and Its Role in the Regulation of Pyruvate,Pi Dikinase. *Arch. Biochem. Biophys.* **1984**, *231*, 175–182. [CrossRef]

26. Eymann, C.; Dreisbach, A.; Albrecht, D.; Bernhardt, J.; Becher, D.; Gentner, S.; Tam, L.T.; Büttner, K.; Buurman, G.; Scharf, C.; et al. A Comprehensive Proteome Map of Growing *Bacillus subtilis* Cells. *Proteomics* **2004**, *4*, 2849–2876. [CrossRef]

27. Mader, U.; Schmeisky, A.G.; Florez, L.A.; Stulke, J. SubtiWiki—A Comprehensive Community Resource for the Model Organism *Bacillus subtilis*. *Nucleic Acids Res.* **2012**, *40*, D1278–D1287. [CrossRef]

28. Reiland, S.; Messerli, G.; Baerenfaller, K.; Gerrits, B.; Endler, A.; Grossmann, J.; Gruissem, W.; Baginsky, S. Large-Scale Arabidopsis Phosphoproteome Profiling Reveals Novel Chloroplast Kinase Substrates and Phosphorylation Networks. *Plant Physiol.* **2009**, *150*, 889–903. [CrossRef]

29. Suzuki, K.; Ito, S.; Shimizu-Ibuka, A.; Sakai, H. Crystal Structure of Pyruvate Kinase from *Geobacillus stearothermophilus*. *J. Biochem.* **2008**, *144*, 305–312. [CrossRef]

30. Rannou, O.; Le Chatelier, E.; Larson, M.A.; Nouri, H.; Dalmais, B.; Laughton, C.; Jannière, L.; Soultanas, P. Functional Interplay of DnaE Polymerase, DnaG Primase and DnaC Helicase within a Ternary Complex, and Primase to Polymerase Hand-off during Lagging Strand DNA Replication in *Bacillus subtilis*. *Nucleic Acids Res.* **2013**, *41*, 5303–5320. [CrossRef]

31. Paschalis, V.; Le Chatelier, E.; Green, M.; Képès, F.; Soultanas, P.; Janniere, L. Interactions of the *Bacillus subtilis* DnaE Polymerase with Replisomal Proteins Modulate Its Activity and Fidelity. *Open Biol.* **2017**, *7*, 170146. [CrossRef]

32. Simmons, L.A.; Grossman, A.D.; Walker, G.C. Replication Is Required for the RecA Localization Response to DNA Damage in *Bacillus Subtilis*. *Proc. Natl. Acad. Sci. USA* **2007**, *104*, 1360–1365. [CrossRef]

33. Monahan, L.G.; Hajduk, I.V.; Blaber, S.P.; Charles, I.G.; Harry, E.J. Coordinating Bacterial Cell Division with Nutrient Availability: A Role for Glycolysis. *mBio* **2014**, *5*, e00935-14. [CrossRef]

34. Weart, R.B.; Lee, A.H.; Chien, A.-C.; Haeusser, D.P.; Hill, N.S.; Levin, P.A. A Metabolic Sensor Governing Cell Size in Bacteria. *Cell* **2007**, *130*, 335–347. [CrossRef]

35. Sanders, G.M.; Dallmann, H.G.; McHenry, C.S. Reconstitution of the *B. subtilis* Replisome with 13 Proteins Including Two Distinct Replicases. *Mol. Cell* **2010**, *37*, 273–281. [CrossRef]

36. Li, Y.; Chen, Z.; Matthews, L.A.; Simmons, L.A.; Biteen, J.S. Dynamic Exchange of Two Essential DNA Polymerases during Replication and after Fork Arrest. *Biophys. J.* **2019**, *116*, 684–693. [CrossRef]

37. Hernández-Tamayo, R.; Oviedo-Bocanegra, L.M.; Fritz, G.; Graumann, P.L. Symmetric Activity of DNA Polymerases at and Recruitment of Exonuclease ExoR and of PolA to the *Bacillus subtilis* Replication Forks. *Nucleic Acids Res.* **2019**, *47*, 8521–8536. [CrossRef]

38. Murray, H.; Koh, A. Multiple Regulatory Systems Coordinate DNA Replication with Cell Growth in *Bacillus subtilis*. *PLoS Genet.* **2014**, *10*, e1004731. [CrossRef]

39. Maciąg, M.; Nowicki, D.; Janniere, L.; Szalewska-Pałasz, A.; Węgrzyn, G. Genetic Response to Metabolic Fluctuations: Correlation between Central Carbon Metabolism and DNA Replication in *Escherichia coli*. *Microb. Cell Factories* **2011**, *10*, 19. [CrossRef]

40. Maciąg-Dorszyńska, M.; Ignatowska, M.; Janniere, L.; Węgrzyn, G.; Szalewska-Pałasz, A. Mutations in Central Carbon Metabolism Genes Suppress Defects in Nucleoid Position and Cell Division of Replication Mutants in *Escherichia coli*. *Gene* **2012**, *503*, 31–35. [CrossRef]

41. Krause, K.; Maciąg-Dorszyńska, M.; Wosinski, A.; Gaffke, L.; Morcinek-Orłowska, J.; Rintz, E.; Bielańska, P.; Szalewska-Pałasz, A.; Muskhelishvili, G.; Węgrzyn, G. The Role of Metabolites in the Link between DNA Replication and Central Carbon Metabolism in *Escherichia Coli*. *Genes* **2020**, *11*, 447. [CrossRef] [PubMed]

42. Bergé, M.; Pezzatti, J.; González-Ruiz, V.; Degeorges, L.; Mottet-Osman, G.; Rudaz, S.; Viollier, P.H. Bacterial Cell Cycle Control by Citrate Synthase Independent of Enzymatic Activity. *eLife* **2020**, *9*, e52272. [CrossRef] [PubMed]

43. Hughes, P.; Landoulsi, A.; Kohiyama, M. A Novel Role for CAMP in the Control of the Activity of the *E. coli* Chromosome Replication Initiator Protein, DnaA. *Cell* **1988**, *55*, 343–350. [CrossRef] [PubMed]

44. Zhang, Q.; Zhou, A.; Li, S.; Ni, J.; Tao, J.; Lu, J.; Wan, B.; Li, S.; Zhang, J.; Zhao, S.; et al. Reversible Lysine Acetylation Is Involved in DNA Replication Initiation by Regulating Activities of Initiator DnaA in *Escherichia coli*. *Sci. Rep.* **2016**, *6*, 30837. [CrossRef]

45. Vogelauer, M.; Rubbi, L.; Lucas, I.; Brewer, B.J.; Grunstein, M. Histone Acetylation Regulates the Time of Replication Origin Firing. *Mol. Cell* **2002**, *10*, 1223–1233. [CrossRef]

46. Cai, L.; Sutter, B.M.; Li, B.; Tu, B.P. Acetyl-CoA Induces Cell Growth and Proliferation by Promoting the Acetylation of Histones at Growth Genes. *Mol. Cell* **2011**, *42*, 426–437. [CrossRef]

47. Sutendra, G.; Kinnaird, A.; Dromparis, P.; Paulin, R.; Stenson, T.H.; Haromy, A.; Hashimoto, K.; Zhang, N.; Flaim, E.; Michelakis, E.D. A Nuclear Pyruvate Dehydrogenase Complex Is Important for the Generation of Acetyl-CoA and Histone Acetylation. *Cell* **2014**, *158*, 84–97. [CrossRef]

48. Wellen, K.E.; Hatzivassiliou, G.; Sachdeva, U.M.; Bui, T.V.; Cross, J.R.; Thompson, C.B. ATP-Citrate Lyase Links Cellular Metabolism to Histone Acetylation. *Science* **2009**, *324*, 1076–1080. [CrossRef]

49. Zheng, L.; Roeder, R.G.; Luo, Y. S Phase Activation of the Histone H2B Promoter by OCA-S, a Coactivator Complex That Contains GAPDH as a Key Component. *Cell* **2003**, *114*, 255–266. [CrossRef]
50. Dai, R.-P.; Yu, F.-X.; Goh, S.-R.; Chng, H.-W.; Tan, Y.-L.; Fu, J.-L.; Zheng, L.; Luo, Y. Histone 2B (H2B) Expression Is Confined to a Proper NAD+/NADH Redox Status. *J. Biol. Chem.* **2008**, *283*, 26894–26901. [CrossRef]
51. Li, X.; Qian, X.; Jiang, H.; Xia, Y.; Zheng, Y.; Li, J.; Huang, B.-J.; Fang, J.; Qian, C.-N.; Jiang, T.; et al. Nuclear PGK1 Alleviates ADP-Dependent Inhibition of CDC7 to Promote DNA Replication. *Mol. Cell* **2018**, *72*, 650–660.e8. [CrossRef]
52. Konieczna, A.; Szczepańska, A.; Sawiuk, K.; Węgrzyn, G.; Łyżeń, R. Effects of Partial Silencing of Genes Coding for Enzymes Involved in Glycolysis and Tricarboxylic Acid Cycle on the Enterance of Human Fibroblasts to the S Phase. *BMC Cell Biol.* **2015**, *16*, 16. [CrossRef]
53. Fornalewicz, K.; Wieczorek, A.; Węgrzyn, G.; Łyżeń, R. Silencing of the Pentose Phosphate Pathway Genes Influences DNA Replication in Human Fibroblasts. *Gene* **2017**, *635*, 33–38. [CrossRef]
54. Wieczorek, A.; Fornalewicz, K.; Mocarski, Ł.; Łyżeń, R.; Węgrzyn, G. Double Silencing of Relevant Genes Suggests the Existence of the Direct Link between DNA Replication/Repair and Central Carbon Metabolism in Human Fibroblasts. *Gene* **2018**, *650*, 1–6. [CrossRef]
55. Grosse, F.; Nasheuer, H.-P.; Scholtissek, S.; Schomburg, U. Lactate Dehydrogenase and Glyceraldehyde-Phosphate Dehydrogenase Are Single-Stranded DNA-Binding Proteins That Affect the DNA-Polymerase-Alpha-Primase Complex. *Eur. J. Biochem.* **1986**, *160*, 459–467. [CrossRef]
56. Popanda, O.; Fox, G.; Thielmann, H.W. Modulation of DNA Polymerases α, δ and ε by Lactate Dehydrogenase and 3-Phosphoglycerate Kinase. *Biochim. Biophys. Acta BBA—Gene Struct. Expr.* **1998**, *1397*, 102–117. [CrossRef]
57. Jindal, H.K.; Vishwanatha, J.K. Functional Identity of a Primer Recognition Protein as Phosphoglycerate Kinase. *J. Biol. Chem.* **1990**, *265*, 6540–6543. [CrossRef]
58. Schaechter, M.; MaalOe, O.; Kjeldgaard, N.O. Dependency on Medium and Temperature of Cell Size and Chemical Composition during Balanced Growth of *Salmonella typhimurium*. *J. Gen. Microbiol.* **1958**, *19*, 592–606. [CrossRef]
59. Stokke, C.; Flåtten, I.; Skarstad, K. An Easy-To-Use Simulation Program Demonstrates Variations in Bacterial Cell Cycle Parameters Depending on Medium and Temperature. *PLoS ONE* **2012**, *7*, e30981. [CrossRef]
60. Helmstetter, C.E. Timing of Synthetic Activities in the Cell Cycle. In *Escherichia coli and Salmonella: Cellular and Molecular Biology*; ASM Press: Washington, DC, USA, 1996; pp. 1627–1639.
61. Sharpe, M.E.; Hauser, P.M.; Sharpe, R.G.; Errington, J. *Bacillus subtilis* Cell Cycle as Studied by Fluorescence Microscopy: Constancy of Cell Length at Initiation of DNA Replication and Evidence for Active Nucleoid Partitioning. *J. Bacteriol.* **1998**, *180*, 547–555. [CrossRef]

Review

Molecular Cytology of 'Little Animals': Personal Recollections of *Escherichia coli* (and *Bacillus subtilis*)

Nanne Nanninga

Molecular Cytology, Swammerdam Institute for Life Sciences (SILS), University of Amsterdam,
1098 XH Amsterdam, The Netherlands; nannenanninga@gmail.com

Abstract: This article relates personal recollections and starts with the origin of electron microscopy in the sixties of the previous century at the University of Amsterdam. Novel fixation and embedding techniques marked the discovery of the internal bacterial structures not visible by light microscopy. A special status became reserved for the freeze-fracture technique. By freeze-fracturing chemically fixed cells, it proved possible to examine the morphological effects of fixation. From there on, the focus switched from bacterial structure as such to their cell cycle. This invoked bacterial physiology and steady-state growth combined with electron microscopy. Electron-microscopic autoradiography with pulses of [^{3}H] Dap revealed that segregation of replicating DNA cannot proceed according to a model of zonal growth (with envelope-attached DNA). This stimulated us to further investigate the sacculus, the peptidoglycan macromolecule. In particular, we focused on the involvement of penicillin-binding proteins such as PBP2 and PBP3, and their role in division. Adding aztreonam (an inhibitor of PBP3) blocked ongoing divisions but not the initiation of new ones. A PBP3-independent peptidoglycan synthesis (PIPS) appeared to precede a PBP3-dependent step. The possible chemical nature of PIPS is discussed.

Keywords: electron microscopy; confocal microscopy; image processing; bacteria, divisome; elongasome; PIPS

Citation: Nanninga, N. Molecular Cytology of 'Little Animals': Personal Recollections of *Escherichia coli* (and *Bacillus subtilis*). *Life* **2023**, *13*, 1782. https://doi.org/10.3390/life13081782

Academic Editor: Lluís Ribas de Pouplana

Received: 18 July 2023
Revised: 9 August 2023
Accepted: 17 August 2023
Published: 21 August 2023

1. Introduction

In the early sixties of the previous century, I was involved in photosynthetic experiments with chloroplasts isolated from succulents. The results deviated from those obtained with spinach chloroplasts (the spinach was purchased from a nearby greengrocer). One of the explanations which came to my mind was the possibility that the structure of succulent chloroplasts might differ from that of the standard spinach chloroplasts. With this idea, I paid a visit to the Laboratory of Electron Microscopy of the University of Amsterdam. This laboratory was headed by Dr. Woutera van Iterson, who came from the Technical University of Delft, where she pioneered very successfully the use of the electron microscope for the study of bacterial flagella (for instance, *Bacterium herbicola* in Houwink and van Iterson, 1950 [1]; Figure 1). She continued her work on bacterial ultrastructure in Amsterdam.

I do not remember our conversation, but presumably I could not explain to her what kind of difference to expect in chloroplast structure, neither how this might explain my results with chloroplasts of succulent plants. Anyway, she more or less convinced me that it would be good to become familiar with the electron microscope and electron microscopic techniques and to begin with bacteria. And so I started with electron microscopy of wall-less L-forms of *Proteus vulgaris*. The aim was to study the organization of bacterial cytoplasm. The idea was that this would be facilitated by the removal of DNA. I found out that, after chemical fixation with osmium tetroxide, I could still remove its DNA with DNase. Later on (after 1965; see below), I felt that a descriptive approach to cytoplasmic structure should be complemented by an analysis of its components, that is, ribosomes and their subunits. So, I embarked on sucrose gradient centrifugation to isolate 50 S ribosomal subunits. I remember my excitement when I obtained my first purified preparation; however, my recollections in this field are less suitable for the current article.

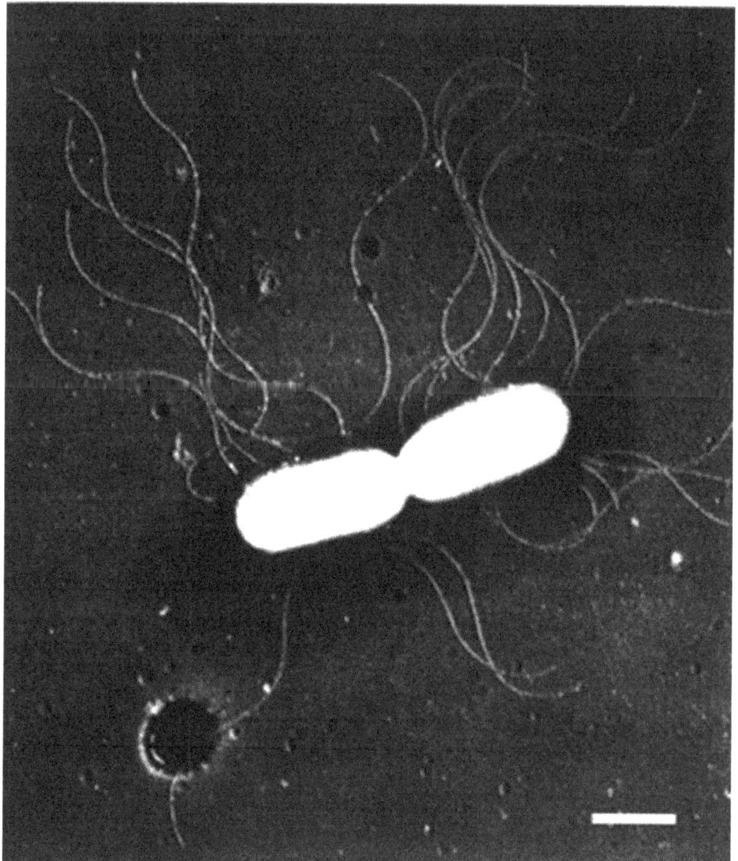

Figure 1. *Bacterium herbicola* with peritrichous flagella after shadowcasting. Bar, 1 μm. Source: Figure 3 from [1]. Copyright © Elsevier Permissions Helpdesk.

In 1965, I completed my biological studies, and at the same time I obtained a position at the Laboratory of Electron Microscopy of the University of Amsterdam. I became heavily involved in fixation and embedding techniques, and I also learned more and more about possible pitfalls in microscopy. A recurrent theme was the concern about the reliability of the structures observed in the electron microscope. Two examples, which I shall indicate briefly here, will be shown later on. The first example refers to the significance of mesosomes (internal membranous structures visible in thin sections of Gram-positive bacteria) and the second refers to the variability of the shape of the nucleoid (the bacterial chromosome) in relation to the fixation technique employed. This latter problem led to a search for a microscopic technique to bridge the resolution gap between light and electron microscopy. It resulted in the construction of one of the first (if not the first) operational confocal scanning light microscope(s). This latter topic will be very briefly dealt with at the end of this article.

In 1970, I finished my Ph.D. study, entitled Dissecting a Bacterium (Promotor: Prof. Dr. D. Stegwee; Coreferent: Dr. W. van Iterson) [2]. In those times, the approach to bacterial anatomy was static. For instance, the question was: what does *Escherichia coli* look like after freeze-fracturing? Then, it was a valid question, because freeze-fracturing was a very novel technique and its application could provide new structural information (Nanninga, 1970 [3] and van Gool and Nanninga, 1971 [4]). However, it was quickly realized that a bacterial cell is a dynamic structure. Cells may elongate and they might start with division. In addition, their genetic material (DNA) has to replicate and segregate during the division cycle. What could electron microscopy contribute to these problems? Furthermore, a dynamic structure required handling of bacterial physiology (see below).

Before embarking on these questions, I will briefly make some remarks on early electron microscopy and then proceed with a general framework revealing the importance of electron microscopy for the study of bacterial DNA segregation and cell division. For instance, is the term mitosis applicable, and do bacteria posses a structure equivalent to the eukaryotic spindle apparatus? In fact, this was the framework in the beginning which guided us (partly in hindsight) towards our future research.

2. Early Electron Microscopy

2.1. From the Outside

Electron microscopic preparations require the presence of heavy elements to facilitate image formation by electrons. Initially, a frequently employed technique consisted of the evaporation of a heavy element (e.g., platinum) in a vacuum chamber onto a small specimen. Generally, the evaporated metal covered the specimen at a defined angle. In this way, no metal was deposited behind an object, revealing an electron-transparent shadow. Reversing the contrast of the recorded images in the darkroom produced a more natural appearance of the objects because of the black shadows, hence the term shadowcasting. Thus, for the first time, viruses could be seen and, in the case of bacteria, appendages such as flagella could be easily distinguished (Figure 1). These results clearly demonstrated the advantage of the higher resolution of the electron microscope as compared with the light microscope.

2.2. From Outside to Inside

However, what can be found in the interior of a bacterial cell? How does a bacterial cell compare to a eukaryotic cell? For instance, is there a nuclear membrane, an endoplasmic reticulum, or a mitochondrion or equivalents of them, and is chromosome segregation based on mitosis? Today, we largely know the answers to these questions, but in the middle of the twentieth century, very little was known.

A breakthrough was a refinement of fixation and embedding procedures to obtain reproducible images by ultrathin sections for electron microscopy (Ryter et al., 1958 [5]). Colloquially, it was denoted as R-K fixation. In addition, embedding techniques and microtomes were improved to allow very thin sections for the electron microscope. This development permitted the visualization of the bacterial cell's interior at high resolution, thus going from outside (shadowcasting) to inside. A wealth of new information was obtained by electron microscopists studying bacterial anatomy. It is beyond the scope of this paper to provide details. However, a notable observation was the demonstration of a membranous body in the middle of the Gram-positive *Bacillus megaterium* cell and it was therefore termed the mesosome (Fitz-James, 1960 [6]). Earlier, they had been denoted as peripheral bodies by Chapman and Hillier (1953) [7]. Van Iterson (1961) [8], in a classic study, referred to them as chondrioids in the supposition that they were the bacterial equivalents of eukaryotic mitochondria (van Iterson, 1961 [8]; van Iterson, 1965 [9]; Figure 2).

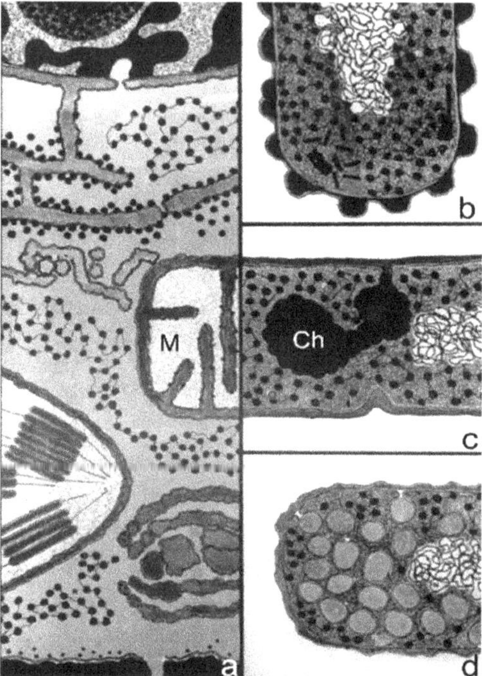

Figure 2. Comparison of a chloroplast-containing eukaryote (**a**) with three prokaryotes: (**b**) A Gram-negative bacterium; (**c**) A Gram-positive bacterium; and (**d**) A photosynthetic bacterium. Note the comparison of a mitochondrion M in (**a**) with a chondrioid or mesosome Ch in (**c**). Source: Modified diagram 1 from [9]. See original [9] for a more detailed description. Copyright © American Society for Microbiology.

2.3. Freeze Fracturing

A completely different approach was a method without chemical fixation: freeze-fracturing. The freeze-fracturing technique was invented by Hans Moor and coworkers from the Swiss Federal Institute of Technology, Zürich, Switzerland (Moore et al., 1961 [10]). Briefly, living specimens are frozen and thereafter fractured in vacuum. The resulting fractures could be replicated by shadowcasting and, after cleaning, the replicas could be studied in the electron microscope. It revealed insights in cellular structure by images which had never been seen before. The technique was further elaborated by a study on the "Fine structure of frozen-etched yeast cells" (Moor and Mühlethaler, 1963 [11]). An interesting observation was the finding that frozen yeast cells could continue their growth after thawing, implying that replicas were made of living cells. The interpretation appeared complicated, however, notably concerning the fracture faces of frozen membranes.

Originally, Moor and Mühlethaler (1963) [11] entertained the idea that a fractured membrane results in two faces, implying two fracture planes, that is, one fracture plane located between the cytoplasm and the external cytoplasmic face of a membrane (concave fracture face) and the other fracture plane between the outside of a membrane and its surroundings (convex fracture face). When I started with freeze-fracturing, I adopted the above interpretation and I remember how difficult (but exciting) it was to understand the novel images. There was no reference point. However, the interpretation of Moor and Mühlethaler (1963) was challenged by Daniel Branton (Branton, 1966 [12]) by providing evidence that, regarding membranes, there is only one fracture plane, which splits the membranes' hydrophobic interior. In other words, the convex and concave appearance are from one and the same fracture. It took some time before the vision of Branton became accepted, partly, because it was an opinion that differed from that of the authority, that is, the inventors of the technique.

What about bacteria? I delved into the topic by two approaches. Firstly, I made complementary replicas in such a way that I retained the fractured material, and made replicas of the original and of the material normally thrown away. To make a long story short, I obtained two replicas, each on its own electron microscopic grid, of one and the same fractured bacterium. How to find them?

In the course of time, this turned out to be possible in a matter of minutes, though it did not, the first time. I take the liberty to recall this first attempt. Firstly, each grid with a replica was photographed in the electron microscope. Secondly, the two grids were printed, and the prints were placed and arranged on the floor of the library. Each grid had a surface of about 10 m^2 (bacteria are small). I walked on them by foot with a pointing stick in my hand. It took some time before I found one and the same bacterium on the two different replicas (Figure 3; Nanninga, 1971 [13]). Another approach was to freeze-fracture fixed cells and subsequently embed them after thawing to be followed by thin sectioning. In this way, I could study fractured cells and inspect the fractures by classical electron microscopy. These results confirmed the interpretation of Branton (1966) [12].

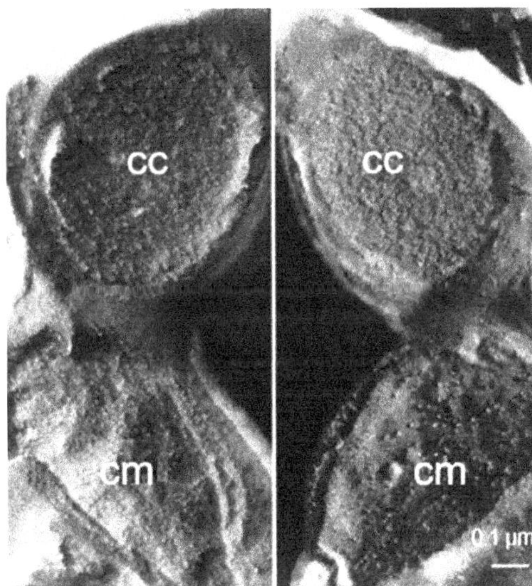

Figure 3. Complementary replicas of freeze-fractured *B. subtilis*. Note the convex and concave surfaces of the cell membrane (cm) at the lower left and the lower right, respectively. Above, the cell content (cc) has been cross-fractured with no clear-cut distinction of the nucleoid. Bar, 0.1 µm. Source: Modified Figure 1a,b from [13]. © 1971, Nanninga, N., Originally published in Journal of Cell Biology [13].

In addition, I revisited *E. coli* with the late August van Gool and offered a different interpretation of the fracture faces (van Gool and Nanninga, 1971 [4]). What the membranous fracture faces demonstrated in particular was the asymmetric appearance of the cell membrane and the outer membrane, thus emphasizing the dramatic differences in the chemical compositions of the respective membrane faces.

3. Mechanism of DNA Segregation

3.1. DNA–Membrane Attachment and Envelope Growth

A seminal paper also inspired by structures visible in the electron microscope dealt with "the regulation of DNA replication in bacteria" (Jacob et al., 1963 [14]). In early electron micrographs, DNA-containing regions could be recognized by the presence of a clew composed of more or less parallel electron-dense threads (presumably, coagulated DNA strands). The clews were largely located in the cell center and not surrounded by a membrane, hence the term nucleoid (nucleus-like). Extensive internal membranous structures could be seen in contact with the nucleoid in *Bacillus subtilis*, a Gram-positive bacterium. These structures seemed to arise from the enveloping cell membrane (Figure 4a; [15]) and they fed the notion that DNA is attached to a membrane. It was not a big step to connect DNA replication and segregation with envelope growth. In fact, DNA–membrane attachment became a popular research theme for decades, as witnessed by countless research papers.

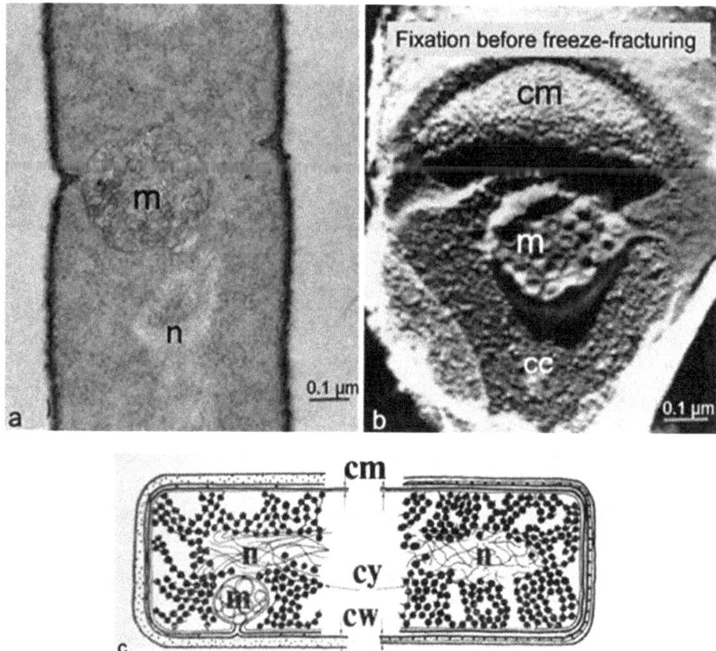

Figure 4. Thin section after RK-fixation of *B. subtilis* (**a**) with mesosome (m) and nucleoid (n). The mesosome is in contact with the nucleoid. To the right (**b**), a freeze-fracture image after chemical fixation. cc, cell content; cm, cell membrane; m, mesosome. Bar, 0.1 µm. Source: Modified Figures 2 (left) and 3 (right) from [15] © 1971, Nanninga, N., Originally published in Journal of Cell Biology [15]. (**c**) Schematic representation of the structures observed after sectioning of chemically fixed bacteria. The major components of a Gram-positive (*B. subtilis*; left), and a Gram-negative (*E. coli*; right) bacterium are shown. cm, cell membrane; cw, cell wall; cy, cytoplasm; m, mesosome; n, nucleoplasm. The cytoplasm is depicted as a network of ribosomes. In these prokaryotic cells, nucleoplasm and cytoplasm are not sharply separated from each other. Note the multilayered structure of the *E. coli* cell wall. The diameter of these bacteria is about 600 nm. Source: My thesis in 1970 [2].

3.2. The Jacob Model

I have discussed the model proposed by Jacob et al. (1963) [14] before, under the heading of "Pictures Considered #32: A Model for DNA Segregation in 1963" in Elio Schaechter's blog Small Things Considered (29 November 2015). I will follow the description of the model (Figure 5) as elucidated in the above contribution. In A of Figure 5, two circular replicons, the bacterial chromosome (Chr.) and a plasmid (F), are attached to "the equatorial perimeter", which represents the "unit of segregation". The circularity was deduced in genetic experiments from the linear sequences of genes in overlapping DNA fragments and by lightmicroscopic autoradiography (Cairns, 1963 [16]). In B, the "cell surface is assumed to transmit to the replicons the signal initiating their replication". In C, D, and E, "Elements of the bacterial membrane are assumed to grow between the two planes of attachment of the daughter replicons putting them progressively apart". (I have shortened the original legend, and wordings of the text of the paper of Jacob et al., 1963 [14] are between quotation marks).

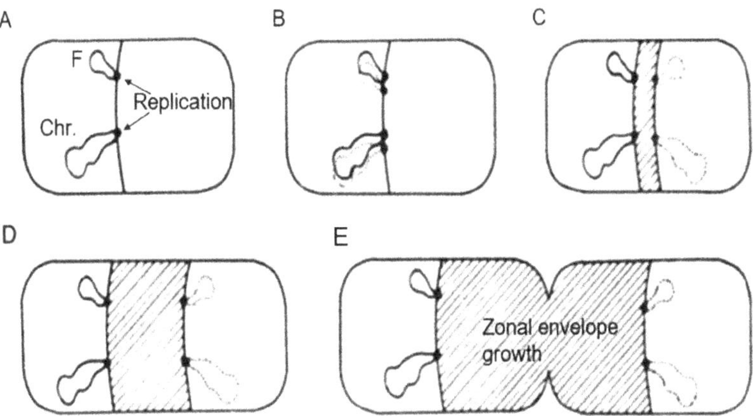

Figure 5. Model for the replication and subsequent segregation of a bacterial chromosome (Chr.) and an F factor (F). The circular molecules are membrane-attached, as is the replication apparatus. Replication takes place before segregation (**A**,**B**). Segregation occurs through zonal envelope growth (hatched areas in (**C–E**)). Source: Adapted Figure 10 from [14].

For the present purpose, I will select three aspects of the model. Firstly, DNA is membrane-attached throughout the cell cycle. Secondly, DNA replication has been completed before segregation (B), in line with eukaryotic mitosis, where already-replicated chromosomes become separated. Thirdly, DNA segregation (C, D, E) is carried out by zonal growth of the cell envelope. Thus, the growing cell envelope with DNA attached to it would serve as an equivalent of the mitotic apparatus. As a matter of course, I should also mention here the pioneering work of Y. Hirota on bacterial thermosensitive mutants of DNA replication and of cell division (Hirota et al., 1968 [17]). He contributed to the development of the field of molecular genetics regarding the above topics.

3.3. DNA–Membrane Attachment

An important consideration leading to the role of membranes in DNA segregation was the observation that the nucleoids in the Gram-positive *B. subtilis* appeared to be in contact with mesosomes (Figure 4a). Serial sections of *B. subtilis* for electron microscopy by Antoinette Ryter led to a model where mesosomes seem to mediate attachment of nucleoids to the cell membrane (A in Figure 6). Moreover, mesosomes divide in two, enabling separation of duplicated nucleoids (C, D and E). It should be noted that *E. coli* does not possess *B. subtilis*-like mesosomes, suggesting that its nucleoid is directly attached to the cell membrane (Ryter et al., 1968 [18]).

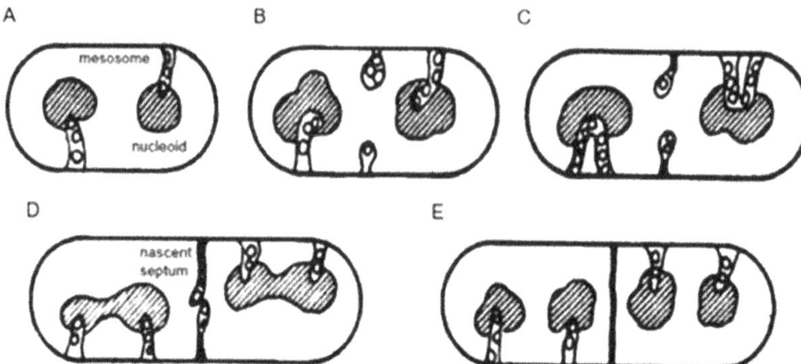

Figure 6. Model with the nucleoid attached to mesosomes as based on serial sectioning. Nucleoids divide, as do the mesosomes and septum arises with the help of mesosomes. (**A**) Mesosomes attached to nucleoids, (**B**) Septum formation with mesosomes, (**C**) Mesosome division and nucleoid growth, (**D**) Nucleoid and mesosome segregation, (**E**) Completion of nucleoid segregation enabled by mesosomes. Source: Modified Figure 23 [18]. Copyright © American Society for Microbiology.

In 1963, the term membrane (cell membrane, cytoplasmic membrane, plasma membrane) had not the same meaning as it did in 1972, when the fluid mosaic model (FMM) was conceived (Singer and Nicolson, 1972 [19]). Similarly, there was not a consistent term for the nucleoid (nucleoplasm, bacterial chromosome, nucleus). Regarding the separation of nucleoids, the fluidity of the bacterial cell membrane precluded a conceptual function of the latter in this process. In the course of time, the focus shifted to the combination of cell membrane and peptidoglycan or murein layer (sacculus). In particular, the growing sacculus was seen as a component instrumental in cell elongation and, therefore, as an agent in DNA segregation. However, I will first return to the presumed role of mesosomes (Figure 4a).

3.4. Mesosomes

Around 1967, I started to use the freeze-fracture technique to study the ultrastructure of *B. subtilis*. Eventually, I came to the conclusion that mesosomes could not be shown in freeze-fractured cells. However, mesosomes became visible after freeze-fracturing when cells were first fixed with osmium tetroxide (R-K fixation) before freezing (Figure 4b), implying that mesosomes might be artifacts of preparation (Nanninga, 1969 [20]; Nanninga, 1971 [15]; Nanninga, 1973 [21]). I quote here the last sentence of my 1971 paper: "Finally, it can be said that some care is needed in drawing conclusions concerning the structure of mesosomes in chemically fixed material" (Nanninga, 1971 [19]).

Presenting my observations on conferences produced disbelief and sometimes outspoken hostility. I learned that rationality and science do not always go together. My final sentence quoted above was an attempt to adopt a more or less neutral position (while softening my real opinion), though the accompanying published scheme left no doubt (Figure 7). For a more detailed discussion on freeze-fracturing of mesosomes, see Nanninga (1973) [21]. Extensive electron microscopic and biochemical experiments on the variability of mesosome structure after chemical fixation have been carried out by Silva et al. (1976) [22]. In the case of the nucleoid, lack of its clear-cut discernibility was not due to its absence (Nanninga, 1969 [19]). I will elaborate on this later on.

 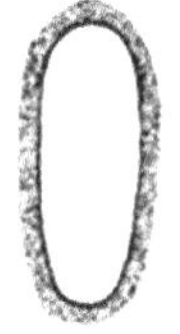 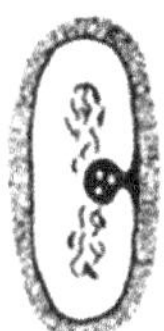

Figure 7. Diagram to depict the effect of chemical fixation on the appearance of mesosomes after freeze-fracturing. Mesosomes are not visible after freeze-fracturing without fixation. Source: Adapted Figure 7 from [20]. © 1971, Nanninga, N., Originally published in Journal of Cell Biology [20].

However, quite unexpectedly, mesosomes became fashionable again from the early nineties of the previous century onwards. They ended up in the field of epistemology. Some paper titles: "Facts, artifacts, and mesosomes: Practicing epistemology with the electron microscope" (Rasmussen, 1993 [23]), "Mesosomes and scientific methodology" (Hudson, 2003 [24]) and "The very reproducible (but illusory) mesosome" (Allchin, 2022 [25]). These are only a few examples. I found out that epistemologists disagree about thinking about science. Fortunately, one of them concluded that I was a real scientist: "For example, referring again to the mesosome episode, if I were to suggest that either Chapman or Hillier or Fitz-James or Nanninga is not in fact a scientist, then I would have a difficult time making my case " (Hudson, 2003 [24]).

3.5. Envelope Structure

In the foregoing, I referred to the Gram-positive cell envelope of *B. subtilis*, i.e., the cell membrane and its attached relatively thick cell wall composed of peptidoglycan and teichoic acids. This was the image visible in the electron microscope after thin sectioning (Figure 4a). By contrast, the envelope of the Gram-negative *E. coli* proved quite different, revealing a more layered appearance after thin sectioning (de Petris, 1967 [26]; Figure 4c). In a pioneering study, Weidel et al. (1960) [27] had already isolated from *E. coli* B what has been termed the sacculus. Its isolation required various chemical treatments resulting most notably in a covalently linked macromolecule in the shape of the original bacterium. The sacculus had been visualized by shadowcasting. The envelope, as already mentioned above, revealed a layered appearance after thin sectioning (de Petris, 1967 [26]).

De Petris also employed various enzymatic and chemical treatments of intact cells and their fragments. Though using a different terminology, de Petris contributed to the concept of a three-layered structure of the *E. coli* cell envelope: outer membrane, peptidoglycan layer, and cell membrane (though the periplasmic space acquired a special consideration because the size of its volume appeared controversial (Graham et al., 1991 [28])).

4. Size Distributions by Electron Microscopy

4.1. Steady-State Growth and Agar Filtration Technique

Our change of thinking, already referred to above, from the static to the dynamic aspects of bacterial structure involved a shift from the individual cell as such to the individuals that are members of a population. What is needed, we (Conrad Woldringh and myself) realized, is not a random collection of cells, but a population with reproducible parameters in time. One such parameter is cell length, which one wishes to correlate with a cell cycle event in time. It dawned on us that an exponentially growing bacterial culture is not well defined and that steady-state growth is required (Maaloe and Kjeldgaard, 1966 [29]). In practice, to create steady-state growth, a small inoculum is used to start a batch culture, after which periodical dilutions with prewarmed growth medium are carried out to achieve a continuous exponential growth (cf. Fishov et al., 1995 [30] and references therein).

An important method to visualize populations was the agar filtration technique (Kellenberger and Arber, 1957 [31]). It has been widely used to make length distributions of bacterial cultures for electron microscopy and to screen for appropriate markers. In addition, it is easy to distinguish dividing from non-dividing cells. Methodically, the cells are fixed with 1% osmium tetroxide to a final concentration of 0.1%. The fixed cells are then applied to an agar filter and processed according to the filtration technique (Kellenberger and Arber, 1957 [31]). Cell length is measured on electron micrographs. In this way, steady-state culture conditions are combined with electron microscopy. (In steady-state, the measured length distributions are constant in time). Below follows an early example that refers to nucleoid separation in *E. coli* (Woldringh, 1976 [32]).

4.2. Length Distributions

In Figure 8, two *E. coli* length distributions are shown, wherein the dividing cells are indicated by hatched areas. The mean length of dividing cells (m_d) can be calculated from the relationship $m_d = \frac{1}{2} \, l_{max} + l_{min}$, as deduced by Harvey et al. (1967) [33]. The minimal cell length (l_{min}) and the maximal cell length (l_{max}) are taken from the length distribution. The mean length of the newborn cells (m_n) = $1/2 \, m_d$. The distance between m_n and m_d represents the duration of the doubling time of the steady state cultures of 32 and 60 min, respectively. It is generally assumed that the relationship between length and age is exponential (for discussion, see Koppes and Nanninga, 1980 [34]). The technique has been helpful in establishing the timing of nucleoid separation in thin-sectioned cells, as described below. Though we did focus on cell length, cell diameter can also be interesting. The latter has been previewed by Zaritsky in 1975 [35] and studied extensively by Trueba and Woldringh (1980) [36].

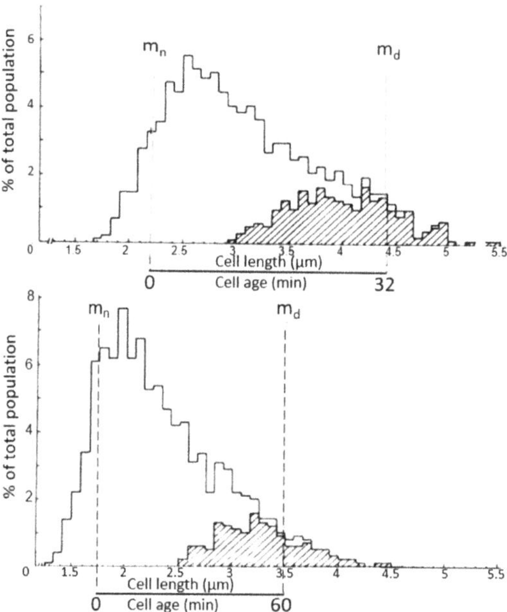

Figure 8. Length distribution of OsO$_4$-fixed cells prepared by agar filtration. Above, *E. coli* B/r growing with a doubling time of 32 min. Below, doubling time 60 min. Hatched areas: Length distribution of cells in the process of constriction. Mean length of newborn (m_n) and dividing cells (m_d). Source: Figure 1 of [32]. Copyright © American Society for Microbiology.

4.3. Measuring Techniques

An important asset for us was the presence and dedication of Norbert O. E. Vischer for helping with and developing measuring tools for analyzing images of bacteria. One example is the software program ObjectJ, which he modified and improved for decades. Details are in the references: (Vischer et al., 1994 [37]; Vischer et al., 2015 [38]). The recent tool Coli-Inspector allowed the presentation of maps of fluorescent profiles of thousands of individually labeled cells. An example and technical details can be found in Figure 27 to be discussed later. The combination of bacterial physiology (steady-state), electron microscopy, and image analysis has been termed the Amsterdam School by Arthur Koch on several occasions. Its scientific context has been elaborated on by Arieh Zaritsky in a personal perspective on chromosome replication, cell growth, division, and shape (Zaritsky and Woldringh, 2015 [39]). Since my retirement, Tanneke den Blaauwen has continued and improved our work with the fullest dedication.

5. DNA Replication and DNA Segregation Go Hand in Hand

5.1. Nucleoid Size and Cell Length

In the model of Jacob et al. (1963 [14]; Figure 5), DNA segregation follows after the completion of DNA replication. The question arose, as mentioned before, whether this notion could be corroborated by electron microscopy. With this in mind, Woldringh (1976) [32] studied the morphological appearance of the nucleoid (or nucleoplasm) during cell elongation, i.e., from birth to division. For this purpose, he followed serially sectioned *E. coli* B/r A cells classified according to length. To allow an estimate of the relation between cell length and cell age, length distributions were made with the agar filtration technique of steady-state-grown cells. B/r strains were employed that had previously been used by Cooper and Helmstetter (1968) [40] to determine the timing of termination of DNA replication in synchronized cells. In this way, the cell length at which DNA replication terminates in the length distributions could be inferred. Visible separation of the nucleoids in Woldringh's images appeared to coincide with the termination of DNA replication, as previously determined by Cooper and Helmstetter (1968) [40]. Before termination of DNA replication, the area occupied by the nucleoplasm in the sectioned cells increased in size with cell length. Earlier work by Cooper and Helmstetter (1968) [40] and Helmstetter and Cooper (1968) [41], complemented by the morphological studies of Woldringh (1976) [32], indicated that DNA replication and DNA segregation go hand in hand. If so, it means that classical eukaryotic mitosis does not apply to *E. coli* and presumably not to prokaryotic cells in general. So, how should prokaryotes and eukaryotes be compared? I have dealt with this topic separately elsewhere (Nanninga, 2001 [42]).

5.2. A Course in Electron Microscopy

In the meantime, we obtained obligations in teaching. One example was an introductory course in electron microscopy. The course was very popular and always attracted a very diverse population of all ages from Amsterdam. The included photograph is from 1978 and depicts a final stage: diplomas have been awarded (Figure 9). Two familiar persons at the left are Frank Trueba and Ronald Verwer. Also present are Conrad Woldringh (to the far right) and the writer of this article.

Figure 9. Diploma awarding after a Course in Electron Microscopy (1978). From left to right: F.T, Frank Trueba; R.V, Ronald Verwer; N.N., Nanne Nanninga; C.L.W., Conrad Woldringh.

6. Peptidoglycan Assembly, Arrangement of Glycan Chains, and the Thickness of the Sacculus

6.1. Peptidoglycan Assembly

In the early sixties of the previous century, the peptidoglycan layer (murein layer or sacculus) was isolated from *E. coli* (Weidel et al., 1960 [27]; Weidel and Pelzer, 1964 [43]). In the electron microsope, as mentioned already above, it was visible as a flattened structure with the outline of the bacterium it came from. The sacculus, which was supposed to maintain cell shape, is composed of glycan chains interconnected with short peptide chains. It is a covalent structure, and as such, it represents a single macromolecule (Nanninga, 1998 [44] and references therein; Figure 10).

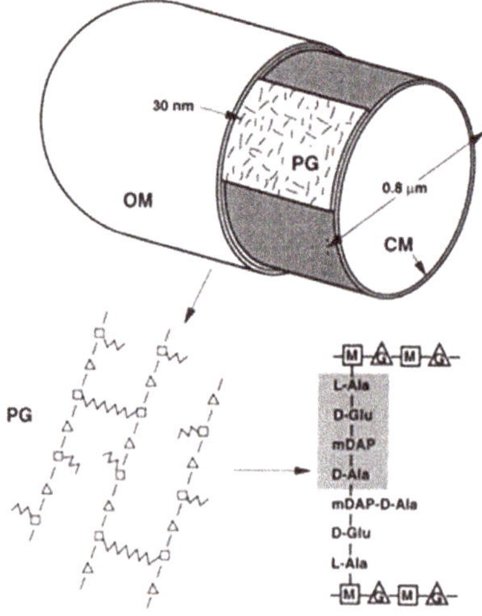

Figure 10. Arrangement of glycan chains in the peptidoglycan layer, the schematic structure of peptidoglycan, and the making of a cross-link. CM, cytoplasmic membrane; G, *N*-acetyl-glucosamine; OM, outer membrane; PG, peptidoglycan. Source: Figure 2 of [44]. Copyright © American Society for Microbiology and Figure 4 of Nanninga, N. et al., 1992. Symp. Soc. Gen. Microbiol. 47: 185–221. Copyright © Cambridge University Press.

Its assembly begins in the cytoplasm, is continued by membrane proteins (in particular PBPs), and is completed as a constituent of the cell envelope (Nanninga, 1998 [44]; Figure 11). Briefly, UDP-MurNAc-pentapeptide in the cytoplasm is bound to undecaprenyl phosphate in the cell membrane by MraY (a translocase) creating lipid I. MraY contains many membrane spanning sequences. Lipid II is made by adding lipid I to UDP-Glc-NAc by MurG (a transferase). MurG is located at the cytoplasmic side of the cell membrane. Next, the prenylated disaccharide pentapeptide has to be presented to the periplasm by a flippase. In 1998, its identity was not yet known (see further on). In the periplasm, the main actors are the penicillin-binding proteins (PBPs; Spratt, 1975 [45]). Here, I will not give further particulars.

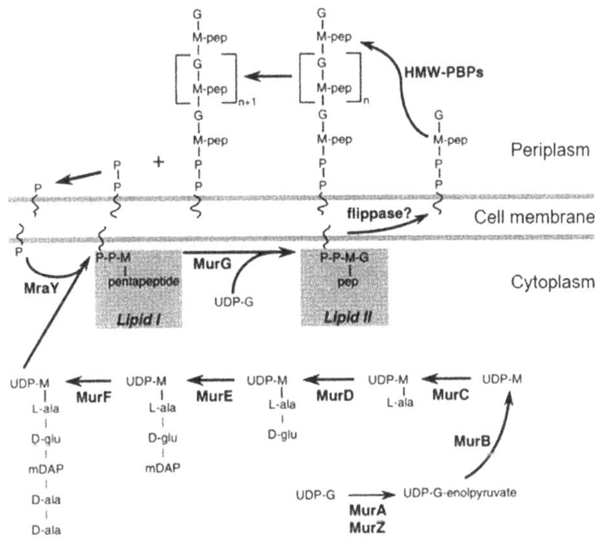

Figure 11. Peptidoglycan assembly. The cytoplasmic steps leading to the prenylated pentapeptide (lipid II) are shown. Through the action of a hypothetical flippase, the disaccharide pentapeptide becomes exposed to the periplasm. The membrane-bound disaccharide peptide becomes attached to a polymer that is also membrane-bound. Glycan chain elongation occurs by transglycosylation. Elongation can occur at the nonreducing end of the glycan chain (this figure) or at the reducing end (not shown in this figure). G, *N*-acetylglucosamine; M, *N*-acetylmuramic acid; P, undecaprenylphosphate; PP, undecaprenyl biphosphate; pep, pentapeptide. Source: Figure 3 of [44]. Copyright © American Society for Microbiology.

The peptidoglycan layer, though a single macromolecule, is not a static structure. Dynamics include rearrangements, turnover, recycling, degradation, and autolysis (van Heijenoort, 1992; personal communication). For more extensive information on peptidoglycan assembly at the end of the nineties of the previous century, see van Heijenoort, 1996 [46] and Höltje (1998) [47]. For a recent excellent review, cf. Egan et al., 2020 [48].

Finally, note that many genes relevant for peptidoglycan synthesis and cell division are located in the 2-min region of the *E. coli* chromosome, which leads to the question of how and to what extent their expression is coordinated (Rothfield and Garcia-Lara, 1996 [49]). This will be elaborated later on.

6.2. Arrangement of Glycan Chains

In collaboration with the group of the late Uli Schwarz in Tübingen, we isolated purified sacculi from *E. coli* and incubated them with enzymes that disrupt peptidoglycan. Thereafter, they were prepared for electron microscopy. Sacculi became fully disrupted after incubation with *E. coli* transglycosylase or egg white lysozyme. These enzymes hydrolyze the glycan chains. However, after hydrolysis of the peptide bridges which link the glycan chains with an *E. coli* endopeptidase, a preferential orientation of the remaining glycan chains more or less perpendicular to the length axis of the cell was observed. This suggests that cell elongation involves insertion of new glycan chains in the same orientation, thereby keeping cell diameter constant (Verwer et al., 1978 [50]).

6.3. Thickness of the Peptidoglycan Layer

In a later stage, there reemerged the thickness of the peptidoglycan layer (Wientjes et al., 1991 [51]), in particular, through the "make before break" concept of the late Arthur Koch (1984) [52]. That is, growth of a single-layered covalently closed peptidoglycan should not lead to rupture. Hence, new glycan chains are first cross-linked to existing peptidoglycan before old cross-links are broken. Next, the new chains are pulled into the stress-bearing plane of the peptidoglycan by osmotic pressure (Koch, 1984 [52], Höltje, 1998 [47]). I also would like to acknowledge here our fruitful contacts with the late Jochen Höltje and coworkers, the successor of Uli Schwarz in Tübingen.

In the meantime, the sacculus grew thicker and thicker (I will refrain from references), so we thought it appropriate to attempt to measure its thickness again. Firstly, a radiochemical method, as outlined by Wientjes et al. (1991) [51], was employed. This method is based on the steady-state incorporation of [meso-^{3}H] diaminopimelic acid ([^{3}H] Dap) during several generations. One can calculate the number of Dap molecules per sacculus from the cell concentration and the specific activity of the [^{3}H] Dap. Secondly, one can measure the Dap content chemically in sacculi isolated from a known number of cells. With both methods, a value of about 3.5×10^6 Dap molecules per sacculus was obtained. Combined with electron microscopic measurements of the surface area of the cells, the data indicate an average surface area per disaccharide unit of ca. 2.5 nm^2. This suggests again that the peptidoglycan is organized in a monolayered structure.

However, there remained the question of how the peptidoglycan layer is growing. The answer has been sought by application of electron microscopic autoradiography of cells pulse-labeled with [^{3}H] Dap. We measured the position of silver grains on electron microscopic images of non-dividing and dividing cells.

7. Zonal or Diffuse Growth of the Peptidoglycan Layer?

Early pioneering autoradiographic experiments on the insertion of [^{3}H] Dap into peptidoglycan in *E. coli* seemed to show the expected central zone of envelope growth in the electron microscope (Schwarz et al., 1975 [53]). However, diffuse incorporation in the lateral wall was also observed. From several subsequent autoradiographic experiments, it became clear that insertion of labeled [^{3}H] Dap in elongating cells occurs diffusely over the entire cell envelope (Verwer and Nanninga, 1980 [54]; Burman et al., 1983 [55]; Woldringh et al., 1987 [56]; Wientjes and Nanninga, 1989 [57]). By contrast, sharply incorporated [^{3}H] Dap could be seen in constricting cells. These experiments thus indicated that envelope growth in between the postulated DNA–membrane attachment points (zonal growth in Figure 5) does not function as part of a prokaryotic equivalent of a mitotic apparatus. So, how are replicated nucleoids segregated without help from the cell envelope and how does replication take place, that is, before or during segregation? These problems have been addressed by Conrad Woldringh (Woldringh, 2023 [58]).

8. Zonal or Diffuse Growth of the Outer Membrane?

With respect to the outer membrane, there was the same discussion as with the peptidoglycan layer. Is the insertion of outer membrane proteins during elongation zonal or diffuse? We were fortunate, being able to collaborate with the late inspiring Bernard Witholt and coworkers of the University of Groningen. We could combine his experience with the induction of outer membrane proteins of *E. coli* with our electron microscopic analytics. Induction was roughly for about 5 min and therefore comparable to a radioactive pulse. Thus, we embarked on the topography of LamB (Vos-Scheperkeuter et al., 1984 [59]) and of lipoprotein (Lpp; Hiemstra et al., 1987 [60]).

8.1. Induction of LamB

LambB is an outer membrane component (a transmembrane porin) as well as a phage λ receptor. Wild-type cells were induced with maltose and cyclic-AMP and a lac-lamB fusion strain with isopropyl-ß-D-thiogalactopyranoside (IPTG). Gold-labeled antibodies were bound to LamB proteins exposed at the surface and their topography analyzed. It was concluded that insertion of LamB proteins into the cell envelope occurred at multiple sites over the entire cell surface, and also at the site of constriction.

8.2. Induction of Lipoprotein (Lpp)

Lpp, which is covalently linked to the peptidoglycan layer (Braun and Rehn, 1969 [61]), is not exposed at the cell surface, thus prohibiting its direct labeling. Therefore, we had to treat cells with Tris-EDTA to make the Lpp accessible for labeling with a protein-A gold probe. Lpp was briefly induced with IPTG in cells carrying lac-lpp on a low-copy-number plasmid in an *E. coli lpp* host. In addition, in this case, the topography of Lpp was diffuse and not zonal, and at the constriction site [60].

For a more recent paper on LamB, see Gibbs et al. (2004) [62]. Sheng et al. (2023) [63] confirmed the tight localization of existing Lpp on the sacculus with atomic force microscopy, thus confirming earlier results with respect to the existing distribution of Lpp. The mode of insertion by induction was not carried out.

9. Centrifugal Elutriation and Peptidoglycan Synthesis in *E. coli*

In a later stage, we wished to compare [^{3}H] Dap incorporation in dividing and non-dividing cells. For this, two approaches have been employed. It required methods to distinguish between the two populations. Firstly, we employed synchronized cells, which allowed the application of biochemical methods, and secondly, we employed electron microscopic autoradiography, which was applied to steady-state-grown cultures in the presence of [^{3}H] Dap. In the latter case, we could easily identify dividing and non-dividing cells in the electron microscope.

9.1. Selection of Small Cells by Centrifugal Elutriation

To allow analysis of peptidoglycan synthesis during the cell cycle, one needs an enrichment procedure for dividing and non-dividing cells. A pioneering method for *E. coli* B/r strains was one where cells were attached by filtration to a membrane whereafter the latter was turned upside down and growth medium was poured through the filter. Under these conditions, the stuck cells continued their growth (and the division process) so that newborn cells left the filter specifically after cell separation. Newborn cells were collected in the cold to arrest their growth. Regrowth of the "babies" produced a synchronous culture (Helmstetter and Cummings, 1964 [64]; Helmstetter, 2023 [65]), invaluable for the experimental background of the Cooper-Helmstetter model (Cooper and Helmstetter, 1968 [40]). See also Helmstetter (2023) [65].

However, initially, not all *E. coli* strains appeared suitable for the above synchronization procedure, presumably because attachment to the membrane filter is influenced by the chemical composition of the *E. coli* cell wall and its appendages. In our case, we wished to study aspects of peptidoglycan synthesis during the cell cycle in *E. coli* by using specific K-12 (non-B/r) strains. This led us to search for an alternative cell separation method to allow synchronous growth. In collaboration with the late W. S. Bont of the Netherlands Cancer Institute (Amsterdam), we embarked on the adaptation of centrifugal elutriation for the separation of bacterial cells (Figdor et al., 1981 [66] and references therein). In centrifugal elutriation, controlled particle flow is in a direction opposite to that of sedimentation. Under appropriate conditions, the smallest particles (in our case, the smallest *E. coli* cells) can be washed out and collected. Elutriated cells were, before regrowth, collected in the cold as with the "baby machine".

The above two cell separation methods are similar in the sense that small cells are selected to allow embarkment on synchronous growth. The selection criteria are, however, essentially different: age (baby machine) versus size (elutriation). Nevertheless, both approaches can produce separation of dividing and non-dividing cells.

9.2. Peptidoglycan Synthesis in Non-Dividing and Dividing Cells

Peptidoglycan synthesis was determined by following the incorporation of [³H] Dap by pulse-labeling of *E.coli* MC4100 *lysA*, either by using synchronized cultures (see above) or by autoradiography.

Small cells were separated by centrifugal elutriation and, as a control, asynchronous cultures were put through the elutriator as well (Creanor and Mitchison, 1982 [67]; Wientjes and Nanninga, 1989 [57]). Regrowth of the cooled samples at 37 °C was for two division cycles and the percentages of constricting cells varied between about 2 and 80% (Figure 12). The rate of [³H] Dap incorporation was measured in samples of synchronized and of exponentially growing cells. The incorporation ratios of the two types of cultures did not change during the experiment (Figure 12). This suggests that during the division cycle, no noticeable variation(s) in [³H] Dap incorporation occurs. On the other hand, it could not be excluded that a percentage of 80 for dividing cells might be too low to detect changes in incorporation.

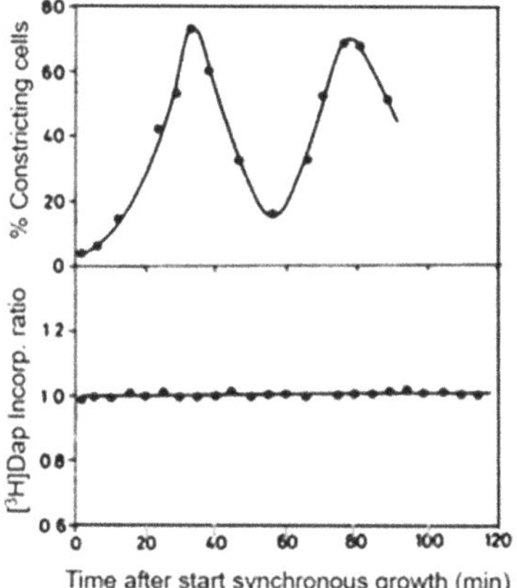

Figure 12. Percentages of cells showing a visible constriction after centrifugal elutriation, as judged by electron microscopy of whole-mount preparations (**top**). Pulse-labeling with [³H] Dap (**bottom**). The ratios of the pulse values of the synchronous and the control culture are given. For details, see [57]. Incorp., Incorporation. Source: Figure 1b,c from [57]. Copyright © American Society for Microbiology.

To further assess this problem, we used again electron microscopic autoradiography on steady-state cultures of the same strain. As indicated before, electron microscopy allows an almost complete distinction of dividing and non-dividing subpopulations. It appeared that dividing cells of the same normalized length as the non-dividing ones produced more silver grains in the cell center (for details, see Wientjes and Nanninga, 1989 [57]). By comparing the topography of [³H] Dap incorporation in the two cases, it can be seen that incorporation at the site of division proceeds at the expense of lateral incorporation (Figure 13; uninterrupted line: dividing cells. Interrupted line: non-dividing cells). An additional interpretation of this figure can be found below.

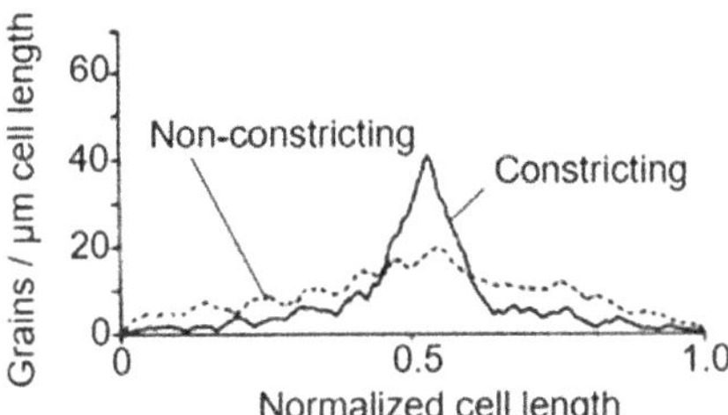

Figure 13. Comparison of the topography of [³H] Dap incorporation in constricting (-) and nonconstricting (--) cells of *E. coli* MC4100*lysA*. The grain distributions over the cells are plotted as grain densities (number of grains per micrometer of cell length) versus normalized cell length. For deeply constricting cells, 399 grains were counted on 23 cells and the average cell dimensions were 2.64 by 0.75 μm. The cells are positioned with their longest half to the left. For nonconstricting cells 867 grains were counted on 94 cells and the average cell dimensions were 1.72 by 0.75 μm. Source: Modified Figure 4 of [57]. Copyright © American Society for Microbiology.

Electron microscopy also allows the distinction of the topography of [³H] Dap incorporation in cells with slight, moderate, and deep constrictions (Figure 14). In these cases, the central peak did not widen during progression of constriction, implying that surface synthesis is not evenly distributed over the nascent polar cap. It suggested incorporation at the leading edge of the constriction. For this interpretation, we also considered that in plasmolyzed *Salmonella typhimurium* cells (similar to *E. coli*), the leading edge of constriction resisted separation of the various envelope layers (MacAlister et al., 1987 [68]). Presumably, this particular tightness at the leading edge reflects local peptidoglycan synthesis, as in Figure 14.

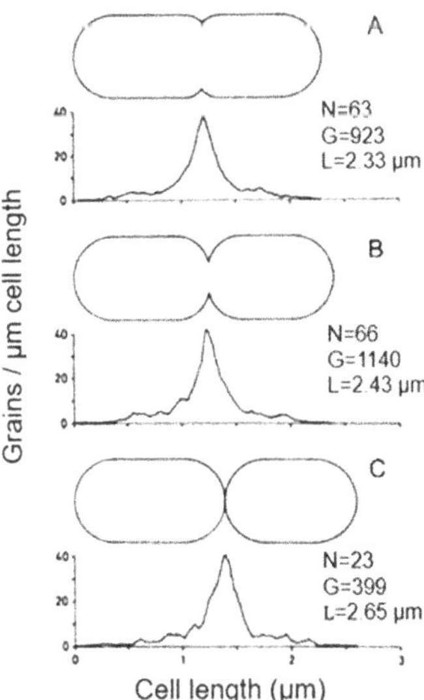

Figure 14. Silver grain distribution over cells in progressing stages of constriction (cells with slight (**A**), medium (**B**), and deep (**C**) constrictions). For labeling conditions, see Wientjes and Nanninga (1989). The grain distributions are plotted as grain densities versus cell length. The cells are positioned with the longest cell half to the left. Drawings of the cells of the different length classes are shown above the distributions, and the number of cells (*N*) and grains (*G*), as well as the average length of the cells (*L*), are given. Source: Modified Figure 5 [57]. Copyright © American Society for Microbiology.

10. Further Details of Peptidoglycan Synthesis: Composition and Mode of Insertion

At the Department of Microbiology, the group of Jan T. M. Wouters acquired expertise in operating the chemostat under various conditions. One study pertained to the influence of various growth conditions on peptidoglycan structure as analyzed by high pressure liquid chromatography (HPLC; Driehuis and Wouters, 1987 [69]). Notably, it was found that the composition of the growth medium dominated cell shape with respect to peptidoglycan structure. Cell shape refers to long cells (relatively little polar cap material) and short cells (relatively much polar cap material).

However, it remained possible that chemical differences existed between the lateral wall and polar cap. So, we turned to peptidoglycan structure in synchronized cultures, asking what products were being formed after incorporation of [^{3}H] Dap after pulslabeling. Table 1 gives an impression of what kind of peptidoglycan digestion products can be found in exponentially growing cells by HPLC (de Jonge et al., 1989 [70]). Note, the ubiquitous presence of Tet and Tet-Tet (bis-disaccharide tetratetra) compounds.

Table 1. Muropeptide composition of an exponentially growing *E. coli* MC4100 culture.

Muropeptide	Fraction of Incorporated [^{3}H]-Dap
Tri	0.4(0.1)
Tet	54.4(0.9)
Pen	1.0(0.1)
Tri-Lys	3.1(0.1)
Tet-Tri-DAP	0.4(0.1)
Tet-Tri	2.4(0.1)
Tet-Tet	28.8(0.4)
Tet-Anh	1.0(0.3)
Tet-Pen	2.0(0.0)
Tet-Tri-DAP-Lys	0.3(0.0)
Tet-Tet-Tri	0.4(0.0)
Tet-Tri-Lys	0.8(0.2)
Tet-Tet-Tet	1.6(0.1)
Tet-Tri-Anh	0.4(0.1)
Tet-Tet-Tet-Tet	0.1(0.0)
Tet-Tet-Tri-Lys	0.3(0.2)
Tet-Tet-Anh	1.2(0.1)
Tet-Tet-Tri-Anh	0.2(0.1)
Tet-Tet-Tet-Anh	0.4(0.1)
Dap-Dap	Degree of cross-linking 0.4(0.0)
	Disaccharide peptides (% of total) containing lipoprotein 4.1(0.3)

After applying the centrifugal elutriation procedure, fractions containing a high proportion of dividing cells were compared to those containing a high proportion of non-dividing cells by HPLC. No difference between the samples was found regarding the peptidoglycan peptides, indicating that division does not require a specific peptide. From the acceptor–donor radioactivity ratio (ADRR) in Tet-Tet, the mode of insertion was derived in dividing and non-dividing cells. In the former, the ratio was high, and in the latter, the ratio was low. This was interpreted to mean that, during cell elongation, the insertion was single-stranded and, during division, multi-stranded ("or sequential single-stranded"). The relationship between ADRR and crosslinking has been discussed by Cooper (1990) [71] and replied by Driehuis et al. (1991) [72], while emphasizing the difference between crosslinking and crosslinkage.

11. Presence of Penicillin-Binding Proteins (PBPs) in Synchronized Cells and Their Cell-Cycle-Dependent Activities

11.1. Presence of PBPs during the Cell Cycle

"Distinct penicillin binding proteins involved in the division, elongation, and shape of *Escherichia coli* K12". This was the informative title given by Brian Spratt for the paper about the detection of penicillin-binding proteins (PBPs; Spratt, 1975 [45]). The PBPs were detected after labeling them with [14C] benzylpenicillin or [14C] mecillinam (FL1060) followed by cell fractionation. Subsequently, they became known as PBP3, PBP2, and PBP1, respectively. PBP3 and PBP2 are transpeptidases, PBP1, now PBP1A and PBP1B, are bifunctional and as such combine transpeptidase and glycosyltransferase activities. The PBPs play various roles in peptidoglycan biosynthesis. Here, I have referred to the main PBPs and I wish here to acknowledge the fruitful collaboration with Brian Spratt on several occasions. For a modern overview, cf. Sauvage and Terrak (2016) [73].

The availability of the centrifugal elutriation technique allowed us to study the binding of [125I] ampicillin to PBPs under various experimental/technical conditions of *E. coli* PA3092 (a K-12 derivative). The main result was that all PBPs became labeled to the same extent, in intact cells as well in membranes derived from them. Figure 15 shows the results with all PBPs (PBP1A/B. PBP1C, PBP2-PBP8) obtained from intact cells. Overall, they appear in a constant ratio during the division cycle, and I quote the general conclusion: "The *E. coli* cells exert their control on shape maintenance and cell wall growth apparently not on the level of concentration of PBPs in the cell but rather on activation of existing components" (Wientjes et al., 1983 [74]). So, what about "activation of existing components"?

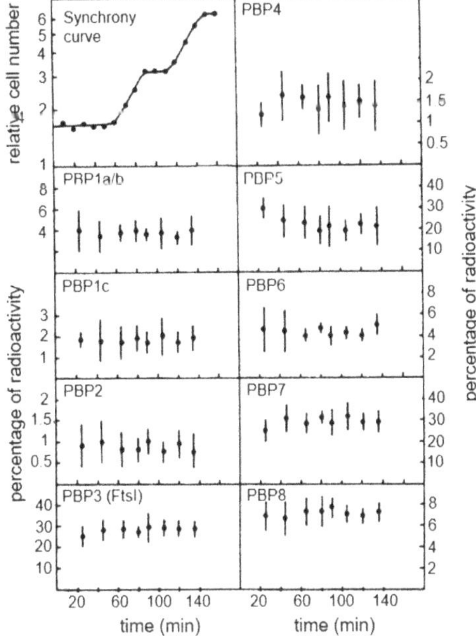

Figure 15. Relative intensities of PBPs after labeling of intact cells with [125I] ampicillin. Intact cells from synchronous cultures (centrifugal elutriation) were labeled with [125I] ampicillin, and the percentage of radioactivity in each PBP band was measured. Average values and standard deviations are shown. PBP1a/b, PBP1c, PBP2, PBP3 (FtsI), PBP$, PBP5, PBP6, PBP7, and PBP8 are indicated. Source: Modified Figure 2 of [74]. Copyright © American Society for Microbiology.

11.2. Inhibition of PBP2 and PBP3 during the Cell Cycle

We tried to answer this question by adding specific antibiotics to synchronized *E. coli* K-12 strain MC4100 *lysA* cells (Wientjes and Nanninga, 1991 [75]). We focused on cell elongation and cell division. In Figure 16, we can observe the rise and fall of the percentage of constricting cells after regrowth of selected small cells. The effect of the antibiotics mecillinam (PBP2) and of cephalexin (PBP3) is plotted as percentage of inhibition of [^{3}H] Dap incorporation. It appears that PBP2 is active all the time (cell elongation), and PBP3 predominantly in dividing cells. Below, I will detail the cellular location of PBP2.

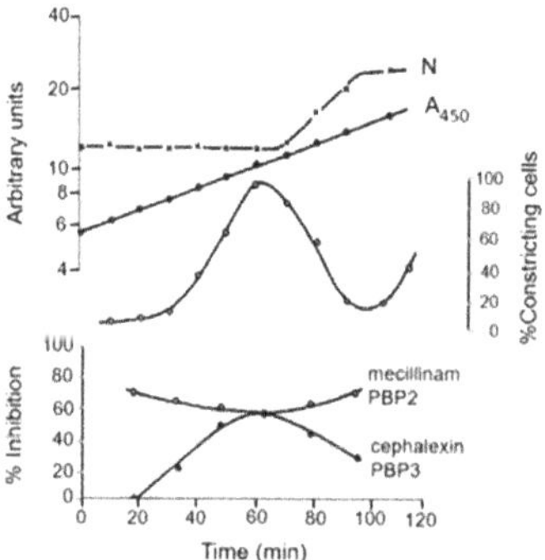

Figure 16. Contribution of PBP2 and PBP3 to peptidoglycan synthesis in relation to the division cycle. The cell number (N) and absorbance (A$_{450}$) of the main culture are shown in the top panel, and the percentage of constricted cells in the middle panel. The lower panel shows the inhibition percentages of [^{3}H] Dap incorporation in the subcultures by mecillinam and cephalexin (for details, see Wientjes and Nanninga, 1991 [75]). Source: Figure 4 of [75]. © Elsevier France.

The roles of PBP2 and PBP3 in cell elongation and cell division, respectively, also became clear by plotting the aspect ratio (L/2R) against the PBP3/PBP2 inhibition ratio. That is, the longer the cell, the smaller the inhibition ratio, and vice versa. For further details, see the legend of Figure 17 (Wientjes and Nanninga, 1991 [75]).

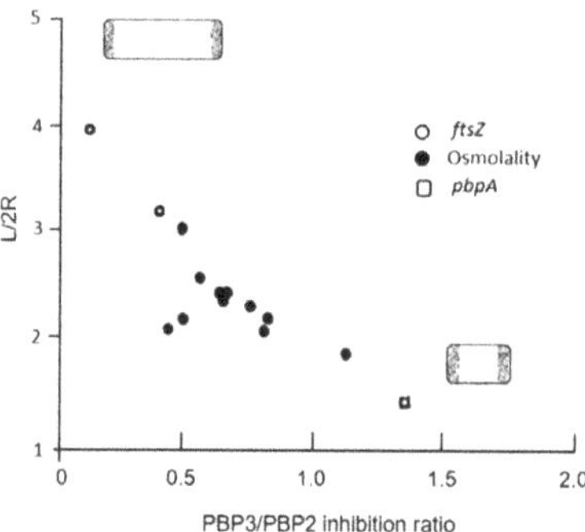

Figure 17. Relation between cell shape and PBP3/PBP2 inhibition ratio. *E. coli* MC4100*lysA* was grown in a phosphate-based minimal medium at 37 °C (doubling time ca. 45 min). The average cell dimensions were measured, and the length/width ratio (L/2R) was calculated. Inhibition of [^{3}H] Dap incorporation was done with cephalexin (10 mm/mL) and by mecillinam (2 mm/mL). The ratio of these inhibition percentages were taken as thePBP3/PBP2 ratio. Further details concerning osmolality and *pbpA* and *ftzs* mutants can be found in [75]. Source: Modified Figure 6 of [75]. © Elsevier France.

12. Localization of Cell Division Proteins

12.1. Making Monoclonal Antibodies

Being microscopists (at least in part), we wished to localize proteins of interest in *E. coli*. This plan was developed before the ubiquitous use of fluorescent markers and the use of green fluorescent proteins. In the latter case, the accompanying loss of resolution was compensated for by the prospect of visualizing fluorescent labels in living cells. It also implied a shift from electron microscopy to fluorescence microscopy (in some cases). However, the detection of FtsZ at the site of constriction by classic immunolabeling of thin sections for the electron microscope was quite exciting (Bi and Lutkenhaus, 1991 [76]), and it stimulated further our project, which started in 1989, to make monoclonal antibodies against cell division proteins. For this, we were happy to collaborate with Arend H. J. Kolk from the N. H. Swellengrebel Laboratory of Tropical Hygiene, Royal Tropical Institute, Amsterdam. Notably, FtsZ (Voskuil et al., 1994 [77]; Koppelman et al., 2004 [78]), PBP1B (den Blaauwen et al. (1989 [79], 1990a [80], 1990b [81]); Zijderveld et al. (1991 [82], 1995a [83], 1995b [84])) and FtsQ (Buddelmeijer et al., 1998 [85]) were studied. These studies involved, amongst others, epitope mappings to chart portions of the respective proteins. Here, I will specifically dwell on the data obtained with FtsQ because they could be placed in a somewhat wider context.

12.2. Location of FtsQ

We found a central location by immunofluorescence microscopy with specific mono-clonal antibodies, as expected, though also in non-central positions. It appeared that the periplasmic part of FtsQ was required for location at the cell center (Buddelmeijer et al., 1998 [85]). Subsequently, it was found that FtsQ occurs in a complex with FtsL and FtsB and that the complex formed before entry into the divisome (Buddelmeijer and Beckwith, 2004 [86]). Further structural and mutational analysis of FtsQ revealed that divisomal positioning of FtsQ and its interaction with Fts L and FtsQ is dependent on two different periplasmic domains (van den Ent et al., 2008 [87]). The first domain is located near the cell membrane, the second domain near the C-terminus. "Both domains act together to accomplish the role of FtsQ in linking upstream and downstream cell division proteins within the divisome" (van den Ent et al., 2008 [87]). This emphasizes the organizational role of FtsQ in divisome construction.

However, "Noncentral FtsQ foci were found in the area of the cell where the nucleoid resides and were therefore assumed to represent sites where the FtsQ protein is synthesized and simultaneously inserted into the cytoplasmic membrane". "An interesting speculation is that the foci might occur where the division and cell wall gene cluster (dcw cluster) resides. To corroborate this intriguing possibility, it will be necessary to combine gene and protein localization studies" (quoted from Buddelmeijer et al., 1998 [85]). And so we did (Roos et al., 2001 [88]).

12.3. Cellular Localization of Cell Division Genes (ftsQAZ Cluster)

The *ftsQ* gene occurs in an *ftsQAZ* gene cluster in the 2-min region of the *E. coli* chromosome. The 2-min region encompasses numerous other genes related to cell division and cell wall synthesis (Wijsman [89]). Hence, as already mentioned above, it has been denoted as the dcw cluster (Figure 18; Ayala et al., 1994 [90]). For further details on dcw, cf. Rothfield and Garcia-Lara (1995 [49]) and Vicente et al. (1998 [91]). Since FtsQ is a transmembrane protein, it was to be expected that the nucleoid, as required by cotranscriptional-cotranslational insertion, is linked through mRNA and ribosomes to the cell membrane (Norris and Madsen, 1995 [92]; Woldringh, 2002 [93]).

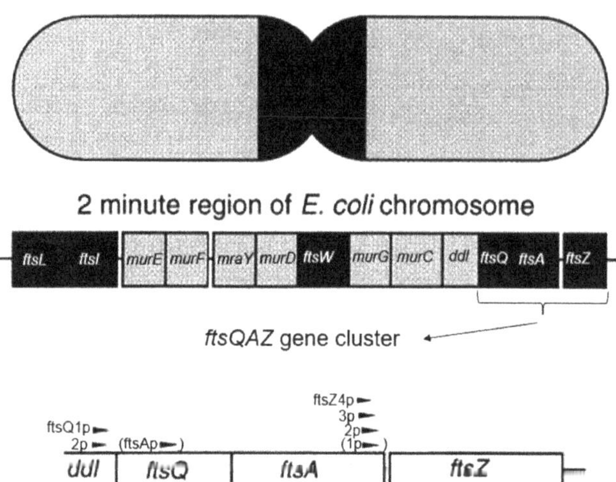

Figure 18. Clustering of genes involved in cell division and cell wall synthesis (dcw cluster) in the 2-min region of the *E. coli* chromosome. Cell division genes are darkly shaded. The genes involved in the production of prenylated disaccharide pentapeptide are lightly shaded (Cf. Figure 11). The *ftsQAZ* gene cluster is depicted at the bottom of the figure. Note the position of the various promoters (see also [90]). Source: Figure 4 of [44] and references therein. Copyright © American Society for Microbiology [44].

A probably naïve supposition was our idea that division genes could be located near the cell center in the vicinity of the nascent divisome during their transcription. This possibility came to our mind before genes or gene regions could be demonstrated by microscopy, the underlying idea being that in bacteria gene and gene product are spatially close together. How to detect genes in the cell?

The microscopic detection of the origin of replication (*oriC*) inside the living cell by fusing gene regions in the neighbourhood of the origin with GFP represented a breakthrough. This was done for *E. coli* (Gordon et al., 1997 [94]) as well as for *B. subtilis* (Webb et al., 1997 [95]). Other genes followed. This took place more or less coincidently with our localization study of FtsQ. In the meantime, we had acquired practical experience (in another context) with fluorescence in situ hybridization (FISH). So, we decided to locate the *ftsQAZ* gene cluster by FISH during the cell cycle of *E. coli* (Roos et al., 2001 [88]). As reference points, we used *oriC* (near the origin) and *minB* (near the terminus). For all genes, plasmid probes were employed for hybridization.

In Figure 19, fluorescent FISH spots are shown together with the DAPI-stained nucleoid. The distance of the center of a fluorescent focus to midcell was measured in a large number of cells. They were grouped according to length so that the position of the foci could be correlated with the known progression of DNA replication. In the interpretive scheme (Figure 20), all three gene regions (*oriC*, *ftsQAZ* and *minB*) are indicated as based on the model of Dingman (1974 [96]). That is, replisomes are located in a replicating apparatus or replication factory wherein DNA strands move apart after replication. In *E. coli* (Koppes et al. 1999 [97]), as well as in *B. subtilis* (Lemon and Grossmann, 1998 [98]), the replication center is in the cell center. The latter authors employed a GFP fusion with the catalytic subunit PolC of DNA polymerase.

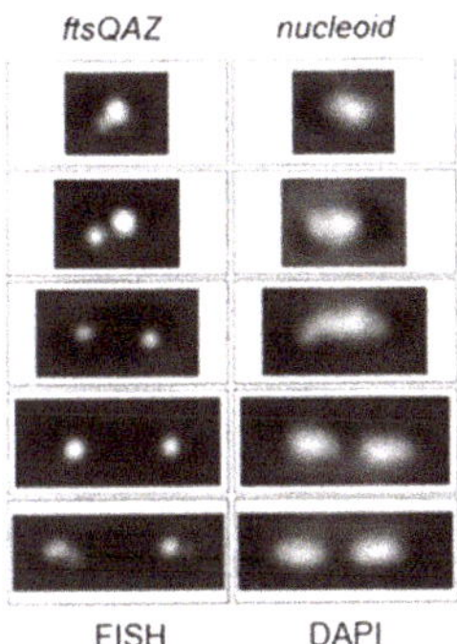

Figure 19. Hybridization foci of *ftsQAZ*- and DAPI-stained nucleoids in fixed cells of increasing length. Cells and foci were photographed in phase contrast and with an Alexa filter (**left**); cells and nucleoids were photographed in phase contrast and with a DAPI filter (**right**). Cells were hybridized with an *ftsQAZ* probe. Source: Modified Figure 2 of [88]. © John Wiley and Sons.

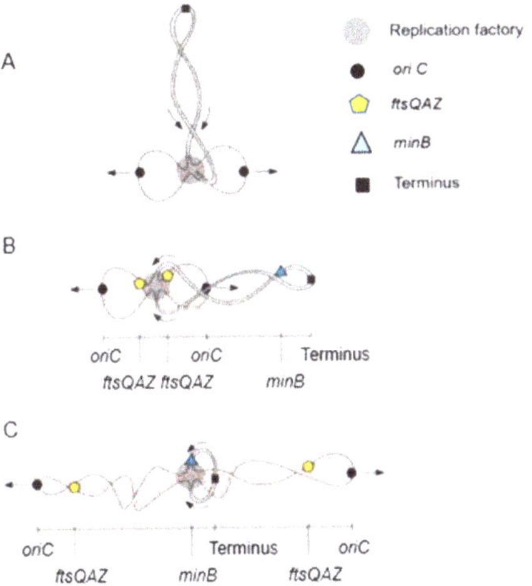

Figure 20. Model for sequential separation of DNA subregions on the *E. coli* chromosome. (**A**) redrawing of the original Dingman figure with the terminus perpendicular to the fixed replisome (replication factory), shortly after initiation of DNA replication [96]. (**B**) Adaptation of (**A**) with the terminus-containing loop in the length axis of the cell. (**C**) Organization of DNA subregions before termination of DNA replication. Arrows indicate the direction of movement of DNA (towards the replisome or away from the replisome in the length axis of the cell). Black circles, *oriC* region; black squares, terminus region; yellow pentagons, *ftsQAZ* region; blue triangles, *minB* region; red dashed line, newly synthesized DNA. Note that, as replication progresses from (**B**) to (**C**), *minB*, moving towards the replisome, passes one of the *oriC*s, which is moving away from the replisome. Source: Modified Figure 5 of [88]. © John Wiley and Sons.

In our case, we pulse-labeled cells with [^{3}H] thymidine for 10 min before preparing for electron microscopic autoradiography. The resulting silver grains were scored in relation to cell length. It was found that incorporation of [^{3}H] thymidine occurred predominantly in the cell center. This indicated that the replisomes are active at a fixed place and that they do not move along the chromosome. However, the model of Dingman with a fixed replication factory might be too simple. Recent considerations have been expressed by Woldringh (2023) [58].

What can be said about gene localization and its expression in the cell? Returning to Figure 20, it emerges that the cellular position of the cell division genes is largely determined by the DNA segregation status of the chromosome. However, this does not reveal when the gene cluster is expressed and to what extent. Moreover, FtsQ, A, and Z are likely produced in different quantities, in the 2-min region. With respect to the *FtsLBQ* cluster, *ftsB*, for instance, is located at 48 min. This also requires coordination regarding its synthesis. Presumably, gene regions are spatially positioned in the cell in an (unknown) dynamic way and their products are able to travel in the plane of the cell membrane, guided or not guided. Anyway, when the products have arrived in the cell membrane, they meet an existing organization. These are only a few thoughts that come to one's mind. Obviously, what lies ahead is a tanglewood of regulations involving a plethora of promoters and concentrations of gene products.

13. Cell Elongation

13.1. Localization of PBP2

Where is PBP2 located in the cell? Localization studies of a GFP-construct of PBP2 led to some surprising results (den Blaauwen et al., 2003 [99]). The PBP2 label was clearly observed at the cell center (not necessarily in a divisome), apart from lateral location (elongasome), as expected. The central label of GFP-PBP2 disappeared upon inhibition of PBP3 with aztreonam, whereas FtsZ remained present (Figure 21). As pointed out before: "This suggests that PBP2 localization at the divisome is dependent on active PBP3" (den Blaauwen et al., 2003 [99]). Furthermore, GFP-PBP2 seemed not to be part of PIPS (see below) in the presence of aztreonam. Nevertheless, what does the central location of GFP-PBP2 mean? I will return to this point later on, after having referred to the association of PBP2 with the actin-like MreB.

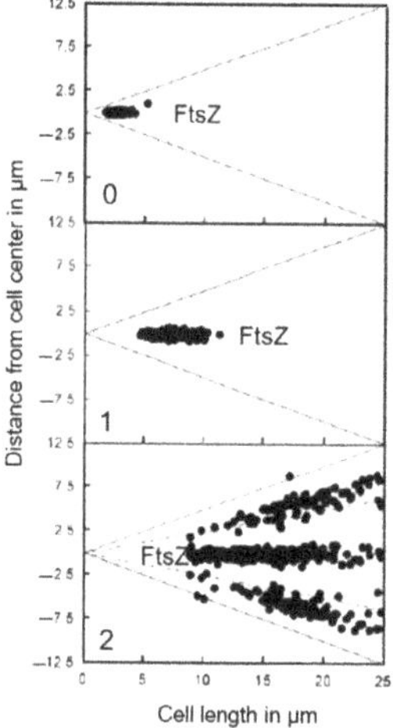

Figure 21. Inhibition of PBP3 by aztreonam. The FtsZ ring is not able to dissociate. The cells were grown to steady state in GB1 medium at 28 °C with a doubling time of 80 min and, subsequently, aztreonam was added (final concentration 1 µg/mL). The culture was sampled just before the addition of aztreonam (0 mass doubling), after one mass doubling (1) and after two mass doublings (2) of growth in the presence of aztreonam, fixed and immunolabelled with Alexa 546-conjugated Fab against FtsZ. The distance from the midcell position of the FtsZ rings is plotted as a function of cell length. Negative values refer to the arbitrary left side of the cell, and positive values refer to the arbitrary right side of the cell. The grey lines represent the border of the cells, and the dotted grey lines represent the one-quarter and three-quarter positions of the cells. Source: Modified Figure 5 of [99]. © John Wiley and Sons.

13.2. Cell Elongation Participants

Actin-like cytoplasmic helical filamentous structures underneath the lateral envelope of MreB in *B. subtilis* (Jones et al., 2001 [100]), and later in *E. coli* (Shi et al., 2003 [101]), revealed the existence of a cytoskeletal element next to the tubulin-like FtsZ. The gene *mreB* and its neighbours *mreC* and *mreD* had already been found in 1987 (Wachi et al., 1987 [102]). Together they form a membrane-bound complex, whereby MreB interacts with MreC, and MreC with MreD (Kruse et al., 2005 [103]). Briefly, they play a role in shape maintenance, together with other components, notably PBP2. PBP2 also interacts with RodA (Stoker et al., 1983 [104]), which is a transglycosylase (Rohs et al., 2018 [105] and references therein).

MreB rotates around the circumference of the cell in coordination with the assembly of peptidoglycan (van Teeffelen et al., 2011 [106]), thus representing the (dynamic) elongasome counterpart of the divisome with FtsZ and PBP3. Substrates (lipid II) for PBP2 are prepared by MurG, a peptidoglycan glycosyltransferase, as mentioned before (Figure 11). MurG occurred all over the envelope, with some accent at the site of division after immunofluorescence microscopy, and it appeared in a complex with MreB and MraY (Mohammadi et al., 2007 [107]). According to the authors, the localized supply of lipid II, as provided by MurG, has to be at the elongasome as well as the divisome. With respect to the divisome, the localization appeared to be PBP3-dependent. This suggests, but does not prove, that peptidoglycan synthesis before PBP3 becomes active, and uses MurG in an MreB environment in the cell center.

Another problem regards the nature of the flippase that translocates lipid II to the periplasm. Confusion has arisen as to whether FtsW or MurJ acts as such. It now appears that FtsW functions as a flippase in vitro (Mohammadi et al., 2014 [108]), whereas this applies for MurJ in vivo (Liu et al., 2018 [109]). MurJ has a location similar to that of MurG. To become located at the divisome, it requires lipid II synthesis and active FtsW; to be located at the elongasome, it possibly requires RodA instead of FtsW (Liu et al., 2018 [109]). Another protein, RodZ, is also involved in shape geometry of *E. coli* (Shiomi et al., 2008 [110]). Its precise function has still to be elucidated. Figure ?? shows a recent version of an elongasome model.

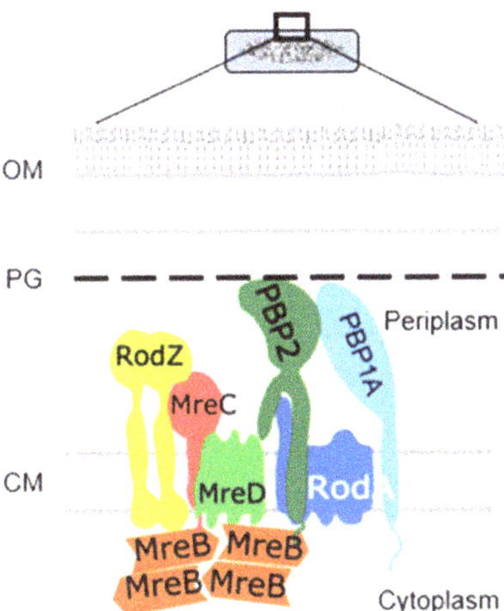

Figure 22. Core elongasome proteins in *E. coli*. Proteins are recruited by MreB. CM, cell membrane; OM, outer membrane; PG, peptidoglycan. Source: Modified Figure 1 of [109]. Open access.

14. PIPS

Inhibition of PBP3 by mutation (a temperature-sensitive *pbpB* mutation) or by furazlocillin, aztreonam, or cephalexin (specific ß-lactam antibiotics for PBP3) produced filaments with blunt constrictions. Upon further growth, the central blunt constriction became flanked by two new normal-looking constrictions (Figure 23; Wientjes and Nanninga, 1989 [57]). I quote: "Inhibition of PBP2 (by mecillinam) and PBPs 1A and 1B (by cefsoludin or moenomycin) does not prevent initiation of division (our unpublished observations). These observations point to the involvement of a penicillin-insensitive peptidoglycan-synthesizing activity (PIPS) during initiation of constriction" (Nanninga, 1991 [111]). I have to make one correction here: "our unpublished observations" are now "my lost observations". I would like to emphasize here that PIPS refers to constricting cells. It appears possible that PIPS still interacts with FtsZ upon inhibition of PBP3, because constrictions are well visible (Figures 21 and 23 and see below). Temperature-sensitive FtsZ filaments are smooth at the non-permissive temperature and no localized peptidoglycan synthesis takes place in that case, as deduced from autoradiograms (Woldringh et al., 1987 [56]).

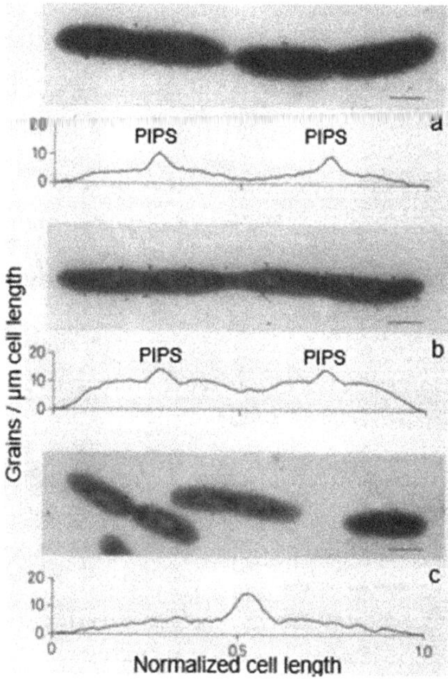

Figure 23. Topography of [³H] Dap incorporation after inhibition of PBP3 and PIPS. (**a**) The temperature-sensitive cell division mutant MC4100 *lysA pbpB* was grown at 28 °C in the presence of 2µg of furazlocillin per mL. Grain distribution of furazlocillin-induced filaments with three constrictions (N = 130; G = 4876; L = 10.68 µm, where N is the number of cells, G is the number of grains, and L = 12.16. (**b**) Cells grown at the restrictive temperature at 37 °C without furazlocillin. Grain distribution of temperature-induced filaments with three constrictions (N = 164; G = 15,506; L = 12.16 µm). (**c**) Grain distribution of normally dividing *pbpB* cells at the permissive temperature (N = 92; G = 1533; L = 4.11 µm). Bar, 1 µm. Source: Modified Figure 6 of [57]. Copyright ©1989, American Society for Microbiology.

With a novel technology, filamentous *E. coli* sacculi obtained from cells with inhibited PBP3 also showed three division sites (a central one and two flanking ones). In this case, cells were grown in the presence of D-cysteine so that sacculi containing modified peptidoglycan with -SH groups could be biotinylated. The latter were detected in the electron microscope by antibiotin antibodies conjugated to gold particles. The two flanking division sites mentioned above, with active peptidoglycan synthesis, became visible by the absence of label after an extensive chase (de Pedro et al., 1997 [112]), thus confirming the presence of PBP3-independent peptidoglycan synthesis at division sites. See, however, below.

15. PIPS and Pre-Septal Peptidoglycan Synthesis (PPS)

15.1. Definition of PIPS

In a recent paper, the following description of PIPS and PPS can be found: "For the purpose of this study, we refer to PIPS as pre-septal PG synthesis" (Pazos et al., 2018 [113]). Also, in an earlier paper, no distinction was made between PIPS and PPS (Potluri et al., 2012 [114]). Pre-septal PG synthesis I have denoted as PPS, for short. I am afraid that combining PIPS and PPS might be confusing. Briefly, the distinction between the two is based on the presence or absence of a constriction, respectively. Below, I will try to elucidate the distinction.

As mentioned before, PIPS arose after inhibition of PBP3. Strictly, it should be defined as PBP3-inhibited peptidoglycan synthesis. Note that [^{3}H] Dap is incorporated, as can be seen in the autoradiogram (Figure 23). Thus, what kind of peptidoglycan synthesis is applicable? Can one maintain that, say, PBP2, is not involved? This appears a valid question because the elongasome has to be replaced by a (nascent) divisome at the cell center. Since PIPS does not require PBP3, will PBP2 carry out the job? Below, this will be adressed further.

15.2. Pre-Septal Peptidoglycan Synthesis (PPS)

I now turn to pre-septal peptide glycan synthesis (PPS). "Pre-septal" means before division (of course), implying that, as soon as a constriction emerges, the term pre-septal does not apply anymore. What unites the activities of PPS and PIPS is that they both require FtsZ, are located at the cell center, and occur before PBP3 comes to the fore (den Blaauwen et al., 2003 [99]). To further delineate PPS and PIPS, it will be helpful to look at their temporal position in the cell cycle (Figure 24; den Blaauwen et al., 1999 [115]). This refers to a steady-state culture of *E. coli* K 12 MC4100 with a doubling time of 85 min at 28 °C. These conditions have been used by us again and again to make experiments comparable which were carried out in the course of time.

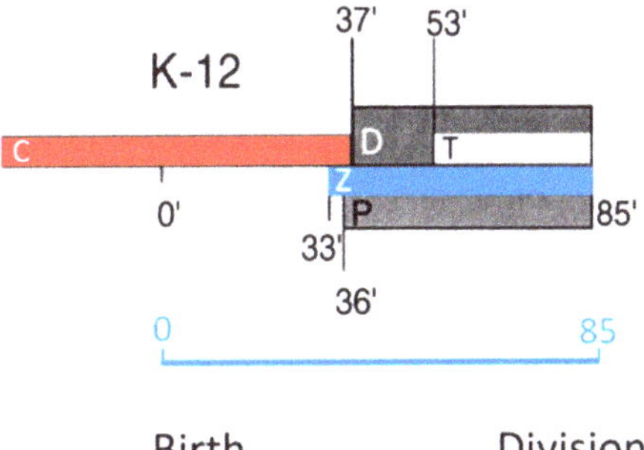

Figure 24. Schematic representation of the cell cycle of *E. coli* K-12 growing with a generation time of 85 min. The C period shows the duration of the DNA replication cycle. Note that C starts before birth. Thus, the cell cycle starts in the middle of C. The D period is the time between the termination of DNA replication and the separation into two daughter cells. The Z period shows the duration of the period in which FtsZ can be detected at midcell. The P period shows the period of peptidoglycan synthesis during constriction [116], and the T period is the duration of the constriction. Source: Modified Figure 7 of [115]. Copyright © 1989, American Society for Microbiology.

Note that, in Figure 24, DNA replication (C-period) starts in the previous cell cycle and finishes at 37 min in the ongoing one. Thus, a newborn cell contains a half-replicated chromosome. The D-period lasts from termination of DNA replication to cell separation, and its duration is thus 85 minus 37 min equaling 48 min. The T-period represents the visible duration of cell division (85 minus 53 min equals 32 min). FtsZ (Z-period) becomes visible at 33 min, i.e., around termination of DNA replication and the start of peptidoglycan synthesis at the cell center (P-period). The duration of the P-period has been taken from Taschner et al. (1988 [116]).

How do PIPS and PPS fit into this scheme? Since PIPS refers to constricting cells, it is located at the beginning of the T-period (the remainder of the T-period refers to PBP3-dependent peptidoglycan synthesis). What about PPS? Visibility of constriction (beginning of the T-period) starts at 53 min, so this is the end of PPS. PPS starts with the beginning of the P-period. The latter more or less coincides with the end of DNA replication (beginning of D) and central positioning of FtsZ (Aarsman et al., 2005 [117]; Pazos et al., 2018 [113]). FtsZ is present during PPS and PIPS.

15.3. A Closer Look at Autoradiography

With this in mind, we can have another look at Figure 14. It depicts silver grain distributions measured in autoradiograms of *E. coli* K-12 cells pulse-labeled with [³H] Dap. Thus, the uninterrupted line represents peptidoglycan synthesis in T-period cells and the interrupted line includes all non-dividing cells. Likely, the central peak in the latter case reflects PPS. We can also have another look at immuno-labeled sacculi mentioned above. My interpretation is that they reflect a combination of PPS and PIPS. There are two arguments for this interpretation. Firstly, PIPS is about pulse-labeling in a local area, that is, what we have termed the leading edge of constriction (Wientjes and Nanninga, 1989 [57]). Secondly, the less-labeled zones visible after immunolabeling (de Pedro et al., 1997 [112]) are rather broad, as compared to a leading edge.

In summary, for the time being, we might distinguish three stages during cell division: PPS, PIPS, and a divisome core activity (Kashämmer et al., 2023 [118]), of which PBP3 is a main participant. Below, I will attempt to further delineate participants of PPS and PIPS.

15.4. Participants of PPS

PPS starts in cells where a functional transition can be expected at the cell center from elongasome to divisome. At the cell center, the bacterial actin filaments of MreB and the tubulin-like filaments of the GTPase FtsZ appear to interact directly (Fenton and Gerdes, 2013 [119]), and the latter authors suggested that MreB bound to FtsZ still stimulated PBP2 and PBP1B activities to produce peptidoglycan. However, other authors stressed the interaction between PBP2 and PBP1A (Banzhaf et al., 2012 [120]) and the interaction between PBP3 and PBP1B (Wientjes and Nanninga, 1991 [75]; Bertsche et al., 2006 [121]; Ranjit et al., 2017 [122]; Boes et al., 2019 [123]). One might speculate that upon the earliest arrival of FtsZ to the cell center, MreB-directed peptidoglycan synthesis, aided by PBP2 and PBP1A and/or PBP1B, synthesizes PPS, at least in part.

15.5. Participants of PPS and PIPS

Earlier work (Figure 24) has shown that FtsZ is positioned at the cell center about 20 min before a constriction becomes visible in an *E. coli* K-12 LMC500 (MC4100*lysA*) strain, thus during PPS. Attachment of FtsZ to the cell membrane is facilitated by FtsA (Ma et al., 1996 [124], 1997 [125]; Wang et al., 1997 [126]; Pichoff and Lutkenhaus, 2002 [127]) and by ZipA (Hale and de Boer, 1997 [128]; Pichoff and Lutkenhaus, 2002 [127]; Potluri et al., 2012 [114]) implying that they are the first components of the nascent or proto-divisome.

According to Potluri et al. (2012) [114], ZipA is required for PPS/PIPS peptidoglycan synthesis. Note that I am using here the term PPS/PIPS as I have argued above. In their comprehensive study, they also investigated other possibilities for peptidoglycan synthesis in the absence of PBP3 and without involving PBP2. They took into account the glycosyltransferases PBP1C (Schiffer and Höltje, 1999 [129]), a peptidoglycan polymerase that catalyzes glycan chain elongation from the lipid-linked precursor MtgA (Di Berardino et al., 1996 [130]) and the penicillin-insensitive L,D-transpeptidases YnhG and YcbB (Magnet et al., 2007 [131], 2008 [132]; Höltje, 1998 [47]; Vollmer and Bertsche, 2008 [133]) and they made a strain with all four mutations. It appeared that so-called PIPS bands were still formed. In addition, not required for PPS/PIPS are AmiC (N-acetylmuramyl-L-alanine amidase), EnvC (murein hydrolase activator), and seven ß-lactamases (PBP1A, PBP4, PBP5, PBP6, PBP7, AmpC (Class C ß-lactamase), and AmpH (Class C ß-lactamase)) in a mutant lacking all of them [114].

Thus, after FtsZ, ZipA, FtsA, and PBP1A and/or PBP1B have taken over, a first step in divisome assembly has been completed (Pazos et al., 2018 [113] and references therein). The role of PBP2 has still to be clarified further (see below).

15.6. Participants of PIPS

With PIPS, the constriction becomes visible in the electron microscope. What does the constriction look like? For this purpose, we can examine recent cryo-sections for the electron microsope (Navarro et al., 2022 [134]). What can be observed is the tight organization of the envelope layers during initial constriction (Figure 25a). Presumably, this is also the case with PIPS. Figure 25 has been selected to address the point I am making here. For numerous other interesting details, the reader should consult the original paper (Navarro et al., 2002 [134]).

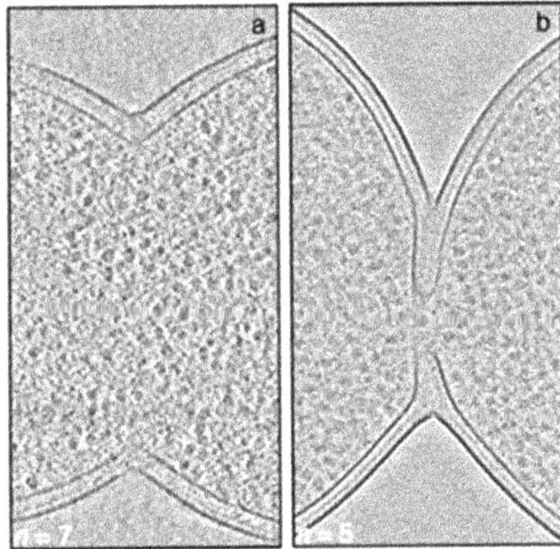

Figure 25. Two different stages of division in wild-type *E. coli*. (**a**) Initiated division. The envelope layers are still together. (**b**) Progression of division with envelope layers disconnected. Summed, projected central slices of cryo-electron tomograms. Source: Figure 1 (partly) of [134]. Open access.

As implied above, PPS is probably facilitated by PBP2 as long as there is no constriction. Nevertheless, does PBP2 have a role for PIPS while PIPS constrictions still occur after inhibition by mecillinam? My speculative answer is based on the experiments of Wientjes and Nanninga (1991) [75], where it was shown that the inhibition by mecillinam was not complete, that is, around 65 percent. In other words, is the remaining 35 percent activity of PBP2 still able to play a role in the initiation of constriction? I will return to this point after having discussed the assembly of the remaining divisomal proteins.

Returning to Figure 25b, it seems that the outer membrane is lagging behind, whereas the peptidoglycan layer is less distinct. Interaction of lipoproteins with the outer membrane seems not very possible here. In particular, would this apply to LpoB (see below) and its partner PBP1B? However, the question can be posed whether, in this stage, PBP3 is still there. Perhaps this is not the case, and are preparations made to separate the daughter cells in the absence of FtsZ (Söderström et al., 2014 [135])?

16. Remaining Divisomal Proteins

So far, I have dealt with the early preparations of the division process, i.e., the assembly of the nascent divisome. How does the composition of the nascent divisome change during the course of preparation and execution of the division process? After FtsA and ZipA, additional proteins become incorporated into the nascent divisome. The sequence is known, as well as the interdependency regarding their interaction (Figure 26; Buddelmeijer and Beckwith, 2002 [86]; Aarsman et al., 2005 [117]). However, this sequence is to some extent misleading because it suggests that most proteins follow one after another while constituting the divisome. In fact, some proteins are grouped independently before they enter the divisome. This applies, for instance, to a trimeric complex of FtsL, FtsB and FtsQ (Buddelmeijer and Beckwith, 2004 [136]).

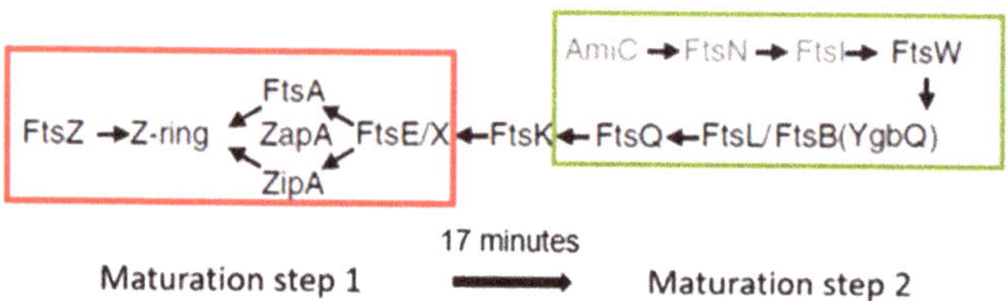

Figure 26. Two maturation steps of divisome assembly. Adapted from [86,117].

So, when do the remaining divisomal proteins arrive at their final destination? To answer this question, the fraction of fluorescently labeled cells at the end of the cell cycle was determined. In this way, the timing of the arrival of divisomal proteins could be calculated from their labeled fractions (Aarsman et al., 2005 [117]). This has been done for FtsQ, FtsW, PBP3, and FtsN. It appeared that, under the growth conditions used (doubling time 85 min at 28 °C), all of them arrived at about the same time at the divisome, i.e., about 17 min after the first positioning of FtsZ. Thus, the authors proposed two maturation steps for divisome assembly (Figure 26).

Since PIPS precedes PBP3 activity, by definition, the continuation of initial constriction is likely based on peptidoglycan synthesis of the PBP3 (transpeptidase)-FtsW (transglycosylase; Taguchi et al., 2019 [137]) couple. A complex of PBP3-FtsW had already been found independent of other divisomal proteins (Fraipont et al., 2011 [138]). Conceptually, this complex resembles that of PBP2 and RodA of the *E. coli* elongasome (Ishino et al., 1986 [139]; Rohs et al., 2018 [105]). Returning again to Figure 25b, one wonders whether the outer membrane is still at its original location, that is, tightly bound to the peptidoglycan layer. Anyway, FtsZ leaves the Z-ring before the end of cell separation, while FtsA, ZipA, FtsQ, FtsL, and PBP3 are still present (Söderström et al., 2014 [135]).

17. Constructing the Core Divisome

Recently, a pentameric complex (FtsWIQBL) composed of the trimeric FtsL, FtsB, and FtsQ and the PBP2-FtsW couple, both from *Pseudomonas aeruginosa*, was purified and frozen. Subsequently, its structure was determined by cryo-electron microscopy at a resolution of 3.7 Å (Kashämmer et al., 2023 [118]). It will be observed that FtsK and FtsN are not yet incorporated. Presumably, this will be a matter of time. As pointed out by the authors: "As FtsWIQBL is central to the divisome, our structure is foundational for the design of future experiments elucidating the precise mechanism of bacterial cell division, an important antibiotic target".

18. A Role for Outer Membrane Proteins LpoA and LpoB

Though peptidoglycan metabolism is generally viewed in the direction from cytoplasm to cell membrane, and from there to the peptidoglycan layer (Figure 11), recent research introduced the direction from outer membrane to peptidoglycan layer and cell membrane (Typas et al., 2010 [140]). This was based on the characterization of two new lipoproteins, called LpoA and LpoB. Briefly, they interact with PBP1A and PBP1B, respectively, to bind new peptidoglycan to the sacculus. Thus, LpoA and LpoB presumably have mainly to do with the elongasome and the divisome, where they interact with the transpeptidase moieties of the respective PBPs1.

19. PPS and PIPS (Preparative Cell Division) and the Core Divisome

Under this heading, I will attempt to integrate the above constituents of the division process. This will be mainly based on the paper of van der Ploeg et al. (2013) [141], entitled "Colocalization and interaction between elongasome and divisome during a preparative cell division phase in *Echerichia coli*". It is, amongst others, an advanced continuation of a paper which appeared one year after my tenure in 2002 (den Blaauwen et al., 2003 [99]). I also wish here to acknowledge the excellence of Tanneke den Blaauwen while pursuing this topic with the persons she assembled around her. I am proud she has continued the philosophy of the Amsterdam School mentioned before.

The above study involved the immunolocalization of MurG, MreB, PBP2, FtsZ, PBP3, and FtsN during the cell cycle (maps of fluorescent profiles) as well as Förster Resonance Energy Transfer (FRET) of PBP2 and PBP3. Remember that MurG catalyzes the transfer of lipid I to lipid II (Figure 11). The maps of fluorescent profiles can be seen in Figure 27 of endogenous elongasome and divisome proteins. A vertical white bar indicates (visually) the central position of a label. Note, that MreB and PBP2 localize more or less together in the cell center, and that FtsZ starts at about the same time (cell cycle fraction) as PBP2. PBP3 enters much later in the midcell and FtsN follows. Also note that the positions of PBP2 and PBP3 overlap, which might explain the above-mentioned FRET data. Their data have been summarized in Figure 28.

Map of fluorescence profiles of LMC500 wild-type cells at 28⁰C

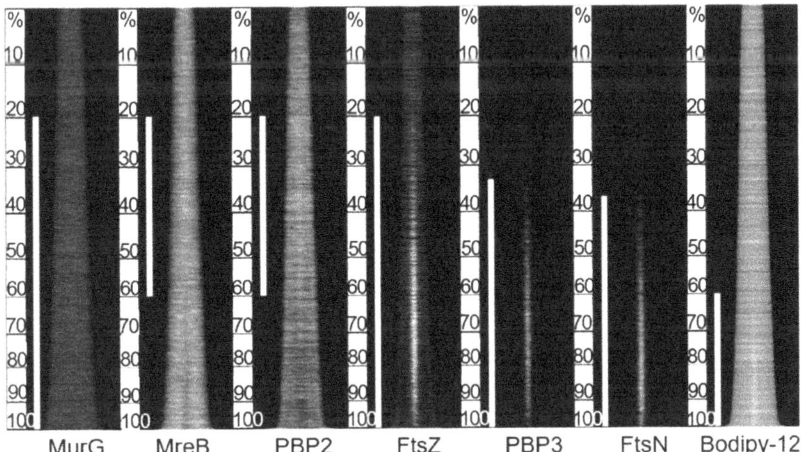

Figure 27. Maps of fluorescent profiles of immunolocalized endogenous elongasome and divisome proteins MurG, MreB, PBP2, FtsZ, PBP3, and FtsN of *E. coli* LMC500 wild-type cells grown to steady state in GB1 medium at 28 °C. The integral fluorescence of each cell is plotted as a function of cell length (*x*-axes) and all cells are plotted with increasing age (*y*-axes). The white bars present midcell localization judged by visual inspection. Bodipy-12 is a general membrane stain. Source: Modified Figure 2 of [141]. © John Wiley and Sons.

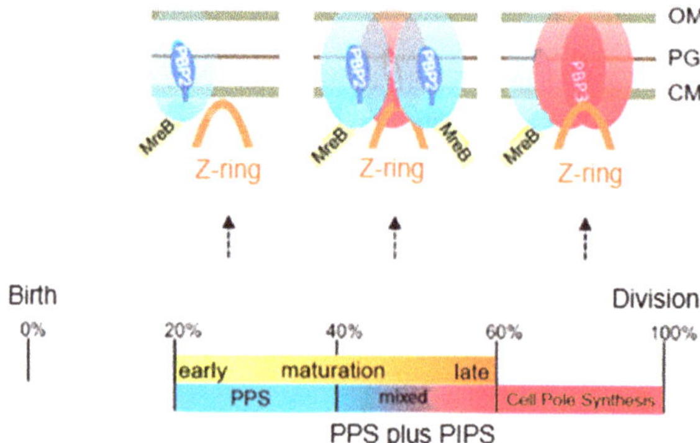

Figure 28. Transition from elongasome (PBP2) to divisome (PBP3) activity at midcell. Growth conditions as in Figure 27. FtsZ is present all the time. During the transition PBP2 and PBP3 are both located at midcell in the presence of MreB and FtsZ. At the onset of cell pole synthesis, PBP2 disappears from the cell center. By contrast, pre-septal peptidoglycan synthesis (PPS) is dependent on PBP2. In between, a mixed situation arises, which I interpret as PPS plus PIPS. PIPS also requires a transglycosylase activity which could be provided by RodA and/or PBP1A. The time scale is divided in percentages to indicate the events occurring from birth to division. CM, cell membrane; OM, outer membrane; PG, peptidoglycan layer. Source: Modified and based on Figure 6 of [141]. © John Wiley and Sons.

The model addresses aspects related to the enigmatic PIPS. To summarize: the FtsZ ring is there all the time, as is MurG. MreB starts to disappear, as does PBP2. The latter is still there when PBP3 becomes involved in cell pole synthesis after the late maturation step of the divisome assembly. We also find PPS commencing the early phase of divisome assembly. After PPS follows a mixed zone which includes PIPS and PBP2, a transpeptidase. Presumably, RodA provides for transglycosylase activity, enabling peptidoglycan synthesis after inhibition of PBP3 with aztreonam. In other words, it is tempting to speculate that PIPS is constituted of PBP2 and RodA.

20. Final Steps

There is a structural continuity between nucleoid and envelope, as mediated by coupled transcription, translation, and insertion of membrane proteins in the cell membrane. Eventually, DNA catenanes in the cell center have to be broken after termination of DNA replication to allow the new nucleoids to find their positions in the daughters. Clearly, DNA should be occluded from the cell center for division to be completed (Mulder and Woldringh, 1989 [142]); hence the later term nucleoid occlusion (NO). One way to occlude fission through the nucleoid would be a division inhibitor that prevents Z-ring assembly "on portions of the membrane surrounding the nucleoid" (Bernhardt and de Boer, 2005 [143]). They found a DNA-associated division inhibitor denoted as SlmA. Another negative regulation is carried out by the Min system (MinC, MinD, and MinE), which prevents the formation of minicells at the poles (Lutkenhaus 2007 [144] and references therein).

A positive regulation mechanism has to do with the DNA-binding protein MatP, which organizes the replication terminus region, the Ter linkage (review: Männik and Bailey, 2014 [145]). MatP can bind to Zap B, which in turn binds to ZapA located on FtsZ (see later). Finally, FtsK, when present in the divisome, can translocate DNA with the aid of ATP when it is in a hexameric assembly (Sheratt et al., 2010 [146]). Presumably, such translocations change the higher-order structure of the nucleoid in such a way that its segregation at the end of DNA replication is facilitated.

The above systems linking DNA segregation to cell division are largely present in cells growing in a rich medium as contrasted to slowly growing cells (Männik and Bailey, 2014 [145]). It would seem that fast-growing cells (presumably, multifork DNA replication) require more coordination to achieve safe cell proliferation. Obviously, according to the authors, the last word has not yet been said on this topic: "Our data, however, is indicative that yet an unidentified, lower fidelity positioning system remains in *E. coli* ΔslmA Δmin cells even without the Ter linkage" [145].

21. What Is the Mechanical Role of FtsZ Polymers during Constriction?

21.1. Role of FtsZ

In the foregoing, I have focused on peptidoglycan assembly and its role in the cell division process. However, what can FtsZ do by itself? Constriction starts with bending of the cell membrane in the presence of the peptidoglycan layer and the outer membrane (Figure 25a). So, what causes membrane bending? Is there a pulling force based on FtsZ as a participant? Next, the constriction proceeds while building two hemispherical domes from the bended surfaces. Is there a shape-maintaining skeleton? Finally, is there a force that results in cell separation? I will start with two opposing models and then try return to the questions formulated above.

Model one starts from the basic observation that FtsZ protofilaments can form upon GTP binding (Mukherjee and Lutkenhaus, 1994 [147]) and that the protofilaments are attached to the cell membrane. Membrane anchors are FtsA and ZipA in the living *E. coli* cell. The C terminus of FtsZ binds FtsA, and the amphipathic helix at the C terminus of FtsA represents its membrane anchor (cf. Osawa et al., 2008 [148]). In a model system, Osawa et al. (2009) [149] used liposomes and, instead of FtsA, a membrane-targeted FtsZ construct called Fts-mts. Inside the liposome tubules, Fts-mts could form constricting Z rings, indicating that the Z rings can generate a force independently of other proteins. Later studies showed that a concave or convex membrane impression could be obtained dependent on whether the mts side was at the C terminus (the normal side) or at the N terminus, respectively (Osawa et al., 2009 [149]; Figure 29A). These in vitro experiments suggest that FtsZ protofilaments can bend liposomal membranes.

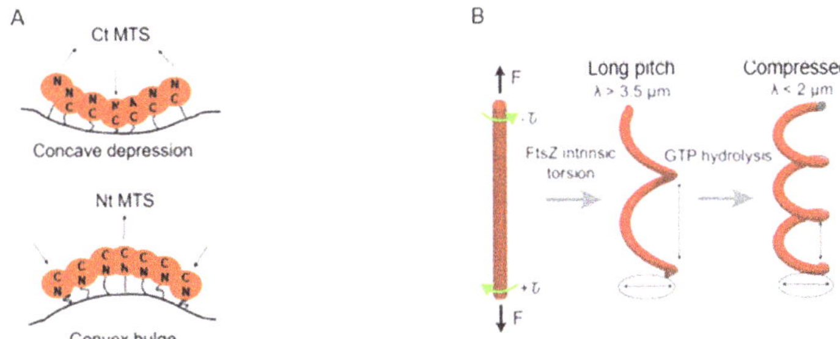

Figure 29. Autonomous bending of FtsZ protofilaments. (**A**) A model of membrane deformations generated by bending force of FtsZ filaments. When the membrane-targeting sequence (mts) is attached at the C- or N-terminus, the bent protofilaments form a concave depression (left panel) or convex bulge (right panel), respectively. The direction of bending to make a concave depression is the same as that of Z-ring constriction. Source: Modified Figure 6B of [149]. (**B**) An elastic rod subjected to a constant force (F) reveals a helical deformation upon twisting. Intrinsic FtsZ torsion rules long-pitch transformations (λ > 3 μm) while GTP enhances further torsion, causing higher pitch states (λ < 2 μm). Source: Modified Figure 3f of [150]. (**A**). © John Wiley and Sons.; (**B**) Open access.

A somewhat different approach was followed by Ramirez-Diaz et al. (2021) [150]. They pulled so-called soft lipid tubes from giant lipid vesicles with optical tweezers. The giant lipid vesicles had been decorated with an FtsZ-mts strain containing a yellow fluorescent protein (YFP) as a label. Remarkably, conformational alterations of the tubes' surfaces occurred by torsional stress after GTPase activity. It has been speculated that torsional stress in FtsZ protofilaments might also contribute to constriction of *E. coli* cells (cf. Figure 29B).

21.2. Role of Peptidoglycan

Model two switches from FtsZ to peptidoglycan as an organizer of constriction (Coltharp et al., 2016 [151]). They showed that the rate of constriction was reduced by a mutation of PBP3 and not so much as by a mutation of FtsZ. The results were interpreted to mean that peptidoglycan remodelling by PBP3 is instrumental for the constriction force. Thus, in contradiction to the above models, the constriction has been supposed to arise from the outside (PBP3) and not from the inside (FtsZ) of the cell.

It appears relevant that, in the first case, we are looking at in vitro models where one always has to find out whether they apply to an in vivo situation. By contrast, the second model is based on presumed peptidoglycan synthesis, though measured in a very indirect (non-biochemical) way. However, it has already been known for a very long time that inhibition of PBP3 leads to impairment of the division process, presumably because PBP3 is a component of the multilayered divisome-encompassing cytoplasm, cell membrane, peptidoglycan layer, and outer membrane. Making peptidoglycan synthesis responsible for constriction seems too simple a conclusion, in particular in light of what is already known about FtsZ protofilaments. It should also be noted that the authors were primarily focused on measuring completion of constriction, and not its initiation (PBP3/PIP2). The PBP3/PIP2 system operates, as I have indicated, prior to the functioning of the core divisome, which includes PBP3 in the presence of FtsZ. The also-present FtsW (transglycosylase) might help in maintaining the shape of the nascent hemispherical domes through interaction with FtsZ. Presumably, the domes are reinforced by nascent peptidoglycan, though the latter might disappear upon last stages of cell separation (Söderström et al., 2014 [135]). Cell separation might be an enzymatic process without the requirement of mechanical assistance.

21.3. Role of Phospholipids

The above dealt with bending of the cell membrane with various external agents. But, what about the cell membranes themselves? For instance, their phospholipid composition, that is, are the hemispherical caps that arise during constriction different from the lateral cell membranes? As an approach, we isolated minicells, assuming that they are constructed of two hemispherical caps (an old one and a new one). *E. coli* LMC500*lysA* was used as the wild-type strain, and *E. coli* LMC1088 was used as a minicell-forming mutant. Cells were grown to steady state with a doubling time of 80 min at 28 °C in glucose minimal medium (our standard conditions). Minicells were separated by centrifugation from LMC1088. Phosphatidylethanolamine (PE), phosphatidylglycerol (PG), and cardiolipin (CL) concentrations were determined in the wild type, in LMC1088, and in minicells. PE was a dominant species in all samples, whereas the overall composition of the wildtype and LMC1088 was very similar. However, minicells contained more Cl and less PG (Koppelman et al., 2001 [152]). These results suggest that cardiolipin might aid in membrane bending. At the moment, this is still a speculation.

22. Divisome Subassemblies
22.1. Subassemblies

In the foregoing, divisome proteins have been treated individually or as interacting individuals. However, a higher order of organization should also be considered. For instance, how many pentameric complexes (FtsWIQBL; Käshammer et al., 2023 [118]) participate in the division process? How are their activities coordinated? Do their numbers change during division (Wientjes and Nanninga, 1989 [57])? As outlined before, most divisomal proteins occur in a limited number of copies per cell as compared with FtsZ, "implying that the FtsZ ring is not saturated with cell division proteins. Therefore, as a tentative model, one can envision that the divisome is composed of subassemblies which are connected by FtsZ polymers" (Figure 30; Nanninga, 1998 [44]).

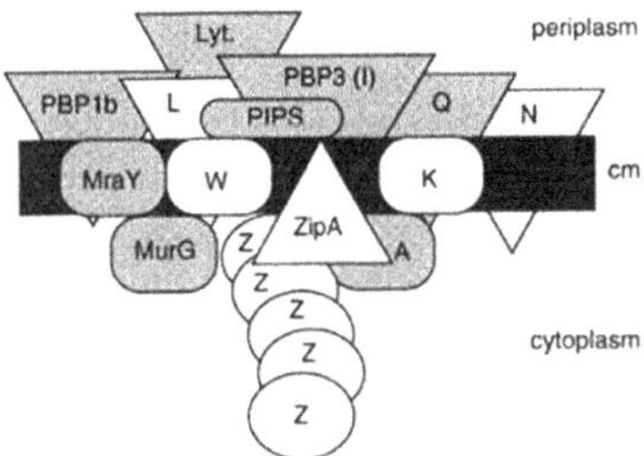

Figure 30. Model of divisome subassembly. It unites cytoplasm and periplasm. The gene products FtsA, FtsI (PBP3), FtsK, FtsL, FtsN, FtsQ, FtsW, and FtsZ have been denoted as A, I, K, L, N, Q, W, and Z, respectively. Lyt. indicates that lytic enzymes must be present; cm, cell membrane. Source: Figure 11 of [44]. Copyright © American Society for Microbiology.

Though the term subassembly might appear appropriate, they are certainly not connected by a continuous FtsZ polymer. The basic architecture of the subassembly encompasses components of the cytoplasm, cell membrane, and periplasm (including the peptidoglycan layer). In particular, components should be present, or nearby, that, as mentioned before, can perform rearrangements, turnover, recycling, degradation, and autolysis of peptidoglycan. Also note the presence of MraY (synthesis of lipid I), MurG (synthesis of lipid II) and a general lytic activity (Lyt.). The identity of the expected flippases (Figure 11) was then not yet known. Last but not least, PIPS is also included as a periplasmic enzymatic activity.

22.2. Stoichiometry of Subassembly Components

The above approach was based on the relative excess number of FtsZ molecules, that is, not all divisome proteins could cover FtsZ. To substantiate this further, more correct information is required about the concentration of divisomal proteins in the cell and in the divisome. Here, I mention two approaches. Earlier in this article, I stressed the importance of measuring cells according to various parameters, as developed by Norbert Vischer. I referred to the program ObjectJ. This was decades ago, and it was also used for decades. Now this expertise has been augmented by a specialized software project, Coli-Inspector, which is used to make fluorescent profile maps of cells arranged according to length and thus to age (Vischer et al., 2015 [38]). It applies to widefield microscopy and immunolabeling of wild-type K-12 strain MC4100 grown to steady state in minimal glucose medium at 28 °C. As stated before, these were our standard growth conditions which we maintained over the course of years.

The amount of fluorescence per cell could be correlated with the concentration of a number of divisome proteins during the cell cycle. Briefly, knowing the number of proteins per cell (Li et al., 2014 [153]) allowed the calibration of fluorescence to the number of proteins per cell (and also at midcell). It can be seen in Figure 31 that the concentrations of ZapA and ZapB remain more or less constant during the division cycle. For FtsZ, PBP3, FtsN, and PBP5, roughly a maximum around 60, 70, 75, and 75 per cent can be observed, respectively, presumably mimicking their developing activities. Note that the authors did not find a concentration change for PBP3 and PBP5, as observed in the past (Figure 15; Wientjes et al., 1983 [74]). Since FtsZ peaks before PBP3, which makes sense, the present data are probably more trustworthy (after forty years).

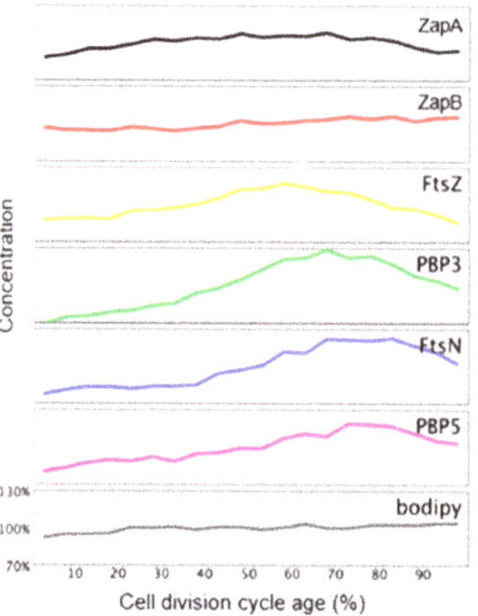

Figure 31. Protein concentration (normalized) as a function of cell age. It applies to ZapA, ZapB, FtsZ, PBP3, FtsN, and PBP5. Bodipy-C12 has been used as a general membrane stain. The concentration of each indicated protein was divided by the average concentration of that protein in the whole population. The individual proteins were plotted against cell division cycle age (%). Source: Figure 6 of [38]. Open access.

Also based on the data of Li et al. (2014) [153], Egan and Vollmer (2015) [154] attempted to determine the stoichiometry of the core divisome proteins (including PBP3). The finding "that most proteins which participate in known multiprotein complexes are synthesized proportional to their stoichiometry" (Li et al., 2014 [153]) led them to derive their stoichiometry from the known protein-protein interactions in the divisome core. In addition, they updated the "make before break" models of Koch (1984) [52] and Höltje (1998) [47].

The above two approaches underpin the notion that groups of divisome proteins are distributed around the circumference of the cell and that their interactions can be studied. We now know that the modern subassembly is a highly dynamic structure (see below) (Du and Lutkenhaus, 2019 [155]).

22.3. Treadmilling of FtsZ Protofilaments

A very remarkable finding is the fact that individual FtsZ filaments are able to treadmill (Loose and Mitchison, 2014 [156]), a phenomenon already known for many years regarding eukaryotic microtubules (remember that the hollow microtubules in a mammalian cell are composed of 13 tubulin filaments). FtsZ treadmilling appeared to be accompanied by peptidoglycan synthesis, thus effecting division (Bisson Filho et al., 2017 [157]). Thus, the subassembly travels along the circumference of the cytoplasmic side of the cell membrane "in both directions with a velocity of about 30 nm/sec". The lifetime of a subassembly is "about 15 s before it disassembles and new clusters appear" (review and references in Du and Lutkenhaus, 2019 [155]).

Treadmilling subassemblies are depicted in Figure 32. Here, peptidoglycan synthesis is carried out by the couple PBP3 (transpeptidase) and FtsW (transglycosylase). FtsK and FtsQLB are involved in recruitment, whereas FtsQLB and FtsN play a role in regulation. The FtsZ protofilament has a very flexible binding to the cell membrane via ZipA, whereas FtsA connects FtsZ to the divisome core (review: Du and Lutkenhaus, 2019 [155]). I may recall a model we made in 1989 (Figure 33; Wientjes and Nanninga, 1989 [57]) and I will quote: "Successive stages in constriction. Since the peptidoglycan-synthesizing activity in the leading is constant during constriction, the concentration of enzymes in the leading edge will increase as the constriction process continues. This is depicted in this drawing as a closer packing of the symbol ◆, which represents PBP3 and possible other enzymes involved in constriction".

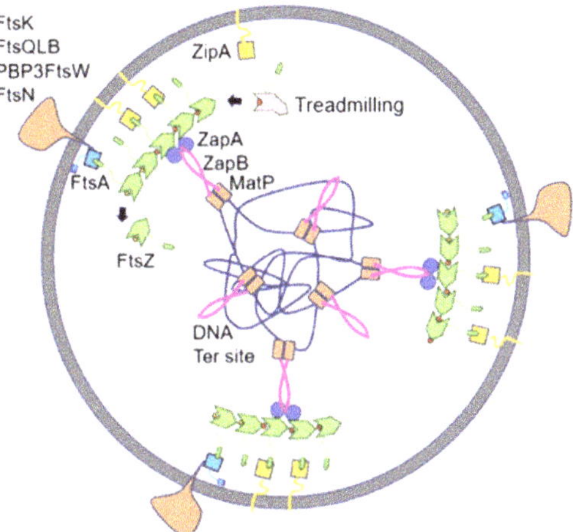

Figure 32. Treadmilling of FtsZ filaments during peptidoglycan synthesis in the direct environment of the terminus of DNA replication. Subassemblies at the cell membrane connected to the peptidoglycan synthetic machinery in the periplasm including FtsK, Fts QLB, FtsWI (PBP3), and FtsN. Internally connected to the terminus of DNA replication (Ter-signal) MatP from FtsZ to MatP via ZapA and ZapB. FtsZ filaments attached to the cell membrane via FtsA and ZipA. The FtsZ filaments are involved in treadmilling in both directions. Source: Modified Figure 4 of [155]. Open access.

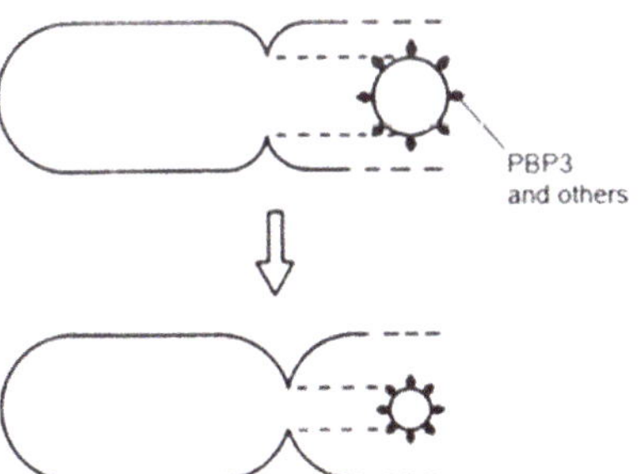

Figure 33. Closer packing of PBP3 and possible other enzymes involved in constriction (◆). As argued in [57], the number of lozenges remains constant. Source: Modified Figure 7 of [57]. Copyright © American Society for Microbiology.

A completely different higher-order organization of the divisome has been proposed by Söderström and co-workers (Söderström et al., 2017 [158]). In their model, the divisome is composed of three concentric rings; the central one spans the cytoplasm and cell membrane and encompasses FtsZ, FtsA, and ZipA, thus representing PPS and PIPS. This ring is sandwiched between a ring composed of core proteins without cytoplasmic partners and a cytoplasmic ring contacting the chromosome.

Since the subassemblies encompass a number of cellular compartments, the required interactions are presumably manyfold. At the moment, it seems too complicated to attempt an answer, unless I have missed something (which is not unlikely).

23. On the Origin of Confocal Scanning Light Microscopy in Amsterdam

23.1. Structure of the Nucleoid

In the beginning of this article, I referred to freeze-fracture experiments with *B. subtilis* with the aim to assess the reliability of chemical fixation (Nanninga, 1969 [20]). This was followed by a focus on mesosomes, with the final conclusion that they arise by chemical fixation (Nanninga, 1971 [15]). In 1969, I was also interested in the appearance of fracture faces of cell membranes after employing prior chemical fixation (RK-fixation; see above). It was shown that cell membranes fractured normally and that their morphology was not visibly altered. However, the nucleoid could not be discerned convincingly (no chemical fixation) unless cells were first fixed with osmium tetroxide. Of course, it could not be concluded that DNA is not there (like mesosomes). Perhaps not surprisingly, the nucleoid's morphology depended strikingly on the chemical fixation method used (Woldringh and Nanninga, 1976 [159]). Though we admired the phase-contrast light microscopic images of Mason and Powelson (1956) [160], the optical resolution was too low to help us with the interpretation of electron microscopic images. So, we felt stuck.

23.2. Origin of Confocal Scanning Laser Microscopy in Amsterdam

Then I came upon a paper of Lemons and Quate (1975) [161], which was about scanning acoustic microscopy and its applications to medical biology. Would this be a method to bridge the gap between light and electron microscopy? I proposed to Fred Brakenhoff to visit them and to have look at the instrument. Why Fred (G.F.) Brakenhoff?

This has to do with the structure of the department in Amsterdam. When at the Technical University of Delft, Woutera van Iterson was accustomed to work with physicists who were involved in designing electron microscopes. With this in mind, I introduced to her a physicist who had just finished his Ph.D. studies, the idea also being not to be fully dependent on a complicated commercial scientific apparatus for our research. So, he became a member of the staff.

Brakenhoff went to Lemons and Quate at the Stanford University, California, USA and worked with the acoustic confocal microscope. Thereafter, he came to a remarkable conclusion: it would be better to make an optical equivalent. The instrument was constructed in our department (Brakenhoff et al., 1979 [162]). It had a scanning table instead of a scanning beam in modern instruments. Initially, the operation of the scanning table was conducted by a spare part of a record player, and the whole instrument rested on an inflated car tyre (these are memories of a biologist). The first prototype is depicted in Figure 34, and a later version can be found in [162].

Figure 34. First prototype of a confocal microscope in Amsterdam. Around 1976. The specimen (S) is present in the common focus of objective 1 and objective 2; hence the term confocal.

Sheppard and Choudhury (1977) [163] calculated that the theoretical gain in optical resolution would be $\sqrt{7}$ and I considered this a significant improvement, but was it useful for our purpose?

In attempting to answer this question, we made an exhaustive study by comparing phase-contrast light microscopy, confocal scanning light microscopy, and electron microscopy of living and chemically fixed *E. coli* cells (Valkenburg et al., 1985 [164]). I will not go into details but show that a higher resolution indeed was obtained. I will give two examples of the above paper. In Figure 35, a phase-contrast image can be compared with a CSLM image of *E. coli* B/r H266 (doubling time 21 min). The latter implies multifork replication, revealing a complex image of the nucleoid, notably in the CSLM cells. The next group of images represent *E. coli* LE316 *gyrB* grown at the non-permissive temperature of 42 °C. Filaments are formed and the nucleoids are not segregated. Compared are chemically fixed cells in Figure 36a and living cells in Figure 36c,d. Of the serial sections, a model was made by cutting the contours (paper) of the nucleoids and they were piled and subsequently photographed (Figure 36b). The latter model (b) of the nucleoids at the non-permissive temperature strikingly resembles those in living cells (Figure 36c,d). So, the gain in resolution, we felt, was demonstrated. Note that the confocal images are based on transmission microscopy and a contrast obtained by light absorption (as contrasted to fluorescence microscopy).

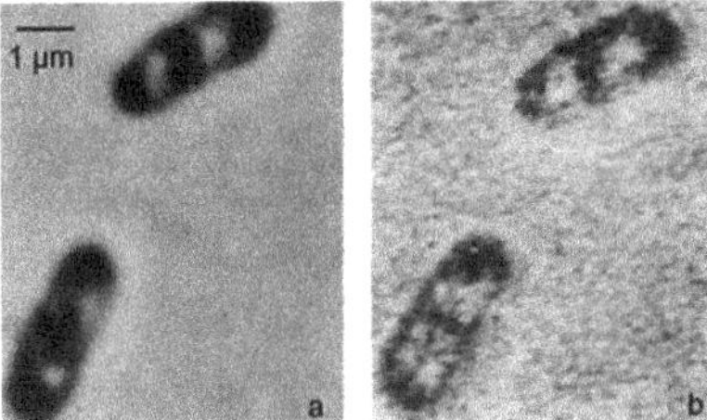

Figure 35. Phase-contrast image (**a**) and confocal transmission image (**b**) of *E. coli* H266. Doubling time 21 min. Bar, 1 µm. Source: Modified Figure 3a,b [164]. Copyright © American Society for Microbiology.

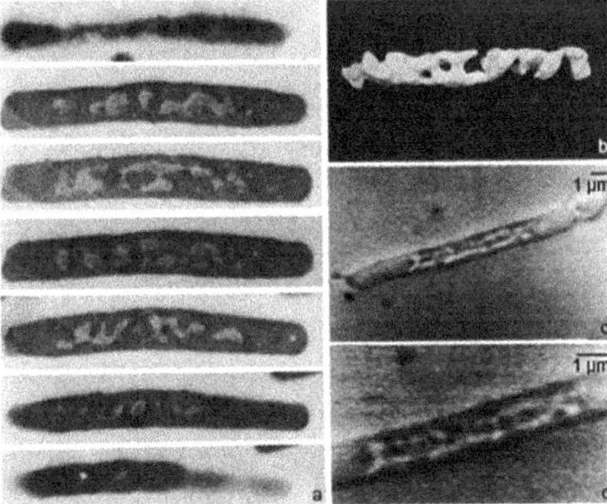

Figure 36. Comparison of electron microscopic thin sections (RK-fixation) (**a**) with a nucleoid reconstruction (**b**) and CSLM transmission image of a living cell of *E. coli* LE316 *gyrB* shifted from 30 to 42 °C and grown for 60 min at 42 °C (**c,d**). The ladder-like nucleoid resembles the model in (**b**). Bar, 1 µm. Source: Modified Figure 2 of [164]. Copyright © American Society for Microbiology.

How to continue from there on? In fact, this was dictated by the emergence of fluorescent probes and lasers and the development of computer technology, three factors that did not yet exist so dominantly when we started with confocal scanning light microscopy in the seventies of the previous century. We therefore altered our approach.

23.3. Three-Dimensional Confocal Scanning Laser Microscopy

The above new possibilities led us to design a computer-controlled confocal microscope (van der Voort et al., 1985 [165]). Essential is the use of pinholes, that is, an illuminated pinhole is focused on to the specimen. Next, the light coming from the specimen is focused on a second pinhole above a photodetector which digitizes the light signal. The importance of a pinhole is illustrated in Figure 37; it shows how the microscope is focused on a thin digital slice by the detection pinhole, thus eliminating out-of-focus information. The mechanical scanning device controlled by a microprocessor produces a series of optical sections (slices). In this way, stacks of images are stored in the computer, and they are thus available for all kinds of image-processing techniques. Screening the stacks from different angles can produce stereo images (Brakenhoff et al., 1985 [166]).

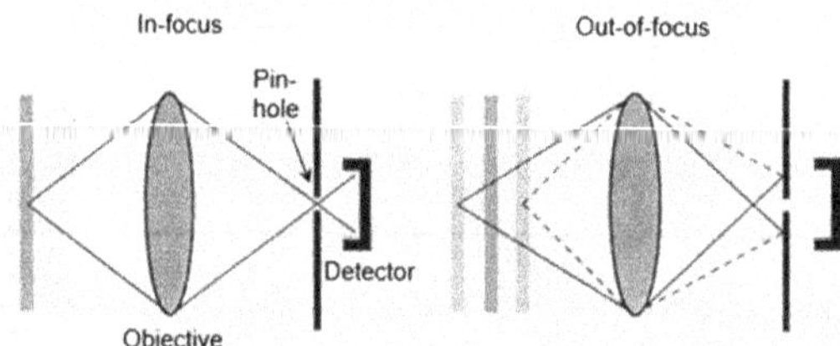

Figure 37. Effect of a pinhole on sharpness of focus. Left, in-focus layer and right, out-of-focus layers that do not contribute to the image. Digital layers arise by scanning, and they can be piled to allow, for instance, a stereo effect.

However, showing such images to an audience appeared very cumbersome. Hence, Hans van der Voort made "clean" solid models of fluorescent objects and added a virtual light source to the whole. The latter created shadows to provide for an impression of depth (van der Voort et al., 1989 [167]). An example is shown in Figure 38 (Oud et al., 1989 [168]). Depicted are *Crepis capillaris* (smooth hawksbead) metaphase chromosomes (2n = 6). We asked the question whether chromosomes have a fixed position towards each other in mitotic anaphase. The answer can be found in the above paper. We embarked on a fruitful project on plant chromosomes with the expert insights of Oof (J. L.) Oud. Also, further optical and image processing techniques, as well as the design of unique fluorescent probes, have been developed. Notably, by my very able successor T. W. J. (Dorus) Gadella, who now occupies the chair of Molecular Cytology at the University of Amsterdam. Unfortunately, this is beyond the scope of this paper.

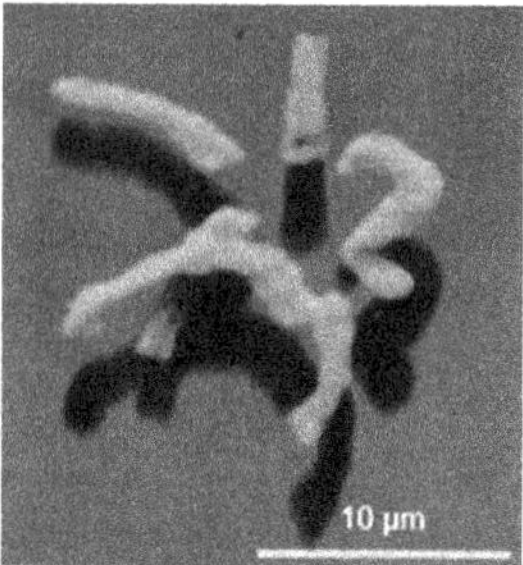

Figure 38. Metaphase chromosomes (2n = 6) with a radial arrangement from a root-tip cell of *Crepis capillaris*. Various filtering procedures, a ray-tracing technique and artificial shadowing were applied to a 3-D reconstruction. Source: Figure 2a of [168]. © Company of Biologists Ltd. [168].

24. Epilogue

Writing this paper has been a journey, and what has been written down is, of course, subjective. I was asked to present my personal recollections, so I can perhaps be forgiven for my subjectivity.

The title of my contribution contains the phrase "little animals", which derives from van Leeuwenhoek (1683). They were little animals because they could move, unlike plants. Actually, the first prokaryote described by van Leeuwenhoek was probably a cyanobacterium. This I have discussed in the blog Small Things Considered: "Van Leeuwenhoek's freshwater microorganisms in 1674. *Spirogyra* or *Anabaena*"? I take the liberty to add as a last figure the cover of my thesis in 1970, which started with the dissecting of the little animals (Figure 39).

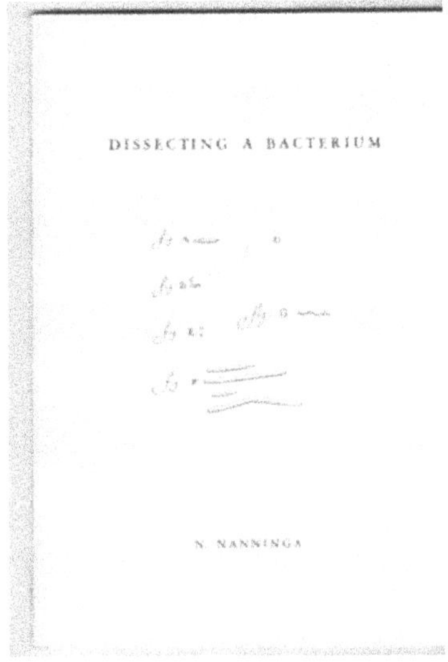

Figure 39. Cover of the Ph.D. thesis of N. Nanninga on 6 May 1970. University of Amsterdam.

Funding: This research received no external funding.

Institutional Review Board Statement: Not applicable.

Informed Consent Statement: Not applicable.

Data Availability Statement: Not applicable.

Acknowledgments: The above was mainly about the past. It is time to concentrate on the future, that is, I wish to dedicate this paper to four of our grandchildren. At the moment, they are studying Life Sciences at three different universities in The Netherlands: Gijs, Nanne, Eva and Dana. The term "our" includes their loving and stimulating grandmother.

Conflicts of Interest: The author declares no conflict of interest.

References

1. Houwink, A.L.; van Iterson, W. Electron microscopical observations on bacterial cytology; a study on flagellation. *Biochim. Biophys. Acta* **1950**, *5*, 10–44. [CrossRef] [PubMed]
2. Nanninga, N. Dissecting a Bacterium. Ph.D. Thesis, University of Amsterdam, Amsterdam, The Netherlands, 1970.
3. Nanninga, N. Ultrastructure of the cell envelope of *Escherichia coli* B after freeze-etching. *J. Bacteriol.* **1970**, *101*, 297–303. [CrossRef] [PubMed]
4. van Gool, A.P.; Nanninga, N. Fracture faces in the cell envelope of *Escherichia coli*. *J. Bacteriol.* **1971**, *108*, 474–481. [CrossRef]

5. Ryter, A.; Kellenberger, E.; Séchaud, J. Electron microscope study of DNA-containing plasms. II. Vegetative and mature phage DNA as compared with normal bacterial nucleoids in different physiological states. *J. Biophys. Biochem. Cytol.* **1958**, *4*, 671–678.

6. Fitz-James, P.C. Participation of the cytoplasmic membrane in the growth and spore formation of bacilli. *J. Biophys. Biochem. Cytol.* **1960**, *8*, 507–528. [CrossRef]

7. Chapman, G.B.; Hillier, J. Electron microscopy of ultra-thin sections of bacteria I. Cellular division in *Bacillus cereus*. *J. Bacteriol.* **1953**, *66*, 362–373. [CrossRef] [PubMed]

8. van Iterson, W. Some features of a remarkable organelle in *Bacillus subtilis*. *J. Biophys. Biochem. Cytol.* **1961**, *9*, 183–192. [CrossRef]

9. van Iterson, W. Symposium on the fine structure and replication of bacteria and their parts. II. Bacterial cytoplasm. *Bacteriol. Rev.* **1965**, *29*, 299–325. [CrossRef]

10. Moor, H.; Mühlethaler, K.; Waldner, H.; Frey-Wyssling, A. A new freezing-ultramicrotome. *J. Biophys. Biochem. Cytol.* **1961**, *10*, 1–13. [CrossRef]

11. Moor, H.; Mühlethaler, K. Fine structure in frozen-etched yeast cells. *J. Cell Biol.* **1963**, *17*, 609–628. [CrossRef]

12. Branton, D. Fracture faces of frozen membranes. *Proc. Natl. Acad. Sci. USA* **1966**, *55*, 1048–1056. [CrossRef] [PubMed]

13. Nanninga, N. Uniqueness and location of the fracture plane in the plasma membrane of *Bacillus subtilis*. *J. Cell Biol.* **1971**, *49*, 564–570. [CrossRef]

14. Jacob, F.; Brenner, S.; Cuzin, F. On the regulation of DNA replication in bacteria. *Cold Spring Harb. Symp. Quant. Biol.* **1963**, *28*, 329–348. [CrossRef]

15. Nanninga, N. The mesosome of *Bacillus subtilis* as affected by chemical and physical fixation. *J. Cell Biol.* **1971**, *48*, 219–224. [CrossRef] [PubMed]

16. Cairns, J. The bacterial chromosome and its manner of replication as seen by autoradiography. *J. Mol. Biol.* **1963**, *6*, 208–213. [CrossRef] [PubMed]

17. Hirota, Y.; Ryter, A.; Jacob, F. Thermosensitive mutants of *E. coli* affected in the processes of DNA synthesis and cellular division. *Cold Spring Harb. Symp. Quant. Biol.* **1968**, *33*, 677–693. [CrossRef]

18. Ryter, A.; Hirota, Y.; Jacob, F. DNA-membrane complex and nuclear segregation in bacteria. *Cold Spring Harb. Quant. Biol.* **1968**, *33*, 669–676. [CrossRef]

19. Singer, S.J.; Nicolson, G.L. The fluid mosaic model of the structure of cell membranes. *Science* **1972**, *175*, 720–731. [CrossRef]

20. Nanninga, N. Preservation of the ultrastructure of *Bacillus subtilis* by chemical fixation as verified by freeze-etching. *J. Cell Biol.* **1969**, *42*, 733–744. [CrossRef]

21. Nanninga, N. Freeze-fracturing of microorganisms. Physical and chemical fixation of *Bacillus subtilis*. In *Freeze-Etching Techniques and Applications*; Benedetti, E.L., Favard, P., Eds.; SFME: Paris, France, 1973; pp. 151–179.

22. Silva, M.T.; Sousa, J.C.F.; Polonia, J.J.; Macedo, M.A.E.; Parente, A.M. Bacterial mesosomes: Real structures or artifacts? *Biochim. Biophys. Acta* **1976**, *443*, 92–105. [CrossRef]

23. Rasmussen, N. Facts, artifacts, and mesosomes: Practising epistemology with the electron microscope. *Stud. Hist. Philos. Sci.* **1993**, *24*, 227–265. [CrossRef]

24. Hudson, R.G. Mesosomes and scientific methodology. *Hist. Philos. Life Sci.* **2003**, *25*, 167–191. [CrossRef]

25. Allchin, D. The very reproducible (but illusory) mesosome. *Am. Biol. Teach.* **2022**, *84*, 321–323. [CrossRef]

26. de Petris, S. Ultrastructure of the cell wall of *Escherichia coli* and chemical nature of its constituent layers. *J. Ultrastruct. Res.* **1967**, *19*, 45–83. [CrossRef] [PubMed]

27. Weidel, W.; Frank, H.; Martin, H.H. The rigid layer of the wall of *Escherichia coli* strain B. *J. Gen. Microbiol.* **1960**, *22*, 158–166. [CrossRef]

28. Graham, L.L.; Beveridge, T.J.; Nanninga, N. Periplasmic space and the concept of the periplasm. *Trends Biochem. Sci.* **1991**, *16*, 328–329. [CrossRef]

29. Maaløe, O.; Kjeldgaard, N.O. *The Control of Macromolecular Synthesis*; W.A. Benjamin Inc.: New York, NY, USA, 1966.

30. Fishov, I.; Zaritsky, A.; Grover, N.B. On microbial states of growth. *Mol. Microbiol.* **1995**, *15*, 789–794. [CrossRef] [PubMed]

31. Kellenberger, E.; Arber, W. Electron microscopical studies of phage multiplication. I. A method for quantitative analysis of particle suspensions. *Virology* **1957**, *3*, 245–255. [CrossRef] [PubMed]

32. Woldringh, C.L. Morphological analysis of nuclear separation and cell division during the life cycle of *Escherichia coli*. *J. Bacteriol.* **1976**, *125*, 248–257. [CrossRef]

33. Harvey, R.J.; Marr, A.G.; Painter, P.R. Kinetics of growth of individual cells of *Escherichia coli* and *Azotobacter agilis*. *J. Bacteriol.* **1967**, *93*, 605–617. [CrossRef]

34. Koppes, L.J.H.; Nanninga, N. Positive correlation between size at initiation of chromosome replication in *Escherichia coli* and size at initiation of cell constriction. *J. Bacteriol.* **1980**, *143*, 89–99.507. [CrossRef]

35. Zaritsky, A. On dimensional determination of rod-shaped bacteria. *J. Theor. Biol.* **1975**, *54*, 243–248. [CrossRef]

36. Trueba, F.J.; Woldringh, C.L. Changes in cell diameter during the division cycle of *Escherichia coli*. *J. Bacteriol.* **1980**, *142*, 869–878. [CrossRef]

37. Vischer, N.O.E.; Huls, P.G.; Woldringh, C.L. Object-image: An interactive image analysis program using structured point collection. *Binary* **1994**, *6*, 160–166.

38. Vischer, N.O.E.; Verheul, J.; Postma, M.; van den Berg van Saparoea, B.; Galli, E.; Natale, P.; Gerdes, K.; Luirink, J.; Vollmer, W.; Vicente, M.; et al. Cell age dependent concentration of *Echerichia coli* divisome proteins analyzed with ImageJ and objectJ. *Front. Microbiol.* **2015**, *6*, 586. [CrossRef]

39. Zaritsky, A.; Woldringh, C.L. Chromosome replication, cell growth, division and shape: A personal perspective. *Front. Microbiol.* **2015**, *6*, 756. [CrossRef]

40. Cooper, S.; Helmstetter, C.E. Chromosome replication and the division cycle of *Escherichia coli* B/r. *J. Mol. Biol.* **1968**, *31*, 519–540. [CrossRef]

41. Helmstetter, C.E.; Cooper, S. DNA synthesis during the division cycle of rapidly growing *Escherichia coli* Br. *J. Mol. Biol.* **1968**, *31*, 507–518. [CrossRef]

42. Nanninga, N. Cytokinesis in prokaryotes and eukaryotes: Common principles and different solutions. *Microbiol. Mol. Biol. Rev.* **2001**, *65*, 319–333. [CrossRef]

43. Weidel, W.; Pelzer, H. Bagshaped macromolecules—A new outlook on bacterial cell walls. *Adv. Enzymol.* **1964**, *26*, 193–232.

44. Nanninga, N. Morphogenesis of *Escherichia coli*. *Microbiol. Mol. Biol. Rev.* **1998**, *62*, 110–129. [CrossRef]

45. Spratt, B.G. Distinct penicillin binding proteins involved in the division, elongation, and shape of *Escherichia coli* K12. *Proc. Natl. Acad. Sci. USA* **1975**, *72*, 2999–3003. [CrossRef]

46. Van Heijenoort, J. Murein synthesis. In *Escherichia coli and Salmonella: Cellular and Molecular Biology*, 2nd ed.; Neidhardt, F.C., Curtiss, R., III, Ingraham, J.L., Lin, E.C.C., Low, K.B., Magasanik, B., Reznikoff, W.S., Riley, M., Schaechter, M., Umbarger, H.E., Eds.; ASM: Washington, DC, USA, 1996; p. 1025.

47. Höltje, J.V. Growth of the stress-bearing and shape-maintaining murein sacculus of *Escherichia coli*. *Microbiol. Mol. Biol. Rev.* **1998**, *62*, 181–203. [CrossRef]

48. Egan, A.J.F.; Errington, J.; Vollmer, W. Regulation of peptidoglycan synthesis and remodelling. *Nat. Rev. Microbiol.* **2020**, *18*, 446–460. [CrossRef]

49. Rothfield, L.I.; Garcia-Lara, J. Regulation of *E. coli* cell division. In *Regulation of Gene Expression in Escherichia coli*; Lin, E.C.C., Lynch, S., Eds.; R.G. Landes: New York, NY, USA, 1996; pp. 547–569.

50. Verwer, R.W.; Nanninga, N.; Keck, W.; Schwarz, U. Arrangement of glycan chains in the sacculus of *Escherichia coli*. *J. Bacteriol.* **1978**, *136*, 723–736. [CrossRef]

51. Wientjes, F.B.; Woldringh, C.L.; Nanninga, N. Amount of peptidoglycan in cell walls of gram-negative bacteria. *J. Bacteriol.* **1991**, *173*, 7684–7691. [CrossRef]

52. Koch, A.L. Shrinkage of growing *Escherichia coli* cells by osmotic challenge. *J. Bacteriol.* **1984**, *159*, 919–924. [CrossRef]

53. Schwarz, U.; Ryter, A.; Rambach, A.; Hellio, R.; Hirota, Y. Process of cellular division in *Escherichia coli*: Differentiation of growth zones in the sacculus. *J. Mol. Biol.* **1975**, *98*, 749–759. [CrossRef]

54. Verwer, R.W.; Nanninga, N. Pattern of *meso*-dl-2,6-diaminopimelic acid incorporation during the division cycle of *Escherichia coli*. *J. Bacteriol.* **1980**, *144*, 327–336. [CrossRef]

55. Burman, L.G.; Raichler, J.; Park, J.T. Evidence for diffuse growth of the cylindrical portion of the *Escherichia coli* murein sacculus. *J. Bacteriol.* **1983**, *155*, 983–988. [CrossRef]

56. Woldringh, C.L.; Huls, P.; Pas, E.; Brakenhoff, G.J.; Nanninga, N. Topography of peptidoglycan synthesis during elongation and polar cap formation in a cell division mutant of *Escherichia coli* MC4100. *J. Gen. Microbiol.* **1987**, *133*, 575–586. [CrossRef]

57. Wientjes, F.B.; Nanninga, N. Rate and topography of peptidoglycan synthesis during cell division in *Escherichia coli*: Concept of a leading edge. *J. Bacteriol.* **1989**, *171*, 3412–3419. [CrossRef]

58. Woldringh, C.L. The bacterial nucleoid: From electron microscopy to polymer physics-a personal recollection. *eLife* **2023**, *13*, 895. [CrossRef]

59. Vos-Scheperkeuter, G.H.; Pas, E.; Brakenhoff, G.J.; Nanninga, N.; Witholt, B. Topography of the insertion of LamB protein into the outer membrane of *Escherichia coli* wild-type and *lac-lamB* cells. *J. Bacteriol.* **1984**, *159*, 440–447. [CrossRef]

60. Hiemstra, H.; Nanninga, N.; Woldringh, C.L.; Inouye, M.; Witholt, B. Distribution of newly synthesized lipoprotein over the outer membrane and the peptidoglycan sacculus of an *Escherichia coli lac-lpp* strain. *J. Bacteriol.* **1987**, *169*, 5434–5444. [CrossRef]

61. Braun, V.; Rehn, K. Chemical characterization, spatial distribution and function of a lipoprotein (murein-lipoprotein) of the *E. coli* cell wall. The specific effect of trypsin on the membrane structure. *Eur. J. Biochem.* **1969**, *10*, 426–438. [CrossRef]

62. Gibbs, K.A.; Isaac, D.D.; Xu, J.; Hendrix, R.W.; Silhavy, T.J.; Theriot, J.A. Complex spatial distribution and dynamics of an abundant *Escherichia coli* outer membrane protein, LamB. *Mol. Microbiol.* **2004**, *53*, 1771–1783. [CrossRef]

63. Sheng, Q.; Zhang, M.Y.; Liu, S.M.; Chen, Z.W.; Yang, P.L.; Zhang, H.S.; Liu, M.-Y.; Li, K.; Zhao, L.-S.; Liu, N.-H.; et al. In situ visualization of Braun's lipoprotein on *E. coli* sacculi. *Sc. Adv.* **2023**, *9*, eadd8659. [CrossRef]

64. Helmstetter, C.E.; Cummings, D.J. Bacterial synchronization by selection of cells at division. *Proc. Natl. Acad. Sci. USA* **1964**, *50*, 767–774. [CrossRef]

65. Helmstetter, C.E. Fifty-five years of research on B, C and D in *Escherichia coli*. *eLife* **2023**, *13*, 977. [CrossRef]

66. Figdor, C.G.; Olijhoek, A.J.M.; Klencke, S.; Nanninga, N.; Bont, W.S. Isolation of small cells from an exponential growing culture of *Escherichia coli* by centrifugal elutriation. *FEMS Microbiol. Lett.* **1981**, *10*, 349–352. [CrossRef]

67. Creanor, J.; Mitchison, J.M. Patterns of protein synthesis during the cell cycle of the fission yeast *Schizosaccharomyces pombe*. *J. Cell Sci.* **1982**, *58*, 263–285. [CrossRef]

68. MacAlister, T.J.; Cook, W.R.; Weigand, R.; Rothfield, L.I. Membrane-murein attachment at the leading edge of the division septum: A second membrane-murein structure associated with morphogenesis of the gram-negative bacterial division septum. *J. Bacteriol.* **1987**, *169*, 3945–3951. [CrossRef]

69. Driehuis, F.; Wouters, J.T.M. Effect of growth rate and cell shape on the peptidoglycan composition of *Escherichia coli*. *J. Bacteriol.* **1987**, *169*, 97–101. [CrossRef] [PubMed]

70. de Jonge, B.L.M.; Wientjes, F.B.; Jurida, I.; Driehuis, F.; Wouters, J.T.M.; Nanninga, N. Peptidoglycan synthesis during the cell cycle of *Escherichia coli*: Composition and mode of insertion. *J. Bacteriol.* **1989**, *171*, 5783–5794. [CrossRef] [PubMed]

71. Cooper, S. Relationship between the acceptor/donor radioactivity ratio and cross-linking in bacterial peptidoglycan: Application to surface synthesis during the division cycle. *J. Bacteriol.* **1990**, *172*, 5506–5510. [CrossRef]

72. Driehuis, F.; de Jonge, B.; Nanninga, N. Cross-linkage and cross-linking of peptidoglycan in *Escherichia coli*: Definition, determination, and implications. *J. Bacteriol.* **1991**, *174*, 2028–2031. [CrossRef]

73. Sauvage, E.; Terrak, M. Glycosyltransferases and transpeptidases/penicillin-binding proteins: Valuable targets for new antibacterials. *Antibiotics* **2016**, *5*, 12. [CrossRef]

74. Wientjes, F.B.; Olijhoek, T.J.; Schwarz, U.; Nanninga, N. Labeling pattern of major penicillin-binding proteins of *Escherichia coli* during the division cycle. *J. Bacteriol.* **1983**, *153*, 1287–1293. [CrossRef]

75. Wientjes, F.B.; Nanninga, N. On the role of the high molecular weight penicillin-binding proteins in the cell cycle of *Escherichia coli*. *Res. Microbiol.* **1991**, *142*, 333–344. [CrossRef] [PubMed]

76. Bi, E.F.; Lutkenhaus, J. FtsZ ring structure associated with division in *Escherichia coli*. *Nature* **1991**, *354*, 161–164. [CrossRef] [PubMed]

77. Voskuil, J.L.; Westerbeek, C.A.; Wu, C.; Kolk, A.H.; Nanninga, N. Epitope mapping of *Escherichia coli* division protein FtsZ with monoclonal antibodies. *J. Bacteriol.* **1994**, *176*, 1886–1893. [CrossRef] [PubMed]

78. Koppelman, C.M.; Aarsman, M.E.G.; Postmus, J.; Pas, E.; Muijsers, O.; Scheffers, D.-J.; Nanninga, N.; den Blaauwen, T. R174 of *Escherichia coli* FtsZ is involved in membrane interaction and protofilament bundling, and is essential for cell division. *Mol. Microbiol.* **2004**, *51*, 645–657. [CrossRef]

79. Den Blaauwen, T.; Wientjes, F.B.; Kolk, A.H.; Spratt, B.G.; Nanninga, N. Preparation and characterization of monoclonal antibodies against native membrane-bound penicillin-binding protein 1B of *Escherichia coli*. *J. Bacteriol.* **1989**, *171*, 1394–1401. [CrossRef]

80. Den Blaauwen, T.; Nanninga, N. Topology of penicillin-binding protein 1b of *Escherichia coli* and topography of four antigenic determinants studied by immunocolabeling electron microscopy. *J. Bacteriol.* **1990**, *172*, 71–79. [CrossRef] [PubMed]

81. Den Blaauwen, T.; Pas, E.; Edelman, A.; Spratt, B.G.; Nanninga, N. Mapping of conformational epitopes of monoclonal antibodies against *Escherichia coli* penicillin-binding protein 1B (PBP1B) by means of hybrid protein analysis: Implication for the tertiary structure of PBP1B. *J. Bacteriol.* **1990**, *172*, 7284–7288. [CrossRef]

82. Zijderveld, C.A.; Aarsman, M.E.; den Blaauwen, T.; Nanninga, N. Penicillin-binding protein 1B of *Escherichia coli* exists in dimeric forms. *J. Bacteriol.* **1991**, *173*, 5740–5746. [CrossRef]

83. Zijderveld, C.A.; Aarsman, M.E.; Nanninga, N. Differences between inner membrane and peptidoglycan-associated PBP1B dimers of *Escherichia coli*. *J. Bacteriol.* **1995**, *177*, 1860–1863. [CrossRef]

84. Zijderveld, C.A.; Waisfisz, Q.; Aarsman, M.E.; Nanninga, N. Hybrid proteins of the transglycosylase and transpeptidase domains of PBP1B and PBP3 of *Escherichia coli*. *J. Bacteriol.* **1995**, *177*, 6290–6293. [CrossRef] [PubMed]

85. Buddelmeijer, N.; Aarsman, M.E.; Kolk, A.H.; Vicente, M.; Nanninga, N. Localization of cell division protein FtsQ by immunofluorescence microscopy in dividing and nondividing cells of *Escherichia coli*. *J. Bacteriol.* **1998**, *180*, 6107–6116. [CrossRef]

86. Buddelmeijer, N.; Beckwith, J. A complex of the *Escherichia coli* cell division proteins FtsL, FtsB and FtsQ forms independently of its localization to the septal region. *Mol. Microbiol.* **2004**, *52*, 1315–1327. [CrossRef]

87. Van den Ent, F.; Vinkenvleugel, T.M.F.; Ind, A.; West, P.; Veprintsev, D.; Nanninga, N.; den Blaauwen, T.; Löwe, J. Structural and mutational analysis of the cell division protein FtsQ. *Mol. Microbiol.* **2008**, *68*, 110–123. [CrossRef]

88. Roos, M.; van Geel, A.B.; Aarsman, M.E.; Veuskens, J.T.; Woldringh, C.L.; Nanninga, N. The replicated *ftsQAZ* and *minB* chromosomal regions of *Escherichia coli* segregate on average in line with nucleoid movement. *Mol. Microbiol.* **2001**, *39*, 633–640. [CrossRef] [PubMed]

89. Wijsman, H.J.W. A genetic map of several mutations affecting the mucopeptide layer of *Escherichia coli*. *Genet. Res.* **1972**, *20*, 65–74. [CrossRef]

90. Ayala, J.A.; Garrido, T.; De Pedro, M.A.; Vicente, M. Chapter 5 Molecular biology of bacterial septation. *New Compr. Biochem.* **1994**, *27*, 73–101.

91. Vicente, M.; Gomez, M.J.; Ayala, J.A. Regulation of transcription of cell division genes in the *Escherichia coli* dcw cluster. *Cell. Mol. Life Sci.* **1998**, *54*, 317–324. [CrossRef] [PubMed]

92. Norris, V.; Madsen, M.S. Autocatalytic gene expression occurs via transertion and membrane domain formation and underlies differentiation in bacteria: A model. *J. Mol. Biol.* **1995**, *253*, 739–748. [CrossRef]

93. Woldringh, C.L. The role of co-transcriptional translation and protein translocation (transertion) in bacterial chromosome segregation. *Mol. Microbiol.* **2002**, *45*, 17–29. [CrossRef]

94. Gordon, G.S.; Sitnikov, D.; Webb, A.; Teleman, A.; Straight, R.; Losick, R.; Murray, A.W.; Wright, A. Chromosome and low copy plasmid segregation in *E. coli*: Visual evidence for distinct mechanisms. *Cell* **1997**, *90*, 1113–1121. [CrossRef]

95. Webb, C.D.; Teleman, A.; Gordon, S.; Straight, A.; Belmont, A.; Lin, D.C.-H.; Grossman, A.D.; Wright, A.; Losick, R. Bipolar localization of the replication origin regions of chromosomes in vegetative and sporulating cells of *B. subtilis*. *Cell* **1997**, *88*, 667–674. [CrossRef]

96. Dingman, C.W. Bidirectional chromosome replication: Some topological considerations. *J. Theor. Biol.* **1974**, *43*, 187–195. [CrossRef]

97. Koppes, L.J.; Woldringh, C.L.; Nanninga, N. *Escherichia coli* contains a DNA replication compartment in the cell center. *Biochimie* **1999**, *81*, 803–810. [CrossRef]

98. Lemon, K.P.; Grossman, A.D. The extrusion-capture model for chromosome partitioning in bacteria. *Genes. Develop.* **2001**, *15*, 2031–2041. [CrossRef]

99. Den Blaauwen, T.; Aarsman, M.E.G.; Vischer, N.O.E.; Nanninga, N. Penicillin-binding protein PBP2 of *Escherichia coli* localizes preferentially in the lateral wall and at midcell in comparison with the old cell pole. *Mol. Microbiol.* **2003**, *47*, 539–547. [CrossRef]

100. Jones, L.J.; Carballido-López, R.; Errington, G. Control of cell shape in bacteria: Helical, actin-like filaments in *Bacillus subtilis*. *Cell* **2001**, *104*, 913–922. [CrossRef] [PubMed]

101. Shih, Y.-L.; Le, T.; Rothfield, L. Division site selection in *Escherichia coli* involves dynamic redistribution of Min proteins within coiled structures that extend between the two cell poles. *Proc. Natl. Acad. Sci. USA* **2003**, *100*, 7865–7870. [CrossRef]

102. Wachi, M.; Doi, M.; Tamaki, S.; Park, W.; Nakajima-Lijima, S.; Matsuhashi, M. Mutant isolation and molecular cloning of *mre* genes, which determine cell shape, sensitivity to mecillinam, and amount of penicillin-binding proteins in *Escherichia coli*. *J. Bacteriol.* **1987**, *169*, 4935–4940. [CrossRef] [PubMed]

103. Kruse, T.; Gerdes, K. Bacterial DNA segregation by the actin-like MreB protein. *Trends Cell Biol.* **2005**, *15*, 343–345. [CrossRef] [PubMed]

104. Stoker, N.G.; Pratt, J.M.; Spratt, B.G. Identification of the RodA gene product of *Escherichia coli*. *J. Bacteriol.* **1983**, *155*, 854–859. [CrossRef]

105. Rohs, P.D.A.; Buss, J.; Sim, S.I.; Squires, G.R.; Srisuknimit, V.; Smith, M.; Cho, H.; Sjodt, M.; Kruse, A.C.; Gamer, E.C.; et al. A central role for PBP2 in the activation peptidoglycan polymerization by the bacterial cell elongation machinery. *PLoS Genet.* **2018**, *14*, e1007726. [CrossRef]

106. Van Teeffelen, S.; Wang, S.; Furchtgott, L.; Huang, K.C.; Wingreen, N.s.; Shaevitz, J.W.; Gitai, Z. The bacterial actin MreB rotates, and depends on cell-wall assembly. *Proc. Natl. Acad. Sci. USA* **2011**, *108*, 15822–15827. [CrossRef]

107. Mohammadi, T.; Karczmarek, A.; Crouvoisier, M.; Bouhss, A.; Mengin-lecreulx, D.; Den Blaauwen, T. The essential peptidoglycan glycosyltransferase MurG forms a complex with proteins involved in lateral envelope growth as well as with proteins involved in cell division in *Escherichia coli*. *Mol. Microbiol.* **2007**, *65*, 1106–1121. [CrossRef]

108. Mohammadi, T.; Sijbrand, R.; Lutters, M.; Verheul, J.; Martin, N.I.; den Blaauwen, T.; de Kruijff, B.; Breukink, E. Specificity of the transport of lipid II by FtsW in *Escherichia coli*. *J. Biol. Chem.* **2014**, *289*, 14707–14718. [CrossRef]

109. Liu, X.; Biboy, J.; Consoli, E.; Vollmer, W.; den Blaauwen, T. MreC and MreD balance the interaction between the elongasome proteins PBP2 and RodA. *PLoS Genet.* **2020**, *16*, e1009276. [CrossRef]

110. Shiomi, D.; Sakai, M.; Niki, H. Determination of bacterial rod shape by a novel cytoskeletal membrane protein. *EMBO J.* **2008**, *27*, 3081–3091. [CrossRef]

111. Nanninga, N. Cell division and peptidoglycan assembly in *Escherichia coli*. *Mol. Microbiol.* **1991**, *5*, 791–795. [CrossRef]

112. de Pedro, M.A.; Quintela, J.C.; Höltje, J.V.; Schwarz, H. Murein segregation in *Escherichia coli*. *J. Bacteriol.* **1997**, *179*, 2823–2834, 7284–7288. [CrossRef]

113. Pazos, M.; Peters, K.; Casanova, M.; Palacios, P.; VanNieuwenhze, M.; Breukink, E.; Vicente, M.; Vollmer, W. Z-ring membrane anchors associate with cell wall synthases to initiate bacterial cell division. *Nat. Commun.* **2018**, *10*, 591. [CrossRef]

114. Potluri, L.-P.; Kannan, S.; Young, K.D. ZipA is required for FtsZ-dependent preseptal peptidoglycan synthesis prior to invagination during cell division. *J. Bacteriol.* **2012**, *194*, 5334–5342. [CrossRef]

115. Den Blaauwen, T.; Buddelmeijer, N.; Aarsman, M.E.; Hameete, C.M.; Nanninga, N. Timing of FtsZ assembly in *Escherichia coli*. *J. Bacteriol.* **1999**, *181*, 5167–5175. [CrossRef]

116. Taschner, P.E.; Ypenburg, N.; Spratt, B.G.; Woldringh, C.L. An amino acid substitution in penicillin-binding protein 3 creates pointed polar caps in *Escherichia coli*. *J. Bacteriol.* **1988**, *170*, 4828–4837. [CrossRef]

117. Aarsman, M.E.; Piette, A.; Fraipont, C.; Vinkenvleugel, T.M.; Nguyen-Disteche, M.; den Blaauwen, T. Maturation of the *Escherichia coli* divisome occurs in two steps. *Mol. Microbiol.* **2005**, *55*, 1631–1645. [CrossRef]

118. Kashämmer, L.; van den Ent, F.; Jeffery, M.; Jean, N.L.; Hale, V.L.; Löwe, J. Cryo-EM structure of the bacterial divisome core complex and antibiotic target FtsWIQBL. *Nat. Microbiol.* **2023**, *8*, 1149–1159. [CrossRef]

119. Fenton, A.K.; Gerdes, K. Direct interaction of FtsZ and MreB is required for septum synthesis and cell division in *Escherichia coli*. *EMBO. J.* **2013**, *32*, 1953–1965. [CrossRef]

120. Banzhaf, M.; van den Berg van Saparoea, B.; Terrak, M.; Fraipont, C.; Egan, A.; Philippe, J.; Breukink, J.; Nguyen-Disteche, M.; den Blaauwen, T.; Vollmer, W. Cooperativity of peptidoglycan synthases active in bacterial cell elongation. *Mol. Microbiol.* **2012**, *85*, 179–194. [CrossRef]

121. Bertsche, U.; Kast, T.; Wolf, B.; Fraipont, C.; Aarsman, M.E.; Kannenberg, K.; von Rechenberg, M.; Nguyen-Distèche, M.; den Blaauwen, T.; Höltje, J.V.; et al. Interaction between two murein (peptidoglycan) synthases, PBP3 and PBP1B, in *Escherichia coli*. *Mol. Microbiol.* **2006**, *61*, 675–690. [CrossRef]

122. Ranjit, D.K.; Jorgenson, M.A.; Young, K.D. PBP1B glycosyltransferase and transpeptidase activities play different essential roles during the *de novo* regeneration of rod morphology in *Escherichia coli*. *J. Bacteriol.* **2017**, *199*, e00612-16. [CrossRef]
123. Boes, A.; Olatunji, S.; Breukink, E.; Terrak, M. Regulation of the peptidoglycan polymerase activity of PBP1b by antagonist actions of the core divisome proteins FtsBLQ and FtsN. *mBio* **2019**, *10*, 1912–1918. [CrossRef]
124. Ma, X.; Ehrhardt, D.W.; Margolin, W. Colocalization of cell division proteins FtsZ and FtsA to cytoskeletal structures in living *Escherichia coli* cells by using green fluorescent protein. *Proc. Natl. Acad. Sci. USA* **1996**, *93*, 12998–30003. [CrossRef]
125. Ma, X.; Sun, Q.; Wang, R.; Singh, G.; Jonietz, E.L.; Margolin, M. Interactions between heterologues FtsA and FtsZ proteins at the FtsZ ring. *J. Bacteriol.* **1997**, *179*, 6788–6797. [CrossRef]
126. Wang, X.; Huang, J.; Mukherjee, A.; Cao, C.; Lutkenhaus, J. Analysis of the interaction of FtsZ with itself, GTP, and FtsA. *J. Bacteriol.* **1997**, *179*, 5551–5559. [CrossRef] [PubMed]
127. Pichoff, S.; Lutkenhaus, J. Unique and overlapping roles for ZipA and FtsA in septal ring assembly in *Escherichia coli*. *EMBO J.* **2002**, *21*, 685–693. [CrossRef] [PubMed]
128. Hale, C.A.; de Boer, P.A.J. Recruitment of ZipA to the septal ring of *Escherichia coli* is dependent on FtsZ and independent of FtsA. *J. Bacteriol.* **1999**, *181*, 167–176. [CrossRef] [PubMed]
129. Schiffer, G.; Höltje, J.-V. Cloning and characterization of PBP1C, a third member of the multimodular class A penicillin-binding proteins of *Escherichia coli*. *J. Biol. Chem.* **1999**, *274*, 32031–32039. [CrossRef]
130. Di Berardino, M.; Dijkstra, A.; Stüber, D.; Keck, W.; Gubler, M. The monofunctional glycosyltransferase of *Escherichia coli* is a member of a new class of peptidoglycan-synthesizing enzymes. *FEBS Lett.* **1996**, *392*, 184–188. [CrossRef]
131. Magnet, S.; Bellais, S.; Dubost, L.; Fourgeaud, M.; Mainardi, J.L.; Petit-Frere, S.; Marie, A.; Mengin-Lecreulx, D.; Arthur, M.; Gutmann, L. Identification of the L,D-transpeptidases responsible for attachment of the Braun lipoprotein to *Escherichia coli* peptidoglycan. *J. Bacteriol.* **2007**, *189*, 3927–3931. [CrossRef]
132. Magnet, S.; Dubost, L.; Marie, A.; Gutmann, L. Identification of the L,D-transpeptidases for peptidoglycan cross-linking in *Escherichia coli*. *J. Bacteriol.* **2008**, *190*, 4782–4785. [CrossRef]
133. Vollmer, W.; Bertsche, U. Murein (peptidoglycan structure, architecture and biosynthesis in *Escherichia coli*. *Biochim. Biophys. Acta* **2008**, *1778*, 1714–1734. [CrossRef]
134. Navarro, P.P.; Vettiger, A.; Ananda, V.Y.; Montero Llopis, P.; Allolio, C.; Bernhardt, T.G.; Chao, L.H. Cell wall synthesis and remodelling dynamics determine division site architecture and cell shape in *Escherichia coli*. *Nat. Microbiol.* **2022**, *7*, 1621–1634. [CrossRef]
135. Söderström, B.; Skoog, K.; Blom, H.; Weiss, D.S.; von Heijne, G.; Daley, D.O. Disassembly of the divisome in *Escherichia coli*: Evidence that FtsZ dissociates before compartmentalization. *Mol. Microbiol.* **2014**, *92*, 1–9. [CrossRef]
136. Buddelmeijer, N.; Beckwith, J. Assembly of cell division proteins at the E. coli cell center. *Curr. Opin. Microbiol.* **2002**, *5*, 553–557. [CrossRef]
137. Taguchi, A.; Welsh, M.A.; Marmont, L.S.; Lee, W.; Sjodt, M.; Kruse, A.C.; Kahne, D.; Bernhardt, T.G.; Walker, S. FtsW is a peptidoglycan polymerase that is functional only in complex with its cognate penicillin-binding protein. *Nat. Microbiol.* **2019**, *4*, 587–594. [CrossRef]
138. Fraipont, C.; Alexaeeva, S.; Wolf, B.; Van der Ploeg, R.; Schloesser, M.; den Blaauwen, T.; Nguyen-Distèche, M. The integral membrane FtsW protein and peptidoglycan synthase form a subcomplex in *Escherichia coli*. *Microbiology* **2011**, *157*, 251–259. [CrossRef] [PubMed]
139. Ishino, F.; Park, W.; Tomioka, S.; Tamaki, S.; Takase, I.; Kunugita, K.; Matsuzawa, H.; Asoh, S.; Ohta, T.; Spratt, B.G.; et al. Peptidoglycan synthetic activities in membranes of Escherichia coli caused by overproduction of penicillin-binding protein 2 and rodA protein. *J. Biol. Chem.* **1986**, *261*, 7024–7031. [CrossRef] [PubMed]
140. Typas, A.; Banzhaf, M.; van den Berg van Saparoea, B.; Verheul, J.; Biboy, J.; Nichols, R.J.; Zietek, M.; Beilharz, K.; Kannenberg, K.; von Rechenberg, M.; et al. Regulation of peptidoglycan synthesis by outer-membrane proteins. *Cell* **2010**, *143*, 1097–1109. [CrossRef] [PubMed]
141. Van der Ploeg, R.; Verheul, J.; Vischer, N.O.E.; Alexeeva, S.; Hoogendoorn, E.; Postma, M.; Banzhaf, M.; Vollmer, W.; den Blaauwen, T. Colocalization and interaction between elongasome and divisome during a preparative cell division phase in *Escherichia coli*. *Mol. Microbiol.* **2013**, *87*, 1074–1087. [CrossRef]
142. Mulder, E.; Woldringh, C.L. Actively replicating nucleoids influence the positioning of division sites in DNA-less cell forming filaments of *Escherichia coli*. *J. Bacteriol.* **1989**, *171*, 4303–4314. [CrossRef]
143. Bernhardt, T.G.; de Boer, P.A.J. SlmA, a nucleoid-associated, FtsZ binding protein required for blocking septal ring assembly over chromosomes in *E. coli*. *Mol. Cell.* **2005**, *18*, 555–564. [CrossRef]
144. Lutkenhaus, J. Assembly dynamics of the bacterial MinCDE system and spatial regulation of the Z ring. *Annu. Rev. Biochem.* **2007**, *76*, 539–562. [CrossRef] [PubMed]
145. Männik, J.; Bailey, M.W. Spatial coordination between chromosomes and cell division proteins in *Escherichia coli*. *Front. Microbiol.* **2015**, *6*, 306. [CrossRef]
146. Sherrat, D.J.; Arciszewska, L.K.; Crozat, E.; Graham, J.E.; Grainge, I. The *Escherichia coli* DNA translocase FtsK. *Biochem. Soc. Trans.* **2010**, *38*, 395–398. [CrossRef]
147. Mukherjee, A.; Lutkenhaus, J. Guanine nucleotide-dependent assembly of FtsZ into filaments. *J. Bacteriol.* **1994**, *176*, 2754–2758. [CrossRef]

148. Osawa, M.; Anderson, D.E.; Erickson, H.P. Reconstitution of contractile FtsZ rings in liposomes. *Science* **2008**, *320*, 792–794. [CrossRef]

149. Osawa, M.; Anderson, D.E.; Erickson, H. P Curved FtsZ protofilaments generate bending forces on liposome membranes. *EMBO J.* **2009**, *28*, 3476–3484. [CrossRef] [PubMed]

150. Ramirez-Diaz, D.A.; Merino-Salomón, A.; Meyer, F.; Heymann, M.; Rivas, G.; Bramkamp, M.; Schwille, P. FtsZ induces membrane deformations via torsional stress upon GTP hydrolysis. *Nat. Commun.* **2021**, *12*, 3310. [CrossRef] [PubMed]

151. Coltharp, C.; Buss, J.; Plumer, T.M.; Xiao, J. Defining the rate-limiting processes of bacterial cytokinesis. *Proc. Natl. Acad. Sci. USA* **2016**, *113*, 1044–1055. [CrossRef]

152. Koppelman, C.M.; Den Blaauwen, T.; Duursma, M.C.; Heeren, R.M.; Nanninga, N. *Escherichia coli* minicell membranes are enriched in cardiolipin. *J. Bacteriol.* **2001**, *183*, 6144–6147. [CrossRef] [PubMed]

153. Li, G.W.; Burkhardt, D.; Gross, C.; Weismann, J.S. Quantifying absolute protein synthesis rates reveals principles underlying allocation of cellular resources. *Cell* **2014**, *157*, 624–635. [CrossRef]

154. Egan, A.J.F.; Vollmer, W. The stoichiometric divisome: A hypothesis. *Front. Microbiol.* **2015**, *6*, 455. [CrossRef]

155. Du, S.; Lutkenhaus, J. At the heart of bacterial cytokinesis: The Z ring. *Trends Microbiol.* **2019**, *27*, 781–791. [CrossRef]

156. Loose, M.; Mitchison, T.J. The bacterial cell division proteins FtsA and FtsZ self-organize into dynamic cytoskeletal patterns. *Nat. Cell Biol.* **2014**, *16*, 38–48. [CrossRef]

157. Bisson-Filho, A.W.; Hsu, Y.-P.; Squyres, G.R.; Kuru, E.; Wu, F.; Jukes, C.; Sun, Y.; Dekker, C.; Holden, S.; VanNieuwenhze, M.S.; et al. Treadmilling by FtsZ filaments drives peptidoglycan synthesis and bacterial cell division. *Science* **2017**, *355*, 739–743. [CrossRef]

158. Söderström, B.; Chan, H.; Shilling, P.J.; Skoglund, U.; Daley, D.O. Spatial separation of FtsZ and FtsN during cell division. *Mol. Microbiol.* **2017**, *107*, 387–401. [CrossRef]

159. Woldringh, C.L.; Nanninga, N. Organization of the nucleoplasm in *Escherichia coli* visualized by phase-contrast light microscopy, freeze-fracturing, and thin sectioning. *J. Bacteriol.* **1976**, *127*, 1455–1464. [CrossRef] [PubMed]

160. Mason, D.J.; Powelson, D.M. Nuclear division as observed in live bacteria by a new technique. *J. Bacteriol* **1956**, *71*, 474–479. [CrossRef]

161. Lemons, R.A.; Quate, C.F. Acoustic microscopy: Biomedical applications. *Science* **1975**, *188*, 905–911. [CrossRef] [PubMed]

162. Brakenhoff, G.J.; Blom, P.; Barends, P.J. Confocal scanning light microscopy with high aperture lenses. *J. Microsc.* **1979**, *117*, 219. [CrossRef]

163. Sheppard, C.J.R.; Choudhury, A. Image formation in the scanning microscope. *Opt. Acta* **1977**, *24*, 1051–1073. [CrossRef]

164. Valkenburg, J.A.C.; Woldringh, C.L.; Brakenhoff, G.J.; van der Voort, H.T.M.; Nanninga, N. Confocal scanning light microscopy of the *Escherichia coli* nucleoid: Comparison with phase-contrast and electron microscope images. *J. Bacteriol* **1985**, *161*, 478–483. [CrossRef]

165. Van der Voort, H.T.M.; Brakenhoff, G.J.; Valkenburg, J.A.C.; Nanninga, N. Design and use of a computer controlled confocal microscope for biological applications. *Scanning* **1985**, *7*, 66–78. [CrossRef]

166. Brakenhoff, G.J.; van der Voort, H.T.M.; van Spronsen, E.A.; Linnemans, W.A.M.; Nanninga, N. Three-dimensional chromatin distribution in neuroblastoma nuclei shown by confocal scanning laser microscopy. *Nature* **1985**, *317*, 748–749. [CrossRef] [PubMed]

167. Van der Voort, H.T.M.; Brakenhoff, G.J.; Baarslag, M.W. Three-dimensional visualization methods for confocal microscopy. *J. Microsc.* **1989**, *153*, 123–132. [CrossRef] [PubMed]

168. Oud, J.L.; Mans, A.; Brakenhoff, G.J.; van der Voort, H.T.M.; van Spronsen, E.A.; Nanninga, N. Three-dimensional chromosome arrangement of *Crepis capillaris* in mitotic prophase and anaphase as studied by confocal laser microscopy. *J. Cell Sci.* **1989**, *92*, 329–339. [CrossRef] [PubMed]

 *life*

Review

Recollections of a Helmstetter Disciple

Alan C. Leonard

Department of Biological Sciences, Florida Institute of Technology, 150 W. University Blvd.,
Melbourne, FL 32952, USA; aleonard@fit.edu

Abstract: Nearly fifty years ago, it became possible to construct *E. coli* minichromosomes using recombinant DNA technology. These very small replicons, comprising the unique replication origin of the chromosome *oriC* coupled to a drug resistance marker, provided new opportunities to study the regulation of bacterial chromosome replication, were key to obtaining the nucleotide sequence information encoded into *oriC* and were essential for the development of a ground-breaking in vitro replication system. However, true authenticity of the minichromosome model system required that they replicate during the cell cycle with chromosome-like timing specificity. I was fortunate enough to have the opportunity to construct *E. coli* minichromosomes in the laboratory of Charles Helmstetter and, for the first time, measure minichromosome cell cycle regulation. In this review, I discuss the evolution of this project along with some additional studies from that time related to the DNA topology and segregation properties of minichromosomes. Despite the significant passage of time, it is clear that large gaps in our understanding of *oriC* regulation still remain. I discuss some specific topics that continue to be worthy of further study.

Keywords: bacterial cell cycle; *oriC*; minichromosomes; DNA supercoiling; chromosome segregation

Citation: Leonard, A.C. Recollections of a Helmstetter Disciple. *Life* **2023**, *13*, 1114. https://doi.org/10.3390/life13051114

Academic Editors: Vic Norris, Arieh Zaritsky and Itzhak Fishov

Received: 31 March 2023
Revised: 25 April 2023
Accepted: 28 April 2023
Published: 30 April 2023

1. Introduction

My meeting with Charles Helmstetter was not planned. As much as I would like to say that it was my life-long dream to work in his lab, nothing could be further from the truth. Our meeting was completely accidental. I arrived at the Roswell Park Cancer Institute in Buffalo, New York wanting to be trained as a cancer research scientist. To me, that meant working with animals, or at the very least, tumor tissue. The idea of spending time in a lab that studied *E. coli* as a model system was unimaginable at the time. With a budding interest in nucleic acids, I sought out a faculty mentor whose laboratory was focused on the study of RNA polymerase activity in leukemic mice. My initial training in the lab was in enzyme purification, specifically the three forms of RNA polymerase from normal and leukemia virus-infected mouse spleens: mostly an experience of learning column chromatography in the cold room.

My limited expertise in the purification and handling of enzymes would turn out to be an important aspect of my introduction to Charles Helmstetter's work and his laboratory group. Another was a fortuitous department reorganization at the institute. The lab I was working in was assigned to a newly organized Experimental Biology department, and while moving into new space was exciting, it also meant dealing with a new boss. The new department head was a well-respected biophysicist known for the exceptional quality of his research, but I knew little about him or his work. A bigger disappointment, at least in my mind, was that he worked on *E. coli*. Why would anyone at a cancer research institute study bacterial cells? With little interest in his research, I only hoped he was a benevolent leader, and I did my best to stay out of his way. It turned out that I was not successful at remaining anonymous for very long. Within the first year, our new head, Charles Helmstetter, came looking for some assistance and my faculty advisor volunteered me to lend a hand. This was not exactly a dream come true for me given my bad attitude, but at the time it seemed politically wise to provide the assistance, and perhaps I could quickly train one of Helmstetter's grad students to replace me.

As initially described to me at a lab meeting with his group, the technical aspects of the project were straightforward: clone the replication origin (*oriC*) from the *E. coli* chromosome and construct an autonomously replicating miniature derivative: a minichromosome. Since studies of *oriC* on the chromosome were limited, particularly because it was an essential region, minichromosomes would be a useful tool. At the time, the location of *oriC* was known with respect to its nearby restriction endonuclease cut sites [1,2], and in theory, the pool of restriction fragments derived from the entire genome could be joined randomly at low concentrations with a non-replicating drug resistance marker isolated from a commonly available plasmid [3]. Only when the marker fragment was joined to *oriC* would it be capable of autonomous replication (see Figure 1).

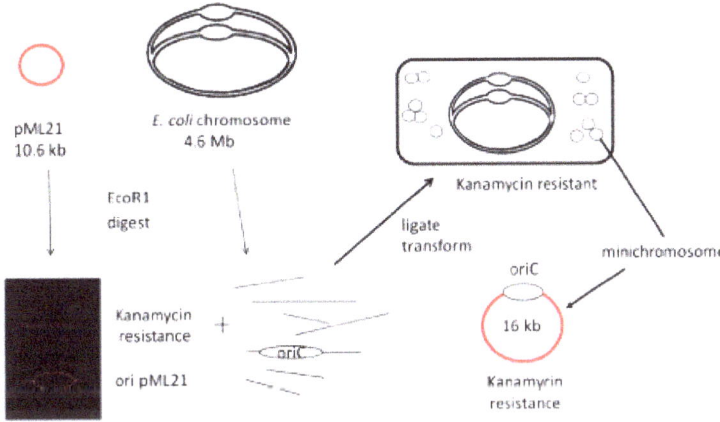

Figure 1. Scheme for construction of minichromosomes. Red color denotes plasmid DNA. Chromosomes are shown in a fast growth configuration used to increase the relative copies of *oriC*. See text for a description of the figure.

Using current technology, this whole project could be completed in about a week, so it may not be obvious why the Helmstetter lab would need any assistance from me. The reader must remember, however, that this work was conducted over forty years ago, prior to the discovery of polymerase chain reaction, and required the brand-new technology of recombinant DNA, using enzymes that at the time were not yet commercially available. The restriction endonucleases and ligases had to be purified from over-producing *E. coli* strains before the project could even be started, and the DNA isolation protocols were not yet established in the lab. I was not intimidated by these roadblocks and remained hopeful that the project could be completed relatively quickly.

I retained my "why would a cancer researcher study *E. coli*" lack of enthusiasm and did not really appreciate why the cloned replication origin was such a prize. Perhaps sensing this, Charles insisted that I meet with him in his office so that he could explain why the project was important. I do not remember all the details of that meeting, nor can I say that I completely understood what he told me. What I do remember was that this was a transformative moment in my career. Charles described his seminal experimental work performed (with Steve Cooper) in the 1960s and explained how these experiments led to the development of the elegant model that describes the bacterial cell cycle: the I + C + D model. Although beyond the scope of this review, this work is beautifully presented in an accessible way in [4], and I urge anyone newly interested in this topic to start with this manuscript. Charles told me the story of the clever technology known as the baby machine, and how "backward" pulse labeling experiments revealed the cell cycle times of G1, S, and G2 (B, C, and D in *E. coli*). He also explained that the bacterial cell cycle was not a cycle at all, but a series of overlapping events; for example see [5,6]. Most impressive to me was that knowing the value of C + D, Charles could tell me exactly when new rounds of chromosome replication would initiate during the cell cycle in cultures growing at any rate. I had never heard anything like this before and the fact that he had worked out the laws for a growing cell (even an *E. coli* cell) was thrilling. I left Charles' office truly enlightened and newly enthusiastic to become part of his lab group. I finally realized why studying *E. coli* was not only appropriate for cancer research, but essential to understand normal cell growth regulation. You could say that I became a true "Helmstetter disciple" from that moment onward.

2. Making *E. coli* Minichromosomes and Cell Cycle Analysis

It turned out that cloning *oriC* was not as simple a task as I had hoped. A big stumbling block was my attempt to purify the non-replicating drug resistant fragment away from the plasmid's own replication origin. This was achieved using gel electrophoresis to separate the two fragments and then purifying out the desired one. It was inefficient and subject to cross contamination. There were no kits available to speed the process, so I failed many times at this step and at the final step of cell transformation with the ligated chromosomal fragments. Unhappily, my repeated failures were noticed by one of Charles' post-docs who, in an effort to be amusing, added a "cloning report" to his weekly tongue-in-cheek newsletter (The Flash) on lab group happenings. While I looked forward to his funny and often clever take on the events of the week, the weekly cloning report of "no progress" was not particularly uplifting (Figure 2).

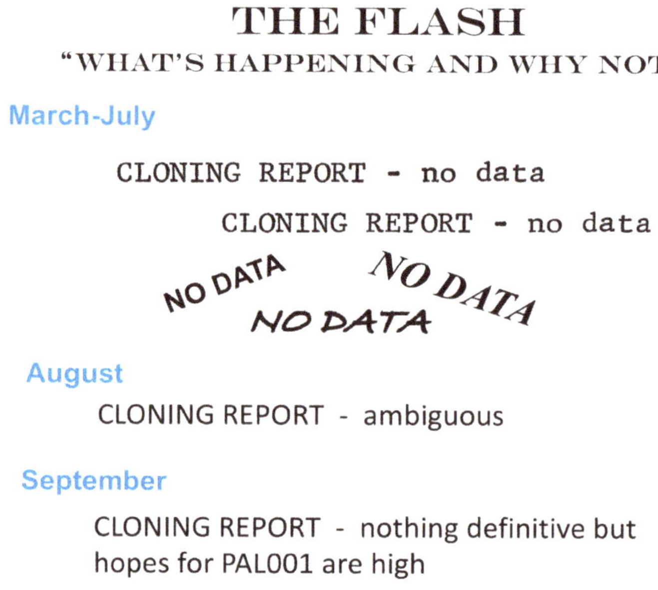

THE FLASH
"WHAT'S HAPPENING AND WHY NOT"

March-July

CLONING REPORT - no data

CLONING REPORT - no data

NO DATA   NO DATA
NO DATA   NO DATA

August

CLONING REPORT - ambiguous

September

CLONING REPORT - nothing definitive but
hopes for PAL001 are high

Figure 2. *OriC* cloning reports from the weekly lab newsletter "The Flash". See text.

When several interesting colonies finally appeared on a transformation plate and were shown to carry a new plasmid of the correct size expected for a minichromosome, I was horrified to observe that these plasmids were highly unstable in my *E. coli* host. This turned out to be due to the carry-over of some genes from the ATP synthetase operon adjacent to *oriC* on the chromosome, but until I was able to construct deletion derivatives, I grew cultures containing my first minichromosome (pAL1) at very high concentrations of antibiotic to kill off the plasmid-less segregants. Using 0.5 mg/mL kanamycin in the media seemed ridiculous at the time, but it allowed me to isolate enough minichromosome DNA for future experiments.

It is important to note that I was not the first to make an *E. coli* minichromosome. That honor went to others in the labs of Yuki Hirota and Walter Messer/Kaspar von Meyenberg, whose ground-breaking work provided the nucleotide sequence of *oriC* and began the difficult process of identifying the protein recognition motifs [7,8]. The enormous efforts of Bob Fuller and Jon Kaguni in Arthur Kornberg's lab then produced an in vitro replication system that was pivotal to dissecting the chromosomal replication machinery [9].

Since our lab was focused on the regulation of DNA replication in living cells, our intent was to measure the replication of minichromosomes during the cell cycle. The obvious question was whether the minimal *oriC* region was sufficient for proper initiation timing during the cell cycle. The answer would not only reveal important features of the regulatory mechanism, but validate the minichromosome model for future studies, particularly those performed using in vitro systems. While we hoped that periodic minichromosome replication was retained, their moderately high copy number (10–20 copies per cell), despite their instability) did not seem compatible with the properly timed, once-per-cycle regulation of chromosomal *oriC*. We anticipated random replication but did not discount the possibility of periodic replication with initiation timed differently than the chromosome.

Rather than working with synchronously growing cells, we based our experimental design on the "backwards" baby machine approach Helmstetter and Cooper had used previously to study chromosome replication [5], since this would minimize artifacts caused by manipulating the cells. Since minichromosomes replicate in a matter of seconds, we believed that a minimally manipulated system would be important in distinguishing cell cycle-specific replication from random replication throughout the cell cycle. In this "backwards" procedure, exponentially growing cultures were pulse-radiolabeled with tritiated thymidine to label any replicating DNA. Then, the entire culture was transferred to a nitrocellulose membrane filter, the unincorporated label was washed out, and baby cells were eluted and collected at 1/10th generation intervals from the dividing cell population on the surface. The eluted cells contained the radiolabel incorporated into minichromosome and chromosomal DNA during the brief pulse label. The next trick was to develop a whole cell lysis protocol that was quantitative and free from the variable recovery artifacts caused when using the multi-step plasmid isolation methods available at the time. By examining many samples eluted over multiple generations of growth, we felt there was a good chance of obtaining a truthful result.

For every cell cycle sample collected over multiple generations of growth, chromosomes and minichromosomes were separated on agarose gels, which were processed, dried, and placed against X-ray film for an extended period of time. I then took the exposed films to a free-standing X-ray developing machine in one of the nearby clinics. The machine would emit many strange noises before the developed film would plop out into a plastic receptacle. I still recall our apprehension as Charles and I would stand there waiting to see each film emerge from the machine. These were often less than beautiful, with missing samples or streaky lanes, but these failures were completely forgotten when the films began to clearly show that minichromosomes were not only cell cycle specific replicons (see Figure 3 and [10]), but also initiated coincidently with the chromosome's *oriC* in successive generations of growth and over a wide variety of growth rates; see Figure 4 [11]. I cannot adequately describe our excitement and how much these findings reshaped our future experiments, as well as our thinking about models for cell cycle regulation of *oriC*. It was also gratifying to see how enthusiastically our findings were accepted by our colleagues in the field. Of course, these studies required many trial and error experiments that extended well beyond my time as a graduate student, and most were conducted after I became a legitimate post-doctoral trainee in the Helmstetter lab. I had not only fallen in love with the science, but also with a doctoral student in the Pharmacology department (Julia Grimwade), who eventually became my wife and ultimately co-investigator in our own lab.

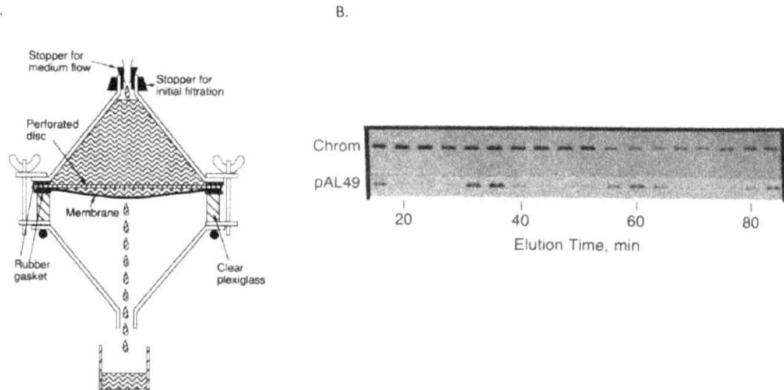

Figure 3. Baby machine analysis of minichromosome cell cycle replication. (**A**). Diagram of baby machine apparatus and sample collection. Cells growing exponentially were labeled with [³H]thymidine for 4 min, bound to a membrane filter, and eluted with glucose/Casamino acids minimal medium. (**B**). Electrophoretic separation of labeled chromosome and pAL49 minichromosome DNA from new daughter cells. Whole-cell lysates of new daughter cells in the effluent were subjected to agarose gel electrophoresis and fluorography. The radioactive bands corresponding to chromosomal and pAL49 DNA are shown for consecutive 4 min samples of the effluent. Exposure times to the x-ray films were 3 h for the chromosomal bands and 10 days for the minichromosome bands. Modified from reference [10].

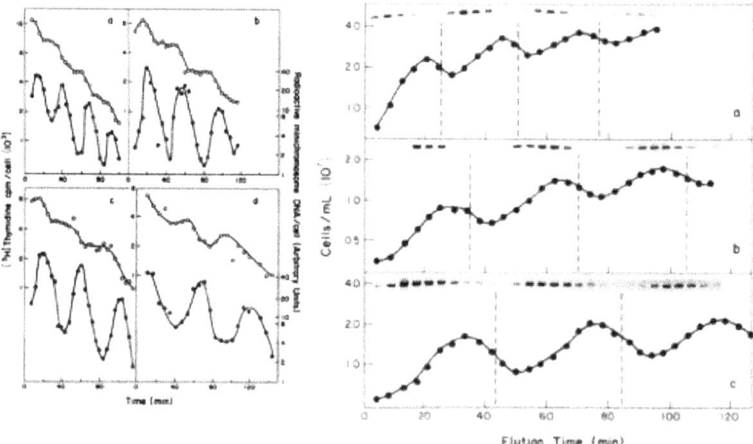

Figure 4. (Left panel) Timing of chromosome and pAL49 minichromosome replication during the division cycle. Exponential-phase cultures of *E. coli* B/r F26(pAL49) growing in glucose plus Casamino Acids (**a**), glucose plus six amino acids (**b**), glucose (**c**), or glycerol (**d**) were pulse-labeled and treated as described in the legend to Figure 3. The radioactivity per cell in minichromosome DNA (closed circles) and total radioactivity per cell (open circles) in newborn cells collected from the effluents of membrane-bound cultures are plotted at the midpoints of the 4 min collection intervals. Abrupt increases in radiolabel (reading right to left) indicate the time of initiation of chromosomal DNA replication. (Right panel). Minichromosome replication during the division cycle of *E. coli* B/r F(pAL49) growing at different rates. Cells growing exponentially in minimal medium containing glucose plus Casamino acids (**a**), glucose plus six amino acids (**b**), or glucose alone (**c**) were pulse-labeled with [³H]thymidine for 4 min, bound to a membrane filter, and eluted with minimal medium

of the same composition. Whole-cell lysates of the newborn cells were treated as in Figure 3. Radioactivity corresponding to closed circular pAL49 minichromosome DNA is shown for consecutive 4 min samples of the effluent at each growth rate. The cell concentrations are also shown, and the vertical interrupted lines indicate the end of each generation of growth on the membrane. Modified from reference [11].

3. Searching for Cell Cycle-Specificity in Plasmid Systems

With the clear cell cycle-specific replication pattern observed for minichromosomes, Charles and I turned our attention to other extrachromosomal replicons in *E. coli*. At the time, plasmid systems were commonly used as surrogates to study chromosome replication, and we were intrigued by the possibility that some plasmids would behave similarly to minichromosomes, particularly the low copy types such as F factors and the many plasmids whose replication origins interact with the chromosomal initiator protein DnaA; some examples are P1 [12], pSC101 [13], R1 [14], and mini F [15]. It seemed obvious that we should use minichromosomes as an internal control for cycle-specific replication, and our studies at the time uniquely included multiple compatible replicons co-inhabiting the same *E. coli* cell. The baby machine procedure to measure minichromosome replication during the cell cycle could be used unaltered, as long as we were careful to use plasmids of sizes that could be resolved from one another on agarose gels. While our efforts were limited to only the most prominent model systems (F, ColE1, pBR322, pSC101, and R1 derivatives), we were unable to identify any plasmid types that showed cell cycle-specific periodicity similar to minichromosomes (for example, see Figure 5, and [16]. Our F plasmid replication data from baby machine experiments was later analyzed using stochastics to reveal that the replication rate function increased monotonically over the cell cycle, with a rapid increase near cell division [17]. The resulting model is consistent with a replication control mechanism that is designed to force most plasmids to replicate before cells undergo division. Extending this model to the case of cell cycle-dependent replication requires additional, as yet unspecified control elements. Later experiments were extended to NR1 and P1 replicons [18]. OriP1 was able to initiate replication at all stages of the cell cycle with a slight periodicity observed in slower growing cells and NR1 plasmid replication was random during the cell cycle.

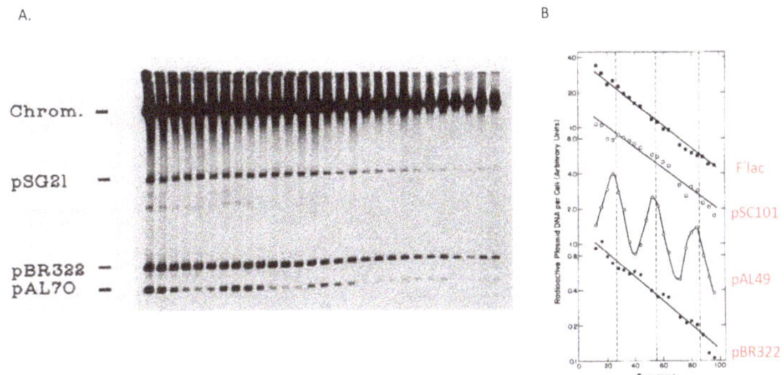

Figure 5. Cell cycle replication of various plasmids. (**A**). Fluorograph of radioactive plasmid DNA in newborn cells from the effluent of a membrane filter-bound culture of *E. coli* B/rF26 containing pSG21 mini-F, pBR322, and pAL70 simultaneously. Cells were grown, pulse-labeled, and prepared as in Figure 3. In this experiment, all lanes contained lysate from the same number of newborn cells. (**B**). Radioactive plasmid DNA in newborn cells from a membrane filter-bound culture of *E. coli* B/r F26 containing F' lac, pSClO1, pAL49, and pBR322 simultaneously. Cells growing exponentially were pulse labeled and treated as described in Figure 3. Modified from [16].

Some controversy about plasmid replication remains unresolved, since later studies from Steve Cooper's lab were consistent with cell cycle specific replication for R6K, P1, F, and mini-F [19–22]. It is not clear why their results contrasted so dramatically with ours. Since all experiments were performed using baby machine selection, any differences in replication patterns must be due to differences in the *E. coli* strains, the post-processing of samples, and/or the assay of radiolabeled DNA.

We were intrigued to find that plasmids whose replication origins were known to interact with DnaA did not use this protein to couple their replication to the cell cycle. For these plasmids, DnaA availability might serve as a monitor of host metabolic activity, acting as either an on or off switch for plasmid replication depending on the plasmid type; for example [23,24]. This function would depend on the free availability of DnaA and not necessarily the ATP-bound state of the initiator. Our inability to identify cell cycle-specific plasmid replicons also raised the question of whether cell-cycle-specific or chromosome-coupled replication is ever beneficial for plasmids. Insights may come from bacteria containing two heterologous chromosomes (see Discussion).

4. A Sidestep into the Role of DNA Supercoiling in Minichromosome Regulation

During my time studying minichromosome replication, rapid advancements were being made in the study of bacterial regulation of DNA supercoiling; for example see reviews [25,26]. These studies intrigued me, and I began discussing DNA supercoiling regulation with Karl Drlica, who at the time was at the University of Rochester, just a short drive away from our lab in Buffalo. There were a number of mutant strains available with defective DNA gyrase (*gyrB* mutants) and topoisomerase 1 (*topA*), and examining the behavior of minichromosomes in increased and decreased supercoiling strains seemed like an interesting way to assess the supercoiling requirements for *oriC* function.

We observed that minichromosomes were sensitive to DNA supercoiling activity and were very unstable in decreased supercoiling strains, in contrast to a variety of other plasmid types whose replication was unperturbed in these strains [27]. We also observed that minichromosomes had significantly lower superhelical density compared to other commonly studied plasmids (Figure 6). However, the stability of minichromosome replication was modulated by the arrangement of active transcriptional promoters on the plasmid (Figure 6).

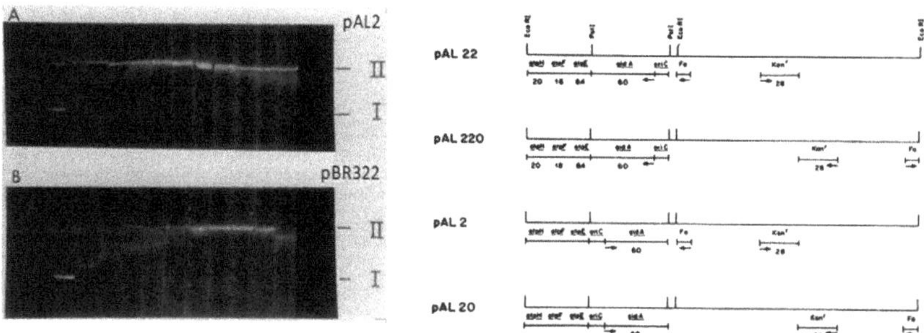

Figure 6. (Left panel) Dye titrations of pAL2 and pBR322 closed circular DNA. pAL2 (**A**) and pBR322 (**B**) DNA was isolated from JTT1 recA grown at 37 °C and electrophoresed in gels containing increasing concentrations of ethidium bromide. pAL2 and pBR322 were electrophoresed through 0.6 and 0.8% agarose, respectively. The concentrations of EtBr (in hundredths of micrograms per milliliter) from left to right are 0, 1, 2, 3, 4, 5, 6, 7, 8, 9, 10 and 11. (I) and (II) are supercoiled and relaxed-nicked circular DNA, respectively. (Right panel) Positions of promoter sequences and the direction of transcription on minichromosomes pAL2, pAL20, pAL22, and pAL220 (indicated by arrows). Only pAL20 and pAL22 were able to replicate in *E. coli* strains (*topA*, *gyrB*) with decreased supercoiling. Modified from [27].

The roles of supercoiling and transcription-driven supercoiling activation of *oriC* were further characterized by others [28–30], and the relationship between transcription, supercoiling, and genome structure remains an active area of study; for example see [31–33]. However, it has yet to be determined whether the molecular requirements for *oriC* function differ under high or low supercoiling conditions in vivo.

5. Studying the Non-Random Segregation of Minichromosomes

E. coli chromosome segregation is nonrandom, despite the presence of an equipartition mechanism that ensures both daughter cells inherit complete genomes. In studies using the "backward" baby machine method, measurement of pulse-radiolabeled chromosomal DNA among progeny cells revealed that label does not segregate with the expected 50–50 distribution, but rather displays a distinctive, growth-media dependent, non-random distribution in successive generations; for more detail see [34–36]. It was difficult to envision a model that explained this mode of segregation, but our best ideas at the time were based on a mechanism determined by cell geometry, with chromosomes behaving as though they were restricted to particular cell locations. The simplest model to impart these restrictions was to envision an intracellular distribution of attachment sites for *oriC* that did not include the existing cell poles (e.g., the poles are dead for chromosome attachment); see Figure 7 and [37] for more detail. Lateral cell envelope growth would provide new attachment sites, but their distribution would remain asymmetric. The non-randomness of segregation would also be dependent on the size of the poles as well as the growth rate, consistent with experimental observations [35,36,38].

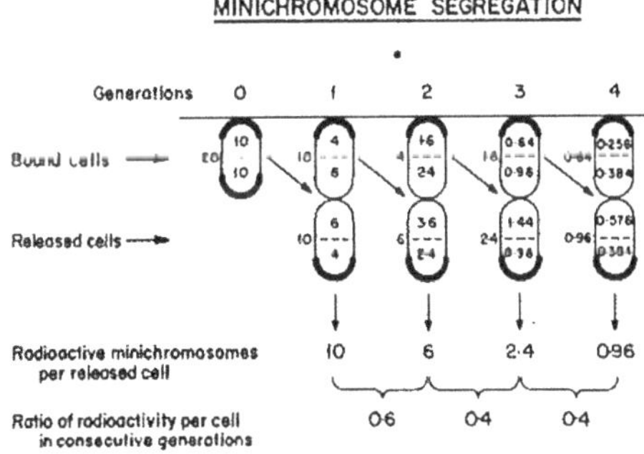

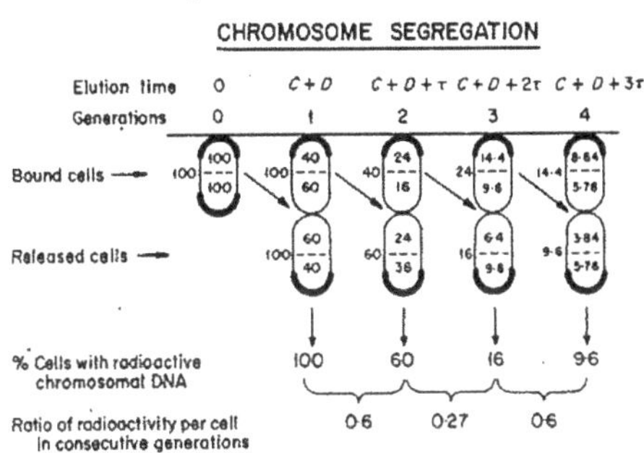

Figure 7. Segregation of minichromosomes and chromosomes during baby machine analysis. Theoretical segregation patterns of radiolabeled DNA are shown for bound (on the horizontal line) and

released cells growing with a generation time of C + D minutes. Four successive generations are shown. Old poles lacking attachment sites are shown by thick lines. (Upper panel) Minichromosome segregation is shown with cells that initially contain 20 copies. Average copy numbers of labeled molecules are shown to the left of each cell and the internal distribution of copies is shown above and below the dotted line representing the division septum. The radiolabeled copies in released cells are also shown. (Lower panel) Chromosome segregation. Radiolabeled chromosomal strands (assuming 2 chromosomes in the initially bound cell) are shown in a similar fashion to panel A. Modified from [37].

Was minichromosomes segregation (despite the lack of an equipartition mechanism) compatible with a model based on *oriC* attachment? Although the ratios of radiolabel released in consecutive generations were not identical to the chromosome, minichromosomes did indeed segregate non-randomly with a distinct pattern that was compatible with the model; see Figure 7 [37,39]. However, evidence for these hypothetical attachment sites for *oriC* remains sparse [40,41]. The activation of initiator protein DnaA by membrane acidic phospholipids is better understood, reviewed in [42–44], but it remains to be determined whether DnaA plays any role in chromosome segregation (see Discussion).

6. Discussion

While considerable effort was made over the intervening decades to understand the regulatory mechanisms for bacterial chromosome replication and segregation, the projects I describe above remain incomplete, because fundamental questions remain unanswered. This is undoubtably due to the complexity of the mechanisms involved, but also to limitations in technology. However, the recent dramatic shift from batch culture studies to single cell analysis has provided the bacterial cell cycle field with a new path forward and many interesting new models for bacterial size regulation, as well as the relationship between chromosome replication and cell division; for example see [45–52]. It is worth noting that these models are not necessarily in agreement with one another, adding a new level of intrigue to the field.

For studies of the initiation step of chromosome replication, analysis of the molecular machinery in single cells is particularly challenging. New approaches will be needed, but whatever technology is applied, the mechanism must have the following properties: (1) it must be able to accommodate once-only initiation from each copy of *oriC* during each cell cycle, (2) it must have the ability to synchronously initiate replication from many copies of *oriC* as a system that does not count origins, (3) it must trigger DnaA-dependent initiation events at the correct time during the cell cycle over a wide range of growth conditions, and (4) it must accommodate the ordered assembly of DnaA multimeric complexes on each copy of *oriC*, reviewed in [53]. Few models are able to satisfy these requirements, but, in my opinion, the initiation titrator model [54], which is now over 30 years old, still remains the front-runner. Yet, even this gold standard may require some tweaking to accommodate fast as well as slow growth conditions; see [55,56]. Finally, the amazing diversity of *oriC* nucleotide sequences obtained from a wide range of microbial types suggests that, despite conservation of the DnaA initiator protein, there are many ways to assemble a functional initiation complex (orisome). Bacteria may use fundamentally different schemes to regulate chromosome replication as best suits the environment of each organism.

Is plasmid replication coupled to the cell division cycle in any bacterial type? Uncoupled replication control provides the best opportunity for plasmid survival and the plasmid-encoded negative regulatory element(s) required for autonomous replication [57] are not compatible with cell cycle-specific plasmid replication. However, the domestication of plasmids has occurred in bacteria with two heterologous chromosomes [58,59], and studies should reveal how plasmid-derived, secondary chromosome replication is controlled during the cell cycle. While it is too early to know whether multiple mechanisms for plasmid domestication exist, studies of *Vibrio cholerae's* chromosome 2 reveal a highly complex regulation for both cell cycle specificity and copy control that remains to be completely understood [60–64].

The relationship between DNA supercoiling, transcription, and the regulation of chromosomal replication origins in bacteria continues to be an under-explored area of research. However, it was recently demonstrated that during the stringent response (such as during nutrient limitation), global reduction of transcription by ppGpp alters DNA supercoiling sufficiently to prevent replication initiations from *oriC* [32]. This finding suggests that other mechanisms may also regulate the assembly of replication origin nucleoprotein complexes by local or global alterations of chromosome supercoiling; see related discussion in [65]. Dissecting these networks will be a complex task, particularly if the transcriptional activity is also coupled to the architecture of the chromosomes and the density of genes during replication; see reviews [33,66].

The topic of non-random segregation of *E. coli* chromosomes remained essentially dormant for over 20 years, but some recent publications indicate that it has been rediscovered [67–69]. Of particular interest is the finding that MukBEF and MatP proteins are involved in nonrandom segregation [67]. MukBEF is the *E. coli* equivalent of the structural maintenance of chromosomes (SMC) complexes found in all cell types, which organize chromosomes and are required for their faithful segregation, reviewed in [70]. MukBEF complexes have a distinctive folded shape that allows movement of DNA stands for regulation of nucleoid shape and chromosome decatenation [71]. MukBEF interact with multiple binding sites around the chromosome, but MatP is able to displace MukBEF from its DNA binding sites within the terminus region [72,73].

There are several ways that MukBEF and MatP might play roles that are compatible with our aforementioned *oriC* attachment model for nonrandom chromosome segregation. MukBEF sites are prominently clustered around *oriC* [74], and the replication origins interact with MukBEF complexes in a self-organizing system [75,76]. Any viable model must include the dynamic assembly and disassembly of these complexes as the cell cycle proceeds. MatP's ability to displace MukBEF and direct it towards (or away) from *oriC* may provide an opportunity for specifically timed assemblies.

What about attachment of *oriC* to internal surface sites? MukBEF complexes may be part of the direct attachment mechanism, but I prefer a model whereby MukBEF produces a particular structure in the *oriC* region that is necessary for attachment. It is reported that MukBEF is capable of DNA loop extrusion [77], and a MukBEF-produced loop might allow *oriC* to be accessible for surface attachment. A novel MukBEF-dependent mechanism for nonrandom chromosome replication may also exist that does not require any attachment of *oriC* to cell surface sites. Hopefully, this will become a future model to test.

Funding: Previous research described in this publication was supported by National Institute of General Medical Sciences of the National Institutes of Health under award number R01GM26429.

Institutional Review Board Statement: Not applicable.

Informed Consent Statement: Not applicable.

Data Availability Statement: Data sharing is not applicable.

Acknowledgments: I thank Arieh Zaritsky, Vic Norris, Itzhak Fishov, and Conrad Woldringh for their efforts in establishing the Helmstetter prize, organizing the meeting where the first awards were presented, and organizing this Special Issue to commemorate the inaugural prize winners. As always, my unmeasurable gratitude to Charles Helmstetter for his mentorship and friendship for over forty years.

Conflicts of Interest: The author declares no conflict of interest.

References

1. Marsh, R.C.; Worcel, A. A DNA fragment containing the origin of replication of the *Escherichia coli* chromosome. *Proc. Natl. Acad. Sci. USA* **1977**, *74*, 2720–2724. [CrossRef] [PubMed]
2. von Meyenburg, K.; Hansen, F.G.; Nielsen, L.D.; Jørgensen, P. Origin of replication, *oriC*, of the *Escherichia coli* chromosome: Mapping of genes relative to R.EcoRI cleavage sites in the *oriC* region. *Mol. Gen. Genet.* **1977**, *158*, 101–109. [CrossRef] [PubMed]
3. Timmis, K.N.; Cabello, F.; Cohen, S.N. Cloning and characterization of EcoRI and HindIII restriction endonuclease-generated fragments of antibiotic resistance plasmids R6-5 and R6. *Mol. Gen. Genet.* **1978**, *162*, 121–137. [CrossRef]
4. Helmstetter, C.E. A ten-year search for synchronous cells: Obstacles, solutions, and practical applications. *Front. Microbiol.* **2015**, *6*, 238. [CrossRef]
5. Cooper, S.; Helmstetter, C.E. Chromosome replication and the division cycle of *Escherichia coli* B/r. *J. Mol. Biol.* **1968**, *31*, 519–540. [CrossRef]
6. Helmstetter, C.; Cooper, S.; Pierucci, O.; Revelas, E. On the bacterial life sequence. *Cold Spring Harb. Symp. Quant. Biol.* **1968**, *33*, 809–822. [CrossRef] [PubMed]
7. Yasuda, S.; Hirota, Y. Cloning and mapping of the replication origin of *Escherichia coli*. *Proc. Natl. Acad. Sci. USA* **1977**, *74*, 5458–5462. [CrossRef]
8. Meijer, M.; Beck, E.; Hansen, F.G.; Bergmans, H.E.; Messer, W.; von Meyenburg, K.; Schaller, H. Nucleotide sequence of the origin of replication of the *Escherichia coli* K-12 chromosome. *Proc. Natl. Acad. Sci. USA* **1979**, *76*, 580–584. [CrossRef]
9. Fuller, R.S.; Kaguni, J.M.; Kornberg, A. Enzymatic replication of the origin of the *Escherichia coli* chromosome. *Proc. Natl. Acad. Sci. USA* **1981**, *78*, 7370–7374. [CrossRef]
10. Leonard, A.C.; Helmstetter, C.E. Cell cycle-specific replication of *Escherichia coli* minichromosomes. *Proc. Natl. Acad. Sci. USA* **1986**, *83*, 5101–5105. [CrossRef]
11. Helmstetter, C.E.; Leonard, A.C. Coordinate initiation of chromosome and minichromosome replication in *Escherichia coli*. *J. Bacteriol.* **1987**, *169*, 3489–3494. [CrossRef] [PubMed]
12. Abeles, A.L.; Snyder, K.M.; Chattoraj, D.K. P1 plasmid replication: Replicon structure. *J. Mol. Biol.* **1984**, *173*, 307–324. [CrossRef] [PubMed]
13. Vocke, C.; Bastia, D. Primary structure of the essential replicon of the plasmid pSC101. *Proc. Natl. Acad. Sci. USA* **1983**, *80*, 6557–6561. [CrossRef]
14. Masai, H.; Arai, K. RepA and DnaA proteins are required for initiation of R1 plasmid replication in vitro and interact with the oriR sequence. *Proc. Natl. Acad. Sci. USA* **1987**, *84*, 4781–4785. [CrossRef]
15. Murakami, Y.; Ohmori, H.; Yura, T.; Nagata, T. Requirement of the *Escherichia coli* dnaA gene function for ori-2-dependent mini-F plasmid replication. *J. Bacteriol.* **1987**, *169*, 1724–1730. [CrossRef]
16. Leonard, A.C.; Helmstetter, C.E. Replication patterns of multiple plasmids coexisting in *Escherichia coli*. *J. Bacteriol.* **1988**, *170*, 1380–1383. [CrossRef] [PubMed]
17. Morrison, P.D.; Chattoraj, D.K. Replication of a unit-copy plasmid F in the bacterial cell cycle: A replication rate function analysis. *Plasmid* **2004**, *52*, 13–30. [CrossRef]
18. Bogan, J.A.; Grimwade, J.E.; Thornton, M.; Zhou, P.; Denning, G.D.; Helmstetter, C.E. P1 and NR1 plasmid replication during the cell cycle of *Escherichia coli*. *Plasmid* **2001**, *45*, 200–208. [CrossRef]
19. Keasling, J.D.; Palsson, B.O.; Cooper, S. Replication of the R6K plasmid during the *Escherichia coli* cell cycle. *J. Bacteriol.* **1992**, *174*, 1060–1062. [CrossRef]
20. Keasling, J.D.; Palsson, B.O.; Cooper, S. Replication of prophage P1 is cell-cycle specific. *J. Bacteriol.* **1992**, *174*, 4457–4462. [CrossRef]
21. Keasling, J.D.; Palsson, B.O.; Cooper, S. Cell-cycle-specific F plasmid replication: Regulation by cell size control of initiation. *J. Bacteriol.* **1991**, *173*, 2673–2680. [CrossRef] [PubMed]
22. Keasling, J.D.; Palsson, B.O.; Cooper, S. Replication of mini-F plasmids during the bacterial division cycle. *Res. Microbiol.* **1992**, *143*, 541–548. [CrossRef] [PubMed]
23. Mukhopadhyay, G.; Carr, K.M.; Kaguni, J.M.; Chattoraj, D.K. Open-complex formation by the host initiator, DnaA, at the origin of P1 plasmid replication. *EMBO J.* **1993**, *12*, 4547–4554. [CrossRef] [PubMed]
24. Wu, F.; Levchenko, I.; Filutowicz, M. Binding of DnaA protein to a replication enhancer counteracts the inhibition of plasmid R6K gamma origin replication mediated by elevated levels of R6K pi protein. *J. Bacteriol.* **1994**, *176*, 6795–6801. [CrossRef] [PubMed]
25. Drlica, K. Bacterial topoisomerases and the control of DNA supercoiling. *Trends Genet.* **1990**, *6*, 433–437. [CrossRef] [PubMed]
26. Muskhelishvili, G.; Travers, A. The regulatory role of DNA supercoiling in nucleoprotein complex assembly and genetic activity. *Biophys. Rev.* **2016**, *8*, 5–22. [CrossRef] [PubMed]
27. Leonard, A.C.; Whitford, W.G.; Helmstetter, C.E. Involvement of DNA superhelicity in minichromosome maintenance in *Escherichia coli*. *J. Bacteriol.* **1985**, *161*, 687–695. [CrossRef]
28. von Freiesleben, U.; Rasmussen, K.V. The level of supercoiling affects the regulation of DNA replication in *Escherichia coli*. *Res. Microbiol.* **1992**, *143*, 655–663. [CrossRef]
29. Skarstad, K.; Baker, T.A.; Kornberg, A. Strand separation required for initiation of replication at the chromosomal origin of E.coli is facilitated by a distant RNA–DNA hybrid. *EMBO J.* **1990**, *9*, 2341–2348. [CrossRef]

30. Baker, T.A.; Kornberg, A. Transcriptional activation of initiation of replication from the *E. coli* chromosomal origin: An RNA-DNA hybrid near *oriC*. *Cell* **1988**, *55*, 113–123. [CrossRef]

31. El Houdaigui, B.; Forquet, R.; Hindré, T.; Schneider, D.; Nasser, W.; Reverchon, S.; Meyer, S. Bacterial genome architecture shapes global transcriptional regulation by DNA supercoiling. *Nucleic Acids Res.* **2019**, *47*, 5648–5657. [CrossRef] [PubMed]

32. Kraemer, J.A.; Sanderlin, A.G.; Laub, M.T. The Stringent Response Inhibits DNA Replication Initiation in *E. coli* by Modulating Supercoiling of *oriC*. *mBio* **2019**, *10*, e01330-19. [CrossRef] [PubMed]

33. Dorman, C.J. DNA supercoiling and transcription in bacteria: A two-way street. *BMC Mol. Cell Biol.* **2019**, *20*, 26. [CrossRef] [PubMed]

34. Pierucci, O.; Zuchowski, C. Non-random segregation of DNA strands in *Escherichia coli* B-r. *J. Mol. Biol.* **1973**, *80*, 477–503. [CrossRef]

35. Pierucci, O.; Helmstetter, C.E. Chromosome segregation in *Escherichia coli* B/r at various growth rates. *J. Bacteriol.* **1976**, *128*, 708–716. [CrossRef]

36. Cooper, S.; Weinberger, M. Medium-dependent variation of deoxyribonucleic acid segregation in *Escherichia coli*. *J. Bacteriol.* **1977**, *130*, 118–127. [CrossRef]

37. Helmstetter, C.E.; Leonard, A.C. Mechanism for chromosome and minichromosome segregation in *Escherichia coli*. *J. Mol. Biol.* **1987**, *197*, 195–204. [CrossRef]

38. Helmstetter, C.E.; Leonard, A.C. Involvement of cell shape in the replication and segregation of chromosomes in *Escherichia coli*. *Res. Microbiol.* **1990**, *141*, 30–39. [CrossRef]

39. Schurr, T.; Grover, N.B. Analysis of a model for minichromosome segregation in *Escherichia coli*. *J. Theor. Biol.* **1990**, *146*, 395–406. [CrossRef]

40. Hendrickson, W.G.; Kusano, T.; Yamaki, H.; Balakrishnan, R.; King, M.; Murchie, J.; Schaechter, M. Binding of the origin of replication of *Escherichia coli* to the outer membrane. *Cell* **1982**, *30*, 915–923. [CrossRef]

41. Landoulsi, A.; Malki, A.; Kern, R.; Kohiyama, M.; Hughes, P. The *E. coli* cell surface specifically prevents the initiation of DNA replication at *oriC* on hemimethylated DNA templates. *Cell* **1990**, *63*, 1053–1060. [CrossRef] [PubMed]

42. Boeneman, K.; Crooke, E. Chromosomal replication and the cell membrane. *Curr. Opin. Microbiol.* **2005**, *8*, 143–148. [CrossRef] [PubMed]

43. Saxena, R.; Fingland, N.; Patil, D.; Sharma, A.K.; Crooke, E. Crosstalk between DnaA protein, the initiator of *Escherichia coli* chromosomal replication, and acidic phospholipids present in bacterial membranes. *Int. J. Mol. Sci* **2013**, *14*, 8517–8537. [CrossRef] [PubMed]

44. Regev, T.; Meyers, N.; Zarivach, R.; Fishov, I. Association of the chromosome replication initiator DnaA with the *Escherichia coli* inner membrane in vivo: Quantity and mode of binding. *PLoS ONE* **2012**, *7*, e36441. [CrossRef]

45. Lee, S.; Wu, L.J.; Errington, J. Microfluidic time-lapse analysis and reevaluation of the Bacillus subtilis cell cycle. *Microbiologyopen* **2019**, *8*, e876. [CrossRef]

46. Zhang, Q.; Zhang, Z.; Shi, H. Cell Size Is Coordinated with Cell Cycle by Regulating Initiator Protein DnaA in *E. coli*. *Biophys. J.* **2020**, *119*, 2537–2557. [CrossRef]

47. Kar, P.; Tiruvadi-Krishnan, S.; Männik, J.; Männik, J.; Amir, A. Using conditional independence tests to elucidate causal links in cell cycle regulation in *Escherichia coli*. *Proc. Natl. Acad. Sci. USA* **2023**, *120*, e2214796120. [CrossRef]

48. Le Treut, G.; Si, F.; Li, D.; Jun, S. Quantitative Examination of Five Stochastic Cell-Cycle and Cell-Size Control Models for *Escherichia coli* and Bacillus subtilis. *Front. Microbiol.* **2021**, *12*, 721899. [CrossRef]

49. Tiruvadi-Krishnan, S.; Männik, J.; Kar, P.; Lin, J.; Amir, A.; Männik, J. Coupling between DNA replication, segregation, and the onset of constriction in *Escherichia coli*. *Cell Rep.* **2022**, *38*, 110539. [CrossRef]

50. Wallden, M.; Fange, D.; Lundius, E.G.; Baltekin, Ö.; Elf, J. The Synchronization of Replication and Division Cycles in Individual *E. coli* Cells. *Cell* **2016**, *166*, 729–739. [CrossRef]

51. Taheri-Araghi, S.; Bradde, S.; Sauls, J.T.; Hill, N.S.; Levin, P.A.; Paulsson, J.; Vergassola, M.; Jun, S. Cell-size control and homeostasis in bacteria. *Curr. Biol.* **2015**, *25*, 385–391. [CrossRef] [PubMed]

52. Witz, G.; van Nimwegen, E.; Julou, T. Initiation of chromosome replication controls both division and replication cycles in *E. coli* through a double-adder mechanism. *Elife* **2019**, *8*, e48063. [CrossRef] [PubMed]

53. Leonard, A.C.; Grimwade, J.E. Regulation of DnaA assembly and activity: Taking directions from the genome. *Annu Rev. Microbiol.* **2011**, *65*, 19–35. [CrossRef] [PubMed]

54. Hansen, F.G.; Christensen, B.B.; Atlung, T. The initiator titration model: Computer simulation of chromosome and minichromosome control. *Res. Microbiol.* **1991**, *142*, 161–167. [CrossRef] [PubMed]

55. Berger, M.; Wolde, P.R.T. Robust replication initiation from coupled homeostatic mechanisms. *Nat. Commun.* **2022**, *13*, 6556. [CrossRef] [PubMed]

56. Zheng, H.; Bai, Y.; Jiang, M.; Tokuyasu, T.A.; Huang, X.; Zhong, F.; Wu, Y.; Fu, X.; Kleckner, N.; Hwa, T.; et al. General quantitative relations linking cell growth and the cell cycle in *Escherichia coli*. *Nat. Microbiol.* **2020**, *5*, 995–1001. [CrossRef]

57. Pritchard, R.H.; Barth, P.T.; Collins, J. Control of DNA synthesis in bacteria. *Symp. Soc. Gen. Microbiol.* **1969**, *19*, 263–297.

58. diCenzo, G.C.; Finan, T.M. The divided bacterial genome: Structure, function, and evolution. *Microbiol. Mol. Biol Rev.* **2017**, *81*, 9–17. [CrossRef]

59. Fournes, F.; Val, M.E.; Skovgaard, O.; Mazel, D. Replicate Once Per Cell Cycle: Replication Control of Secondary Chromosomes. *Front. Microbiol.* **2018**, *9*, 1833. [CrossRef]

60. Ramachandran, R.; Jha, J.; Paulsson, J.; Chattoraj, D. Random versus Cell Cycle-Regulated Replication Initiation in Bacteria: Insights from Studying *Vibrio cholerae* Chromosome 2. *Microbiol. Mol. Biol. Rev.* **2017**, *81*, e00033-16. [CrossRef]

61. Espinosa, E.; Barre, F.X.; Galli, E. Coordination between replication, segregation and cell division in multi-chromosomal bacteria: Lessons from *Vibrio cholerae*. *Int. Microbiol.* **2017**, *20*, 121–129. [PubMed]

62. Venkova-Canova, T.; Chattoraj, D.K. Transition from a plasmid to a chromosomal mode of replication entails additional regulators. *Proc. Natl. Acad. Sci. USA* **2011**, *108*, 6199–6204. [CrossRef] [PubMed]

63. de Lemos Martins, F.; Fournes, F.; Mazzuoli, M.-V.; Mazel, D.; Val, M.-E. *Vibrio cholerae* chromosome 2 copy number is controlled by the methylation-independent binding of its monomeric initiator to the chromosome 1 *crtS* site. *Nucleic Acids Res.* **2018**, *46*, 10145–10156. [CrossRef] [PubMed]

64. Fournes, F.; Niault, T.; Czarnecki, J.; Tissier-Visconti, A.; Mazel, D.; Val, M.-E. The coordinated replication of *Vibrio cholerae's* two chromosomes required the acquisition of a unique domain by the RctB initiator. *Nucleic Acids Res.* **2021**, *49*, 11119–11133. [CrossRef] [PubMed]

65. Muskhelishvili, G.; Sobetzko, P.; Travers, A. Spatiotemporal Coupling of DNA Supercoiling and Genomic Sequence Organization-A Timing Chain for the Bacterial Growth Cycle. *Biomolecules* **2022**, *12*, 831. [CrossRef] [PubMed]

66. Verma, S.C.; Qian, Z.; Adhya, S.L. Architecture of the *Escherichia coli* nucleoid. *PLoS Genet.* **2019**, *15*, e1008456. [CrossRef]

67. Mäkelä, J.; Uphoff, S.; Sherratt, D.J. Nonrandom segregation of sister chromosomes by *Escherichia coli* MukBEF. *Proc. Natl. Acad. Sci. USA* **2021**, *118*, e2022078118. [CrossRef]

68. White, M.A.; Eykelenboom, J.K.; Lopez-Vernaza, M.A.; Wilson, E.; Leach, D.R. Non-random segregation of sister chromosomes in *Escherichia coli*. *Nature* **2008**, *455*, 1248–1250. [CrossRef]

69. Lopez-Vernaza, M.A.; Leach, D.R. Symmetries and asymmetries associated with non-random segregation of sister DNA strands in *Escherichia coli*. *Semin Cell Dev. Biol.* **2013**, *24*, 610–617. [CrossRef]

70. Mäkelä, J.; Sherratt, D. SMC complexes organize the bacterial chromosome by lengthwise compaction. *Curr. Genet.* **2020**, *66*, 895–899. [CrossRef]

71. Bürmann, F.; Lee, B.G.; Than, T.; Sinn, L.; O'Reilly, F.J.; Yatskevich, S.; Rappsilber, J.; Hu, B.; Nasmyth, K.; Löwe, J. A folded conformation of MukBEF and cohesin. *Nat. Struct. Mol. Biol.* **2019**, *26*, 227–236. [CrossRef] [PubMed]

72. Nolivos, S.; Upton, A.L.; Badrinarayanan, A.; Müller, J.; Zawadzka, K.; Wiktor, J.; Gill, A.; Arciszewska, L.; Nicolas, E.; Sherratt, D. MatP regulates the coordinated action of topoisomerase IV and MukBEF in chromosome segregation. *Nat. Commun.* **2016**, *7*, 10466. [CrossRef] [PubMed]

73. Mäkelä, J.; Sherratt, D.J. Organization of the *Escherichia coli* Chromosome by a MukBEF Axial Core. *Mol. Cell* **2020**, *78*, 250–260.e5. [CrossRef] [PubMed]

74. Japaridze, A.; van Wee, R.; Gogou, C.; Kerssemakers, J.W.J.; van den Berg, D.F.; Dekker, C. MukBEF-dependent chromosomal organization in widened *Escherichia coli*. *Front. Microbiol.* **2023**, *14*, 1107093. [CrossRef] [PubMed]

75. Hofmann, A.; Mäkelä, J.; Sherratt, D.J.; Heermann, D.; Murray, S.M. Self-organised segregation of bacterial chromosomal origins. *Elife* **2019**, *8*, e46564. [CrossRef]

76. Marko, J.F.; De Los Rios, P.; Barducci, A.; Gruber, S. DNA-segment-capture model for loop extrusion by structural maintenance of chromosome (SMC) protein complexes. *Nucleic Acids Res.* **2019**, *47*, 6956–6972. [CrossRef]

77. Bürmann, F.; Funke, L.F.H.; Chin, J.W.; Löwe, J. Cryo-EM structure of MukBEF reveals DNA loop entrapment at chromosomal unloading sites. *Mol. Cell* **2021**, *81*, 4891–4906.e8. [CrossRef]

life

MDPI

Opinion

Extending Validity of the Bacterial Cell Cycle Model through Thymine Limitation: A Personal View

Arieh Zaritsky

Faculty of Natural Sciences, Life Sciences Department, Ben-Gurion University of the Negev, Kiryat Bergman, HaShalom St. 1, Be'er-Sheva 8410501, Israel; ariehzar@gmail.com; Tel.: +972-54-595-5670

Abstract: The contemporary view of bacterial physiology was established in 1958 at the "Copenhagen School", culminating a decade later in a detailed description of the cell cycle based on four parameters. This model has been subsequently supported by numerous studies, nicknamed BCD (The Bacterial Cell-Cycle Dogma). It readily explains, quantitatively, the coupling between chromosome replication and cell division, size and DNA content. An important derivative is the number of replication positions n, the ratio between the time C to complete a round of replication and the cell mass doubling time τ; the former is constant at any temperature and the latter is determined by the medium composition. Changes in cell width W are highly correlated to n through the equation for so-called nucleoid complexity NC ($=(2^n - 1)/(\ln 2 \times n)$), the amount of DNA per $terC$ (i.e., chromosome) in genome equivalents. The narrow range of potential n can be dramatically extended using the method of thymine limitation of thymine-requiring mutants, which allows a more rigorous testing of the hypothesis that the nucleoid structure is the primary source of the signal that determines W during cell division. How this putative signal is relayed from the nucleoid to the divisome is still highly enigmatic. The aim of this *Opinion* article is to suggest the possibility of a new signaling function for nucleoid DNA.

Keywords: bacterial physiology; division cycle; cell dimensions; nucleoid complexity; replication position; eclipse

---

Motto-1: *Look; don't touch* (science principle; attributed to Ole Maaløe [1]).

Motto-2: *It makes sense to try clarifying ideas that emerge even if they cannot be tested right away* [2].

1. Brief Historical Highlights

1.1. Pioneering Bacterial Physiology

The conceptual process leading to an understanding of the bacterial cell cycle was hampered by the previously known eukaryotic cycle [G_1-S-G_2-M(-G_0)] because there was an expectation that the cycle would resemble that of eukaryotes. The usually circular bacterial chromosome replicates bidirectionally from a single origin $oriC$, and the unanticipated, surprising reports of reinitiating replication prior to the completion of the previous round at $terC$ [3,4] revolutionized the field.

Major ideas in quantitative microbial physiology [5] were established in the so-called "Copenhagen School" led by Maaløe [2]. The back-to-back pioneering publications from this laboratory, coauthored with Schaechter and Kjeldgaard [6,7], opened the field of bacterial physiology. The soon-after groundbreaking studies of Hanawalt [8], manipulating the "immunity" to thymine-less death (TLD) [9] and using autoradiography, discovered two distinct, seemingly independent stages involved in DNA replication, initiation and elongation.

Citation: Zaritsky, A. Extending Validity of the Bacterial Cell Cycle Model through Thymine Limitation: A Personal View. *Life* **2023**, *13*, 906. https://doi.org/10.3390/life13040906

Academic Editor: Tom Santangelo

Received: 15 February 2023
Revised: 14 March 2023
Accepted: 24 March 2023
Published: 29 March 2023

1.2. The Central Dogma of the Bacterial Cell Cycle (BCD)

The seminal, extensive series of Helmstetter's studies with *Escherichia coli* B/r during the 1960s [1] culminated with the understanding of the basic properties of the relationships between chromosome replication and cell division [10]. The ultimate quantitative description of the bacterial cell cycle includes only four parameters that can define the cell's physiological state (ignoring the variabilities in populations). At a constant temperature under steady-state exponential growth [11], (i) doubling time τ depends on nutritional conditions [6]. The other three, all related to chromosome replication, are relatively constant at a wide range of τs (between ~20 and 70 min) irrespective of the medium composition: (ii) strain-dependent mass per *oriC* when replication initiation occurs, Mi [10,12]; (iii) the time taken for a round of replication to be completed, C; and (iv) the time between replication-termination and cell division, D. In slow growth, when $(C + D) < \tau$, another temporal parameter (albeit not independent) sometimes appears, the time B $(=\tau - (C + D))$ between cell birth (upon splitting its mother) and subsequent replication initiation. This description, which was quantitively consistent with the results obtained at the time, has been repeatedly confirmed by numerous investigations during the following decades and hence may be termed as the Central Dogma of The Bacterial Cell Division Cycle (BCD) [13]

1.3. Average Cell Size and Chromosomal DNA Content

Applying Powell's age distribution function $f(a)$ $(= (\ln2/2) \times 2^{-a})$, where $0 \leq a \leq 1$ [14], the BCD generates, convincingly and realistically so, the average cell size and DNA content, respectively: $<M>$ $(= Mi \times 2^{(C+D)/\tau})$ [12] and $<G>$ $(= (\tau/C\ln2) \times (2^{(C+D)/\tau} - 2^{D/\tau}))$ [10]. Two important parameters were thus derived from BCD: "set number" $(C + D)/\tau$ is "the number of generations between the start of a round and the division at the end of that round" [10], and "the number of replication positions per chromosome" n $(=C/\tau)$, defined earlier [15] as "the position of a set of equivalent [simultaneously initiating] replication points on a chromosome. Replication with more than one position is called 'multi-forked' and replication with less than one position implies the presence of a resting period". Both quantities describe the chromosomal state; the former relates to the whole cell and the latter to the chromosome itself, i.e., to its unique terminus *terC*. The simple parameter n turned out to be key to understand the cell's physiological state as well; by avoiding the value of D, the molecular mechanism of which is still enigmatic (and see in Section 3.3), the amount of DNA per chromosome in genome equivalents NC $(=(2^n - 1)/(\ln2 \times n))$ [15] and the DNA/mass ratio (i.e., "DNA concentration") in the cell $(G/M)_c$ $(=(1 - 2^{-n})/(Mi \times n \times \ln2))$ [16] can be evaluated.

1.4. Cell Dimensions

A cylindrical (rod-shaped) cell such as *E. coli* grows via elongation, with a hardly discernible change in width W during the cell cycle under slow growth rate [17], likely due to systematic variation of " … the internal osmotic pressure … decreased during elongation and increasing again during constriction". Unpredictably, during faster growth (shorter τs) in richer media, the cells are also wider ([6,18], Figure 1, and see details below, in Table 1). With Bob Pritchard, we were the first to suggest a connection of W-change with the parameters of the BCD [19,20], more directly and rigorously so in [21]. The idea prevailing in the 1970s and 1980s, that cell length-growth is bi-linear, coupled to devoted, discrete envelope-growth sites, was inspired by the then-popular, so-called replicon model [22]. With the knowledge that cell mass and volume grow exponentially, W would passively be determined by the active extension of both, leaving no degree of freedom for the mode of its determination. Soon after the idea of "wall growth zones" was precluded experimentally [23], an alternative hypothesis was presented [24,25], that W is " … actively determined by the amount of DNA packed in an individual chromosome", in which case the cell elongates by default at a rate that depends on the other two to preserve a constant mean buoyant cell density [26]. This idea was prompted by our observations [27] that during a nutritional shift up cells elongate temporarily faster than before, overshooting their final steady-state length, and that the new polar caps are the first to widen, resulting in "pear-shaped" cells, before stabilizing the new steady-state dimensions. Thus, a presumed primary signal for both, cell division and *W* determination occurs simultaneously during the action of the divisome in concerted temporal and spatial processes that must be coupled to the nucleoid segregation as well. This notion has been strengthened by our detailed studies with thymine limitation, as described below (at Section 4).

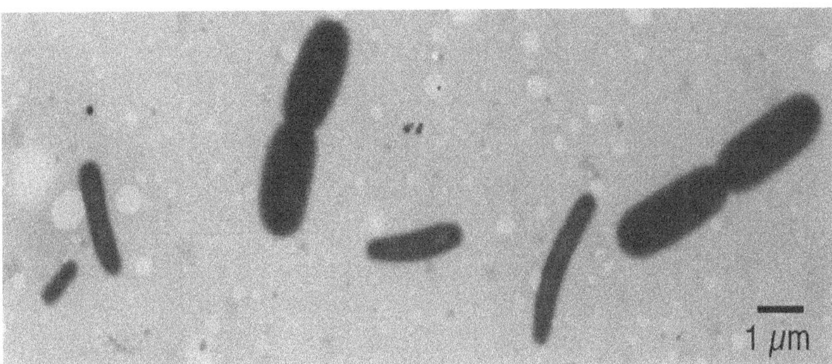

Figure 1. Electron micrograph (taken by Conrad Woldringh) of a mixture of two *E. coli* B/r cultures grown with doubling times τ = 22 min (the bigger cells) and τ = 150 min (the smaller cells).

Table 1. Cell width *W*, nucleoid complexity *NC* and the ratios between them at different doubling times τ. (The *oriC/terC* ratio *o/t* and the ratios between them and *W* are shown [c] for comparisons).

τ (min) [a]	*W* (µm) [a]	*NC* [b]	*W/NC*	*oriC/terC* [c]	*W/(o/t)* [c]
22	1.43	2.01	0.711	3.53	0.405
32	1.22	1.60	0.762	2.38	0.513
60	0.93	1.28	0.727	1.59	0.585
98	0.87	1.18	0.737	1.33	0.654
17.14	0.98	2.77	0.354	5.06	0.205
22.22	0.80	2.15	0.376	3.43	0.248
26.67	0.72	1.76	0.384	2.83	0.254
30.77	0.67	1.64	0.389	2.51	0.282
37.50	0.61	1.48	0.396	2.08	0.481
50.85	0.55	1.33	0.402	1.73	0.324
51.28	0.55	1.33	0.402	1.72	0.319

Adapted from [28]. [a], data from [6,21] (top 4) with *Salmonella typhimuriun* and [29] (bottom 7) with *Escherichia coli*. Values of *W* for *E. coli* were calculated according to $W = 0.41 \times 2^{0.36 \times 60/\tau}$ [29]. [b], $NC = (2^n - 1)/(n \times \ln 2)$, and [c], the ratio $oriC/terC = 2^n$ were calculated assuming C = 40 min.

The coefficient of variation CV of the ratio *W/NC* is 2.9 and 4.4% in *S. typhimurium* and *E. coli*, respectively. Such small CV values for a relationship between two independently measured, seemingly unrelated parameters are rare in biological systems hence re-enforce the suggestion that they are not fortuitous.

2. Dissociation between Rates

The dissociation between the rates of replication C^{-1} and of mass growth τ^{-1} [10] readily explained many then-puzzling observations such as that cells are larger at richer media supporting faster growth [6] and "rate maintenance" of divisions during $(C + D)$ min following a nutritional shift-up transition [7]. The near-constant values of *Mi*, *C* and *D* leaves little freedom to manipulate the cell cycle except τ through medium composition. Conditional lethal mutants in numerous, indispensable genes involved in fixing these parameters enhanced our understanding of the biochemical pathways involved in DNA replication and cell division. For example, modulating expression levels of *dnaA-ts* mutants at intermediate temperatures clarified the mechanism of replication initiation [30]. Such mutants are, however, usually pleiotropic, which affects other pathways and hence often blurs the picture.

As soon as this dissociation was fully understood [10], I arrived at Leicester University's Genetics department to study for a PhD (1969–1971). My mentor Bob (Robert H) Pritchard (1930–2015) [31] immediately realized that the dissociation has two sides, and that some consequences of results in the then-current studies utilizing thymine-requiring mutants to follow DNA replication were often flawed due to the varying concentrations of thymine [T] supplied. To minimize expending radioactive material (commonly ^{3}H- or ^{14}C-labeled), very low [T]s have often been exploited at high specific radioactivities (Cu/gr). I was set to quantitatively test the hypothesis that, under such conditions, the value of C depends on [T]; in a couple of years, we succeeded to extend the validity of this dissociation by elongating C up to about three-fold without any noticeable change in τ (in identical medium) simply by modulating the supply of thymine to *thyA* mutants of *E. coli* [16,32–35]. Thus, the mean number of "replication positions" n, limited in Thy+ strains to ~2 due to the minimum τ_{min} of 20 min (whereas C is constant at 40 min), could only be increased by lengthening C. The least physiologically disturbing way to overcome this limit seems to be by reducing the externally supplied [T].

The Thymine Limitation Tool

Thymine is the only base solely incorporated into DNA but is not used as such in Thy+ strains; their immediate metabolite for DNA synthesis, T-dRib-P-P-P is made from U-dRib-P in a dedicated pathway through a couple of phosphorylations of the resultant T-dRib-P [19,34], indicating that a specific permease for thymine has never evolved. In *thyA* mutants (with an inactive thymidylate synthetase), T-dRib-P is also exclusively exploited for DNA synthesis, but here it is synthesized via a bypass salvage pathway, using supplied thymine entering the cell by diffusion alone [19,34]. Intracellular [T_i] is therefore related to the externally supplied [T]. Hence, n ($=C/\tau$) can be manipulated in two ways, by varying C at a constant τ in *thyA* mutants or varying τ at a constant C in Thy+ strains. The systematic manipulation of both (which has never yet rigorously studied, to the best of my knowledge) is anticipated to dramatically affect NC [20] (later defined as "Nucleoid Complexity" [34]; see below for explicit definition also in Table 1, [28], and Equation (6) at [25]).

3. Repercussions

3.1. Valuable Uses of Thymine Limitation

Validating the hypothesis that C inversely varies with [T] by several means [16,32,33] opened the way to discover additional aspects of cell physiology at a wider range of n exploiting the powerful thymine limitation leverage (mostly summarized and referenced in [34]). Several attributes were studied by scientists in laboratories from all around the world, e.g., the bidirectionality of chromosome replication, kinetics of mutagenicity, control of plasmid replication, and localizing replication forks by SeqA foci distribution. Some studies were performed as follow-up at Leicester after my departure: the dependence of constitutive gene output at different DNA concentrations and relative gene dosages, dependence of D on C, and changes in cell dimensions and shape [36]. The latter two were also among the subjects that I pursued at Ben-Gurion University of the Negev (BGU) during 50 years of my tenure there, being naturally attracted by the highly effective tool of thymine limitation, mostly in cooperation with colleagues and trainees at home and abroad (see below and in [35]).

3.2. The Eclipse

My first years at BGU reflected several subjects related to my studies at Leicester— particularly two that are still at the heart of yet-unsolved questions: lack of a steady state in terms of cell dimensions during fast growth under thymine limitation [20], and the relationship between cell width W and BCD parameters [21]. An explanation of the former emerged in results of the long-term inhibition of DNA replication, envisioned as [33]: "It is postulated that this second replication position (20), which was ready to initiate when thymine was restored, remained "stacked" until the previous one had traversed the presumed minimal distance away from the origin of replication. This hypothesis should not be elaborated further, but should serve as a working hypothesis to be tested by direct means" (and see [2] Motto-2). The existence of a minimal distance possible between two successive replisomes, later termed "Eclipse" [34], was confirmed by several laboratories over-expressing *dnaA* (summed up and discussed in [37]). With a long C, it takes more time for the replisome to reach this distance; if this time is longer than τ, the real start of replication occurs at a larger cell mass than the typical M_i for the said strain, and hence this delay is cumulative [37]. It reaches a maximum before branching, as does NC, readily explaining, at least qualitatively, the lack of steady-state cell dimensions as observed [20] under such conditions (short τ in rich media and long C at low [T]s); this is fully consistent with the concept that W is related to NC [28].

3.3. Dependence of D on C: Contradicting and Enigmatic Results

This dependency under thymine limitation was investigated by employing several methods, with contradictory results: (i) in my hands [38,39], using the "division-rate maintenance" phenomenon [7], when C was longer (at lower [T]), apparent "D" increased proportionately, with a relationship of [38] "D" = 0.83C − 16. On the other hand (ii), Meacock and Pritchard [40], using the "baby machine" method [1,10] and calculating from measurements of cell composition [16] (average size, DNA content and DNA/mass ratio), arrived at the opposite conclusion, that "D" was rather shortened upon the lengthening of C. Their main conclusion, however, that "the time of cell division is determined by termination … " is obvious. Option (i) seems more realistic and makes more sense because at longer C the cells are wider with a larger circumference hence require longer time to assemble the divisome and complete the division process, leading to a longer D, but then, of course, the results of the more extensive baby machine and cell composition studies [40] must be differently explained. A possible reason for this discrepancy is that the latter used cells that grew a shorter period under long C before uploading the baby machine, whereas W takes much longer to change, as was the alternative practice [38,39]. This apparent paradox should be thoroughly looked at again to be resolved. A relationship between these two parameters suggests some sort of coupling between their functions, one that is still to be deciphered in molecular terms or otherwise. In the Discussion of Helmstetter's lecture at the 1968 Cold Spring Harbor Symposium (page 822 in [10]), Dr. GA Herrick suggested that the proportionality between C and D at slow growth rates "supports the implication of membranes in DNA synthesis", an interesting idea that has not been pursued yet, to the best of my knowledge.

4. Are DNA Functions Exhausted?—Bacterial Cell Dimensions

DNA may have been considered as a reservoir of bases due to its monotonic structure. It was recognized as The Essence Molecule of Life only in the mid-20th century [41]. Its essential roles are still not fully clear. The following fundamental functions of DNA have gradually been discovered over 150 years: chromosome-linked genes, store of genetic information, self-replication, coding for proteins, and the regulation of gene expression. Some others are still under intense investigations, e.g., in bacteria: nucleoid structure, segregation, and "vetoing" cell division (i.e., "nucleoid occlusion"), and involvement in the determination of cell dimensions. I am curious to identify what may be the primary signal(s) for the determination of both cell division and dimensions. Such a signal(s) may turn out to be another function of DNA that would be contained, historically, as "a paradox" in "The Dogmatic Phase" at which we are, as described by Stent [42].

Numerous proteins are jointly and coordinately involved in these crucially essential processes, considered to be downstream of a presumed major signal. Is it an elusive signal that stems from another discipline? Inspired by the "Enzyme-Cannot-Make-Enzyme paradox" [41], I recently proposed a "regulator-of-the-regulator paradox" [43]: "The template feature came from another discipline (information science) than chemistry (producing an enzyme); by analogy, triggering cell division may stem from physics—or another discipline that we are not aware of currently rather than the proteins involved in the division process itself. Can the divisome activation be triggered by the nucleoid's complexity or replication status?" Two plausible, potential mechanisms related to DNA are transertion [44,45] and supercoiling; the former has been proposed [46,47] as involved in accurately placing the divisome at mid-cell, whereas the latter [48] has yet to be considered. The antibacterial properties of certain naturally derived drugs have recently been mentioned, but their potential action in affecting cell division per se has not been discussed (see "Sitafloxacin" at [49]). In a recent review [50], hyperstructures formed by combining DNA strand-dependent inheritance of nucleoid-associated proteins (NAP) and topoisomerases are proposed to play a central role during the cell cycle by helping generate daughter cells with different phenotypes (via DNA segregation and cell division) and populations with different, average phenotypes (via different degrees of supercoiling).

Whatever the elusive, primary signal(s) for assigning the divisome to precisely act in time and space will be proposed, a powerful tool to test the idea(s) would be, no doubt, the thymine limitation procedure.

Funding: This investigation was supported by grants from the U.S.-Israel Binational Science Foundation (BSF, No. 2017004) and the ISF-NSFC joint research program (No. 3320/20), and facilities by BGU administration.

Institutional Review Board Statement: Not applicable.

Informed Consent Statement: Not applicable.

Data Availability Statement: Not applicable.

Acknowledgments: Charles Helmstetter, Conrad Woldringh, Vic Norris, Charles Howard, and Galit Yehezkel are gratefully acknowledged for help and constructive comments composing the manuscript.

Conflicts of Interest: The authors declare no conflict of interest.

References

1. Helmstetter, C.E. A ten-year search for synchronous cells: Obstacles, solutions, and practical applications. *Front. Microbiol.* **2015**, *6*, 10. [CrossRef] [PubMed]
2. Maaløe, O.; Kjeldgaard, N.O. *Control of Macromolecular Synthesis*; Benjamin: New York, NY, USA, 1966; p. 296.
3. Pritchard, R.H.; Lark, K.G. Induction of replication by thymine starvation at the chromosome origin in *Escherichia coli*. *J. Mol. Biol.* **1964**, *9*, 288–307. [CrossRef] [PubMed]
4. Yoshikawa, H.; O'Sullivan, A.; Sueoka, N. Sequential replication of the *Bacillus subtilis* chromosome, III. Regulation of initiation. *Proc. Natl. Acad. Sci. USA* **1964**, *52*, 973–980. [CrossRef] [PubMed]
5. Major Ideas in Quantitative Microbial Physiology: Past, Present And Future; Copenhagen. 2022. Available online: https://indico.nbi.ku.dk/event/1592/ (accessed on 14 March 2023).
6. Schaechter, M.; Maaløe, O.; Kjeldgaard, N.O. Dependency on medium and temperature of cell size and chemical composition during balanced growth of *Salmonella typhimurium*. *J. Gen. Microbiol.* **1958**, *19*, 592–606. [CrossRef] [PubMed]
7. Kjeldgaard, N.O.; Maaløe, O.; Schaechter, M. The transition between different physiological states during balanced growth of *Salmonella typhimurium*. *J. Gen. Microbiol.* **1958**, *19*, 607–616. [CrossRef] [PubMed]
8. Hanawalt, P.C.; Maaløe, O.; Cummings, D.J.; Schaechter, M. The normal DNA replication cycle. II. *J. Mol. Biol.* **1961**, *3*, 156–165. [CrossRef]
9. Cohen, S.S.; Barner, H.D. Studies on unbalanced growth in *Escherichia coli*. *Proc. Natl. Acad. Sci. USA* **1954**, *40*, 885–893. [CrossRef] [PubMed]
10. Helmstetter, C.E.; Cooper, S.; Pierucci, O.; Revelas, E. On the bacterial life sequence. *Cold Spring Harb. Symp. Quant. Biol.* **1968**, *33*, 809–822. [CrossRef]
11. Fishov, I.; Zaritsky, A.; Grover, N.B. On microbial states of growth. *Mol. Microbiol.* **1995**, *15*, 789–794. [CrossRef]
12. Donachie, W.D. Relationship between cell size and time of initiation of DNA replication. *Nature* **1968**, *219*, 1077–1079. [CrossRef]
13. Zaritsky, A.; Wang, P.; Vischer, N.O.E. Instructive simulation of the bacterial cell division cycle. *Microbiology* **2011**, *157*, 1876–1885. [CrossRef] [PubMed]

14. Powell, E.O. Growth rate and generation time of bacteria, with special reference to continuous culture. *J. Gen. Microbiol.* **1956**, *15*, 492–511. [CrossRef] [PubMed]
15. Sueoka, N.; Yoshikawa, H. The chromosome of *Bacillus subtilis*. I. Theory of marker frequency analysis. *Genetics* **1965**, *52*, 747–757. [CrossRef] [PubMed]
16. Pritchard, R.H.; Zaritsky, A. Effect of thymine concentration on the replication velocity of DNA in a thymineless mutant of *Escherichia coli*. *Nature* **1970**, *226*, 126–131. [CrossRef]
17. Trueba, F.J.; Woldringh, C.L. changes in cell diameter during the division cycle of *Escherichia coli*. *J. Bacteriol.* **1980**, *142*, 869–878. [CrossRef]
18. Woldringh, C.L.; de Jong, M.A.; van den Berg, W.; Koppes, L. Morphological analysis of the division cycle of two *Escherichia coli* substrains during slow growth. *J. Bacteriol.* **1977**, *131*, 270–279. [CrossRef]
19. Pritchard, R.H. Review lecture on the growth and form of a bacterial cell. *Philos. Trans. R. Soc. Lond. B Biol. Sci.* **1974**, *267*, 303–336. [CrossRef]
20. Zaritsky, A.; Pritchard, R.H. Changes in cell size and shape associated with changes in the replication time of the chromosome of *Escherichia coli*. *J. Bacteriol.* **1973**, *114*, 824–837. [CrossRef]
21. Zaritsky, A. On dimensional determination of rod-shaped bacteria. *J. Theor. Biol.* **1975**, *54*, 243–248. [CrossRef]
22. Jacob, F.; Brenner, S.; Cuzin, F. On the regulation of DNA replication in bacteria. *Cold Spring Harb. Symp. Quant. Biol.* **1963**, *28*, 329–348. [CrossRef]
23. Woldringh, C.L.; Huls, P.; Pas, E.; Brakenhoff, G.J.; Nanninga, N. Topography of peptidoglycan synthesis during elongation and polar cap formation in a cell division mutant of *Escherichia coli* MC4100. *Microbiology* **1987**, *133*, 575–586. [CrossRef]
24. Zaritsky, A.; Woldringh, C.L. *Similarity of Dimensional Rearrangement in Escherichia Coli Cells Responding To Thymine Limitation (Step Down) and to Nutritional Shift-up*; Vicente, M., Ed.; EMBO Workshop: Segovia, Spain, 1987; pp. 78–80.
25. Woldringh, C.L.; Mulder, E.; Valkenburg, J.A.C.; Wientjes, F.B.; Zaritsky, A.; Nanninga, N. Role of the nucleoid in the toporegulation of division. *Res. Microbiol.* **1990**, *141*, 39–49. [CrossRef] [PubMed]
26. Kubitschek, H.E.; Baldwin, W.W.; Schroeter, S.J.; Graetzer, R. Independence of buoyant cell density and growth rate in *Escherichia coli*. *J. Bacteriol.* **1984**, *158*, 296–299. [CrossRef] [PubMed]
27. Woldringh, C.L.; Grover, N.B.; Rosenberger, R.F.; Zaritsky, A. Dimensional rearrangement of rod-shaped bacteria following nutritional shift-up. II. Experiments with *Escherichia coli*. *J. Theor. Biol.* **1980**, *86*, 441–454. [CrossRef]
28. Zaritsky, A.; Rabinovitch, A.; Liu, C.; Woldringh, C.L. Does the eclipse limit bacterial nucleoid complexity and cell width? *Synth. Syst. Biotechnol.* **2017**, *2*, 267–275. [CrossRef] [PubMed]
29. Taheri-Araghi, S.; Bradde, S.; Sauls, J.T.; Hill, N.S.; Levin, P.A.; Paulsson, J.; Vergassola, M.; Jun, S. Cell-size Control and homeostasis in bacteria. *Curr. Biol.* **2015**, *25*, 385–391. [CrossRef]
30. Hansen, F.G.; Rasmussen, K.V. Regulation of the *dnaA* product in *Escherichia coli*. *Mol. Gen. Genet.* **1977**, *155*, 219–225. [CrossRef]
31. Zaritsky, A. Personal recollections of an exciting scientific period (1969–1971 and beyond): A tribute to Bob. In *The Life of Professor Robert Hugh Pritchard*; World Scientific: Singapore, 2017; pp. 61–78. [CrossRef]
32. Zaritsky, A.; Pritchard, R.H. Replication time of the chromosome in thymineless mutants of *Escherichia coli*. *J. Mol. Biol.* **1971**, *60*, 65–74. [CrossRef]
33. Zaritsky, A. Rate stimulation of deoxyribonucleic acid synthesis after inhibition. *J. Bacteriol.* **1975**, *122*, 841–846. [CrossRef]
34. Zaritsky, A.; Woldringh, C.L.; Einav, M.; Alexeeva, S. Use of thymine limitation and thymine starvation to study bacterial physiology and cytology. *J. Bacteriol.* **2006**, *188*, 1667–1679. [CrossRef]
35. Zaritsky, A. List of Peer-Reviewed Publications. Available online: https://orcid.org/0000-0001-9333-6854 (accessed on 14 March 2023).
36. Meacock, P.A.; Pritchard, R.H.; Roberts, E.M. Effect of thymine concentration on cell shape in Thy- *Escherichia coli* B/r. *J. Bacteriol.* **1978**, *133*, 320–328. [CrossRef] [PubMed]
37. Zaritsky, A.; Vischer, N.; Rabinovitch, A. Changes of initiation mass and cell dimensions by the "Eclipse". *Mol. Microbiol.* **2007**, *63*, 15–21. [CrossRef] [PubMed]
38. Zaritsky, A. Chromosome replication and cell division in bacteria. Ph.D. Thesis, Leicester University, Leicester, UK, 1971; pp. 50–55. Available online: http://ariehz.weebly.com/uploads/2/9/6/1/29618953/phd1971_pp_49-62.pdf (accessed on 14 March 2023).
39. Zaritsky, A.; Van Geel, A.; Fishov, I.; Pas, E.; Einav, M.; Woldringh, C.L. Visualizing multiple constrictions in spheroidal *Escherichia coli* cells. *Biochimie* **1999**, *81*, 897–900. [CrossRef] [PubMed]
40. Meacock, P.A.; Pritchard, R.H. Relationship between chromosome replication and cell division in a thymineless mutant of *Escherichia coli* B/r. *J. Bacteriol.* **1975**, *122*, 931–942. [CrossRef] [PubMed]
41. Stent, G.S.; Calendar, R. *Molecular Genetics*, 2nd ed.; WH Freeman and Co.: San-Francisco, CA, USA, 1978.
42. Stent, G.S. That was the molecular biology that was. *Science* **1968**, *160*, 390–395. [CrossRef] [PubMed]
43. Zaritsky, A.; Vollmer, W.; Männik, J.; Liu, C. Does the nucleoid determine cell dimensions in *Escherichia coli*? *Front. Microbiol.* **2019**, *10*, 1–7. [CrossRef]
44. Norris, V.; Madsen, M.S. Autocatalytic Gene expression occurs via transertion and membrane domain Formation and underlies differentiation in bacteria: A model. *J. Mol. Biol.* **1995**, *253*, 739–748. [CrossRef]
45. Woldringh, C.L.; Jensen, P.R.; Westerhoff, H.V. Structure and partitioning of bacterial DNA: Determined by a balance of compaction and expansion forces? *FEMS Microbiol. Lett.* **1995**, *131*, 235–242. [CrossRef]

46. Rabinovitch, A.; Zaritsky, A.; Feingold, M. DNA–membrane interactions can localize bacterial cell center. *J. Theor. Biol.* **2003**, *225*, 493–496. [CrossRef]
47. Woldringh, C.L. The role of co-transcriptional translation and protein translocation (transertion) in bacterial chromosome segregation. *Mol. Microbiol.* **2002**, *45*, 17–29. [CrossRef]
48. Srivastava, A.; Ahmad, R.; Srivastava, A.N. DNA topoisomerases: A review on their functional aspects in cell division. *ERA'S J. Med. Res.* **2018**, *5*, 162–165. [CrossRef]
49. Sitafloxacin. Available online: https://www.chemicalbook.com/ChemicalProductProperty_KR_CB8222468.htm (accessed on 14 February 2023).
50. Norris, V.; Kayser, C.; Muskhelishvili, G.; Konto-Ghiorghi, Y. The roles of nucleoid-associated proteins and topoisomerases in chromosome structure, strand segregation, and the generation of phenotypic heterogeneity in bacteria. *FEMS Microbiol. Rev.* **2022**, fuac049. [CrossRef] [PubMed]

MDPI
St. Alban-Anlage 66
4052 Basel
Switzerland
www.mdpi.com

Life Editorial Office
E-mail: life@mdpi.com
www.mdpi.com/journal/life